The Brain Show—Behind the Scenes

WHAT IS GOING ON INSIDE OUR BRAIN WHILE WE ARE LIVING OUR LIFE

Zeev Nitsan, MD

The Brain Show—Behind the Scenes / Zeev Nitsan MD

© All rights reserved by the author, 2017

No part of this publication may be reproduced, printed, photocopied, presented in public, published, be made into a play or translated in any form or by any means, without prior permission of the copyright owner.

Copyediting of the English edition by Adirondack Editing
Contact information: brainshowbook@gmail.com

ISBN-13: 978-1540722973
ISBN-10: 154072297X

This book is dedicated to all the people upon whose basis of insights this book was built.

Acknowledgements

There is a French saying: "Gratitude is the memory of the heart, and my heart is full of such memories."

Observing the horizon reflected in this book was possible due to the permission granted to my brain to observe the horizons reflected in the ideas of many other brains—and I am grateful to them.

Similar to other books, this book is an eclectic combination of insights born in other brains and insights born in the author's brain.

I owe a great deal to many people. The seeds sprouted in this book originated in various gardens: books, journals, lectures, etc. I allow myself to paraphrase Newton's words and say I managed to look at the secrets of the brain since I relied on many insights of other brains.

Since it is difficult to compress a comprehensive review of such a broad subject into such a short book, I included selected aspects of the reviewed themes and some aphorisms related to them.

A book sometimes enables us to read about ourselves in words written by another person. Some readers will recognize themselves in all of the pages, some will recognize themselves in some of the pages, and some only in a few of the pages (and I hope in no less than that).

A book is a blind date with an anonymous reader, and I hope this book will create an "affair" between my brain and the brains of my honorable readers. I also hope it will bring you pleasure and be a source of knowledge about the impact of the brain's function on our daily lives.

Table of Contents

Acknowledgements — 5

Introduction — 9

Part A: Aspects of Brain Function — 17

  Chapter 1 Selected Aspects of Brain Function — 19

  Chapter 2 The Perception — 83

  Chapter 3 Attention, Consciousness, and Awareness at the Center and Margins of the Attention Beam — 145

  Chapter 4 Aspects of Inner Motivation and Reward — 169

  Chapter 5 The Story of Dopamine — 181

  Chapter 6 Psychonautica — 193

  Chapter 7 Memory Functions — 203

  Chapter 8 Thoughts About Thoughts: Climbing The Mountain of Thoughts — 263

  Chapter 9 A Beautiful Mind — 313

  Chapter 10 Stricken Brain—Dealing with Brain Injuries — 338

  Chapter 11 Memetics—When the Brain Meets an Idea — 360

  Chapter 12 Unity of Contradictions: Two Minds, One Brain—the Hemispheres — 415

Part B: The Seasons of the Brain: Changing of the Brain in Different Periods of Life— How Does Our Brain deal with the Different Stages of Life? 441

    Chapter 13 Patterns of Brain Function in Different Periods of Life 443

    Chapter 14 How the Brain Deals with the Dimension of Time 448

    Chapter 15 Brain Development from Embryo to Adult 460

    Chapter 16 Aspects of Mature Brain Functions 477

    Chapter 17 The Aging Brain 517

Part C: Aspects of Body and Soul 537

    Chapter 18 Body and Soul—Are They the Same? 539

Epilogue 560

Reference 561

Introduction

The Universe Within

Looking from space, Earth seems like a bluish marble, abandoned and solitary, in the gloomy vastness of space, which is colder than any winter known to man.

When one examines the human brain during a neurosurgical operation, the brain seems like a fragile organ whose vulnerable softness is screaming out loud. Examining the naked brain and watching Earth from space are awakening experiences. A possible insight derived from these observations is of a mutual containment of the brain in its physical materialization in planet Earth. On the other hand, there is containment of the earth, a concept in the spiritual perception of the brain. Earth—our collective home—is contained within the brain—our private home—and vice versa.

The Human Brain—an Evolutionary Infant

Earth is four and a half billion years old. Our species, Homo sapiens (modern intelligent human), and our brain, which is unique to us, are two hundred thousand years old—a blink of an eye in evolutionary terms. In a sense, our brain is a living fossil; the contemporary version of our brain has not changed significantly in the last two hundred thousand years. One might say the model of the caveman's brain is still with us in the twenty-first century.

If we compare Earth's age to a twenty-four-hour period, however, it means that our brain and our species, in their contemporary version, have existed for a little less than **the last** four seconds.

Although the size of the human brain has not changed since Homo sapiens first appeared, it does not mean this is the final version of the brain. It is more correct to refer to it as a building in the process of being built, an evolutionary phase. One piece of evidence that the brain is not yet a final product is the lack of integration between certain parts of the brain or, as author Arthur Koestler claimed, a few screws in the human brain are loose.

The Brain as Breacher of Evolutionary Order

A popular perception is that our body is a vessel designed to preserve our genes and make sure they are passed to future generations. In this sense, we are compared to a runner in an evolutionary relay race, who is supposed to pass the stick. The collection of genes represented in our reproductive cells to the next generation is the next runner in the race.

We fulfill the wish of every DNA strand to become two strands and are servants of the dream of every gene to be passed on to a new human being.

Evolution is not about the survival and struggle of an individual but the struggle of all genes the individual is carrying.

The human brain, as a product of evolution, is designed to improve the survival potential of the genes in the body it serves. Thus, the primary purpose of the brain is not researching the phenomena in itself, but researching the phenomena in relation to improving the survival skills of the body that stores the brain and the genes. It seems the human brain avoids the evolutionary order and enables us to examine the roots of certain phenomena, including those that do not serve a direct survival purpose.

Ignorance About the Brain as a Risk Factor

The message that is so popular in different forms in various guidebooks conveys the same meaning: one can live a long life yet live very little. Ignorance with regard to our brain's capabilities and the best usage of them is one of the main causes of a dull life that compromises our potential as human beings. Familiarity with our brain is highly beneficial to us in the practical sense.

We were thrown into life involuntarily, and our brain resembles a map that can help us determine our destination and a compass that guides the direction of the way to it.

At each and every stage of our life, we will be able to select the best steps for us if we have a clearer picture of what is going on inside our brain. This is how we will be able to better use our most important natural treasure more efficiently and have better control of our destiny. Understanding the rules of our personal brain as well as the rules of the universal human brain is a promising pathway to increase our self-awareness and our sense of satisfaction from life. The blessing of knowledge will enable us to maximize our brain's potential. "Know thyself"—the saying inscribed on the top of the Delphi temple and adapted as the first decree by Socrates—means, to a great extent, "Know your brain."

The human culture will better its values when the knowledge basis of the "culture organ" becomes greater and available to more people.

A Multidisciplinary Approach to Understanding the Brain

Brain research, as with research in other complex sciences fields, has known many downfalls. Brain researchers constantly struggle with unfulfilled forecasts and expectations. A known saying describes a scientific tragedy as a "murder of a beautiful theory by an ugly fact." The territory of brain science is drenched in the blood of such beauties; thus, in this sense, it is a very tragic place.

It is said nature "loves to hide," and with respect to the human brain it seems this tendency is even stronger. Mental endurance, which grants the ability to search persistently for what is concealed from the eye of knowledge, is a key ingredient in the recipe for scientific success in this field.

Brain research and mainly – the higher functions of the soul has been traditionally considered part of occultism. Nowadays, the arm of proven facts has turned brain research into an empiric discipline, full of observations and experiments. Nevertheless, the arm of hypotheses that guides research paths also flourishes in the field of brain science. In this book, I will try to share with readers some insights, theories, and guiding insights, all of which are held in the hands connected to these two arms.

Praise for Multidisciplinary Understanding

For many years, brain researchers of various disciplines based their findings exclusively on concepts originated in their disciplines. They would describe the entire phenomenon as it was observed through the narrow keyhole of their specific discipline. Turning this field of research into a multidisciplinary research significantly improved the quality of information related to the brain, as each expert brought his own toolbox, and the gathering of experts from different disciplines will enhance research abilities.

Multidisciplinary study builds bridges. The bridges connect over the wide conceptual gulf separating the disciplines and enable us to move from one kingdom of knowledge to another and intensify our understanding.

Brain research requires an eclectic approach of connecting different disciplines and methodological and conceptual mutual fertilization, which is essential to promoting our understanding. Brain researchers use different types of paintbrushes and draw a picture of the human brain from various angles. Intellectual pluralism is the basis of brain science due to the multidimensionality of the riddle.

"The Theory of Everything" in Brain Science

The human brain is probably the most complex part in the amazing puzzle that is life on planet Earth. Some praise the human brain further and see it as a means through which nature reaches its own consciousness. Nevertheless, cartographers, mapmakers of the human brain, aspire to be able to follow beats of thought and fluctuations in the dial of emotions.

The brain is a multipartite riddle; in addition to being a neuroengineering riddle, well planned and chaotic at the same time, it is a thermodynamic riddle that in its core stands the question of how the rays of meaning and order break through randomly and chaos surrounds our brain.

Similarly to physicists who try to find a formula combining all physical forces, some brain researchers attempt to find the brain version of the "Theory of Everything." In both cases, success has yet to be achieved.

In the modern era, studies of brain science have turned from intellectual desert soil to green fields of insights. Recently we have witnessed the formation of new disciplines that deal with the reciprocal relationship between the brain and various aspects of human function. These disciplines are termed "hyphen sciences" and include neuro-economics, neuro-criminology, neuro-ethics, neuro-philosophy, neuro-linguistics, neuro-aesthetics, neuro-cybernetics, and so on.

There is no doubt that we will soon witness the birth of new disciplines that will expose the brain's role with regard to additional aspects of human behavior.

We are in the midst of a brain research odyssey. Most secrets of the brain are still in the dark, but brain researchers are, little by little, revealing the grammar rules for the language of the brain.

In this book, I have tried to gather selected aspects of new and old pieces of information discovered during the explorations journey of the brain continent.

The Limits of Understanding Our Brain

Contemporary hegemonic insights regarding our brain have not led us yet to the Archimedean point that will allow us to form a model explaining the complexities of the brain function satisfactorily. The continuous path of explanation "from the neuron to consciousness" is full of wide and deep gaps we might not be able to bridge. The skeptics among us might say that such uniform understanding is nothing but wishful thinking.

Certain aspects of brain function nowadays seem as a challenge similar to a kōan riddle, which cannot be solved by means of logical processing and requires intuitive, logical thinking.

Discussions about consciousness are characterized by built-in duality. Consciousness is absorbing a subject as well as the absorbed object, and some believe it reflects a built-in fallacy.

The Freud syndrome, or 'in the splint of era knowledge', is a situation in which the attempt to conceptualize a field of knowledge is made in spite of knowing the available knowledge is partial and limited, so the conceptualization is doomed to be partial and limited as well.

Such was the gloomy situation of neurologist Sigmund Freud, who felt his ability to peek at the obscurity of the human brain and soul, based on the scientific knowledge of the time, was painfully limited. Thus, he was forced to abandon his impartial scientific observations and wander away to the land of subjectiveness. Such a situation is easy to identify with even today, where the keyhole through which the brain is revealed is much wider. The main role of the scientist is to light up the dark, unknown areas, but the ability to do so depends on the intensity of the flashlight provided by the era he lives in.

Acknowledging the fact that our knowledge of the brain is only partial, and realizing there is no such thing as complete knowledge (and that it will probably remain so for good, though we hope to come closer to full understanding), it may shed appropriate light on the riddle of the brain.

The Brain Makes Us Unique

The saying "The unexamined life is not worth living" attributed to Socrates, is well known. On the other hand, we can quote the cynical, humoristic remark of author Kurt Vonnegut: "Life happens too fast for you ever to think about it."

The brain is the element that enables us to live a worthy life in the spirit of Socrates' words. Our reflection as human beings is reflected in the mirror of the human brain. The roulette rounds of our lives' events wire our brain in different patterns. Each one of us resides on a unique planet created in our brain, though we are all residents of Earth. The head cliff rising above our neck carries "us" within it. Everything that we are, as a unique intelligent entity, rises above our neck. In many senses, the limits of our brain are the limits of our world.

In the fifth century BC, Hippocrates, the father of western medicine, wrote, "Humans ought to know that from the brain, and from the brain only, arises our pleasures, joys, laughter and jests, as well as our sorrows, pains, grieving and tears. Through it, in particular, we think, see, hear, and distinguish the ugly from the beautiful, the bad from the good. In these ways, I am of the opinion that the brain exercises the greatest power in the man."[1]

The seventeenth-century philosopher John Locke created a thinking experiment examining the question of self-identity[2]. He described a shoemaker and a prince who exchange their brains. He concluded that when the shoemaker's brain is resting inside the prince's skull, he becomes a shoemaker wearing a royal gown; on the other hand, there is a horrified prince who finds himself walking barefoot for the very first time in his life.

Our body and our brain resemble the ship of Theseus, king of Athens. According to myth, when the beams of the ship became rotten they were gradually replaced, until, as time went by, all of the original beams that the ship was built from were replaced. Therefore we might wonder whether the renovated ship, which does not include any part of the original ship, retains its original identity? It seems that almost every molecule of our young body will be replaced by another as we climb the mountain of the years. In the physical sense, there is almost nothing left from the child we once were and left behind. Nevertheless, we do not conclude that it is a different "self," and our assumption is that our original being is retained.

One might see Franz Kafka's story "Metamorphosis"[3] as an experiment of thought. The protagonist is a man who wakes up in an insect's body but retains his human identity, in spite of the bodily metamorphosis. His memories, temper and thoughts are all retained and reflect the perception that mental continuity, rather than physical continuity, is what makes human essence unique.

Each person has a unique brain-print, yet, within the skull of a certain person, this brain-print is constantly changing as a reflection of the continuous transformation that is taking place in the brain. We are similar and different, cut from the same cloth, but have unique characteristics. We are all made of the same star stuff, as Carl Sagan once said, but each and every one of us is a unique star in the universe's sky. Our brains are the same—they share, as snowflakes, similarity and uniqueness—yet there is not such a thing as two totally identical brains. Each brain is deserve to be called "Atsam Yelash" (an Ethiopian name for girls and boys that means "there is no one like you.")

Part A:
Aspects of Brain Function

Chapter 1
Selected Aspects of Brain Function

Facts About the Brain

The brain—the universe of our being—is also a fluttery chunk of protoplasm that weighs about one thousand and three hundreds grams. It is confined to its bony cage, grooved like a walnut and its texture resembling a ripe avocado. But, as one should not judge a book by its cover, we should not draw conclusions about the essence of the brain merely by focusing on its structure. A fragment of a memory, a line of thought, and a mental image—we cannot assess their weight, although many of them are part of the flesh-and-blood tissue of the brain.

The forefront of contemporary knowledge still leaves behind a broad explanatory gap about the structural and functional aspects of brain function and is still lacking with respect to the formation of thought and spirit. In this sense, the brain is like the philosophers' stone that, according to the alchemists, had the power to turn cheap metal into gold. The brain, which in a way is still unknown to us, knows how to translate biochemical and bioelectrical responses into thoughts and emotions.

Speedy Journey in Brain Kingdom—the Structural Outline

The brain scenery is characterized by a repetitive motif of crevices and ridges that grant it a unique fingerprint, or 'brain-print'. This configuration derives from evolution's attempt to maximize the surface area of the cortex within the skull.

As we age, the crevices encircling the ridges become deeper and wider as a result of the wearing of the ridges, which results from the decrease in the number of neurons that create the ridges.

The external neurons layer, located in the brain mantle, is called the cortex. The thickness of this layer that covers the brain like a handkerchief is between one and a half to four millimeters in different areas of the brain, and its average thickness is three millimeters—as thick as a pack of nine playing cards. The cortex is made of layers of neurons that are organized horizontally and layered one on top of the other, but they also connected to one another in the vertical dimension. Different cortex layers have different features due to the differences in the nature of the cells that form them, their density, and the nature of connections between them.

The number of the layers varies in different areas of the brain. In areas that are "recently acquired" in the sense of evolution (such as the frontal lobes), one can detect six separate layers of neurons, and in more ancient areas of the brain (such as the hippocampus) one can detect three separate layers of neurons.

The cortex area is between 2,000 and 2,400 square centimeters—as small as a tablecloth. Due to the grooved structure, two-thirds of this area are located in the depths of the cortex channels.

The Columns of Neurons—the Pillars of the Thinking Hall

A common communication pattern between the neurons in the brain is in the form of a vertical column called the "neurons column." The six layers characterizing the neocortex are represented in each of the columns. The column penetrates through them and is formed out of their contribution. The number of neurons in a cerebral column varies significantly, from clusters of fifty neurons to clusters of ten thousand neurons. On average, most columns consist of several hundreds of neurons, which rise to the height of the brain cortex—three millimeters on average.

The neurons that form a column are operated in a sequential manner as a response to the same initial stimulus. Lateral connections exist between the vertical columns—hence the crisscross pattern of neurons connections. The neurons column is the basic information processing unit of the brain. According to common estimations, there are about a hundred million columns in the human brain.

The Negative Demography of the Brain Kingdom

Our brain loses neurons in different periods of our life, and in an adult's brain very few neurons are created.

A common assumption among brain researchers is that our brain loses 0.25 percent of its volume every year. In other words, our brain loses 2.5 percent of its volume every decade from the third decade of life. Most of the loss is ascribed to the frontal lobes.

Along with the process of neurons loss, a process of cellular differentiation of stem cells into neurons takes place in a few parts of the brain. The inspiration for the discovery of the formation of new neurons in the human brain partially came from the study of the ornithologist Fernando Nottebohm, who investigated, in 1980, the courting habits of songbirds and discovered that the winged troubadours recompose their songs every season. While studying their brains, Nottebohm discovered that new brain cells grow every season in the specific area of the brain that is in charge of birds singing.

With respect to the brain of an adult person, as far as we know today, we can make a generalization and claim that the demography of neurons in the brain is negative, with the exception of three "districts" as far as we know today. In these districts, new neurons are formed even in the brain of adults. This takes place in the area of the dentate gyrus, which is a substructure of the hippocampus located at the internal part of the temporal lobes, as well as in the subventricular zone of the lateral ventricles and in the olfactory bulb, which is the thickened part of the olfactory nerve located at the bottom of the frontal lobe close to the nasal bridge. Moreover, there are certain reports (yet to be validated) about the creation of new brain cells in the central core of the brain in an area called the caudate nucleus and in other locations in the brain, though these assumptions are still controversial.

The Maps of the Brain

As with the map of Earth, the brain-surface map can be analyzed topographically and mapped according to various functions.

The sensorial representation map of the body, the somatosensory cortex, is located in the cortex at the front of the parietal lobes in the hemisphere opposite to the relevant side of the body. The motor representation map, the motor cortex, is located in the cortex at the rear of the frontal lobes, in the hemisphere opposite to the relevant side of the body. These two representation maps are formed in a somatotopic pattern (which creates correlation between body areas and a unique representation area in the brain). The area that maps the entire body is called "homunculus" ("little man" in Latin). In this area, the representation of the areas in the body is in correlation with the level of its innervations or movement. So, for example, in the sensorial representation map, the hands have a greater area compared to the legs—whose surface area is much bigger—since the sensorial sensitivity of the hand fingers, and in particular of fingertips, is the highest in the body. The representational maps of the brain can be changed with respect to representational relations of the various areas of the body that reflect them.

When a certain organ of the body is damaged, its representational area in the cortex changes accordingly, based on the general brain flexibility rules: the inactive elements will lose their vitality and will vanish gradually. On the other hand, active brain maps will strike their roots deeper into the bed of neurons.

Similar to the popular belief in the spell power of voodoo dolls, stimulation of the homunculus area will create, in a corresponding pattern, stimulation in the specific area of the body represented by it.

The Structure of the Brain

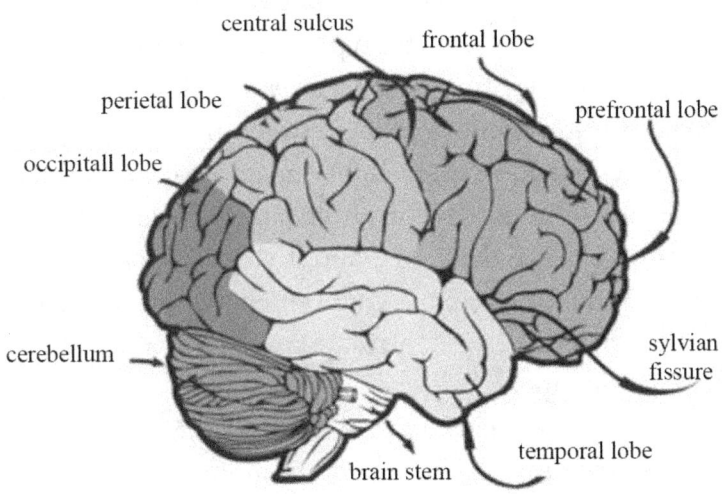

The Algebra of Neurons

There are about one hundred billion neurons in the brain, each of which "chats" regularly with the other ten thousand neurons on average. We have the ability to listen to this chat by means of various technological devices, created by human wisdom, whose purpose is studying this very same wisdom.

The teamwork of neurons is at the basis of the brain activity. It is possible due to mediation of the "runners" who pass on information directly between the various parts of the brain. These runners are actually chemical information molecules, which can be divided into several types: hormones, peptides (small proteins), and neurotransmitters.

Except for the neurons, the other cells that compose the brain are divided into several types and are generically called "glia cells." These cells were traditionally considered to be the cells responsible for the logistics in the brain—some sort of support cells that assist the neurons in performing their role.

The glia cells are divided into three types:

The oligodendrocytes—These cells constitute about three-quarters of the glia cells population. This type of cell constitutes the workshop that creates the coating of the axons, which are the main traffic avenue of passing information between neurons. This coating is called myelin, and its main components are oily.

The astrocytes—These cells constitute one-fifth of the glia cells. New studies emphasize their role, and it seems, in addition to "plain support" and providing a "convenient working environment" for the neurons, they contribute an active component for the thinking functions.

The microglia—These cells constitute between 5 to 10 percent of the glia cells. They serve as the brain's immune protective shield.

There is controversy surrounding the overall number of glia cells in the brain. According to a common estimation, there are nine times more glia cells than neurons. Thus, the total number of cells in our brain is about one trillion (a thousand billion), and the neurons constitute one-tenth of the cells. Another estimation states a similar figure—one hundred billion neurons and glia cells in the brain.

The total length of axons, which connect the neurons in our brain, is estimated at 170,000 kilometers—an overall length that allows them to circle Earth about four and one-quarter times.

Fundamental aspects of our intelligence skills are formed in the cortex. Anatomy researchers estimate that a typical cortex contains about thirty billion neurons. These thirty billion cells are a core infrastructure of our identity as human beings. Some claim they host most of our memories, knowledge, skills, and life experiences. It seems the main fabric of our worldview is woven in the loom of this thin layer of neurons.

The branched connectivity pattern between the neurons in the brain creates a huge potential of neural circuits, which is estimated at the figure 10 followed by a million zeros (for sake of comparison, the estimated number of particles in the known universe is the figure 10 followed by seventy-nine zeros). According to a common assumption, in practice, the number of links between brain cells is about one hundred trillion, similar to the estimated number of leaves in the rain forests of the Amazonas.

This is the inspiration for the common saying that the human brain is the most complex system known to man. This infrastructure is the source of the great gap between the culture developed by human beings and the spiritual heritage of the most intelligent animals on our planet. In other words, this is the course on which the preeminence of human was built. We should not, however, ignore the fact that the basic building blocks of the human brain are also found in animals' brains.

Neural Darwinism

In different periods of life, neural Darwinism exists due to two forces: genetic dictation and development that depends upon environmental stimulus.

A volume unit in the skull is a real-estate asset of great value. In the spirit of the Middle Ages theological question regarding the number of angels that can crowd together on one pinhead, one can claim that if we draw a tiny square whose side is one millimeter long it is enough to mark the location of a hundred thousand neurons. Due to space and energy-saving considerations, neural connections that carry unnecessary or less-necessary information are unraveled.

According to the "usage-dependent existence" rule, brain areas that are frequently used are probably essential to survival from an evolutionary point of view. On the other hand, areas that are not used frequently are doomed to destruction.

Survival of the Fittest—Natural Selection—the Inner-Skull Version

A process of neural Darwinism takes place in our brain. The evolutionary selectivity instrument within the skull box is, in many senses, the replica of the natural selection process that takes place in the kingdom of nature.

The human brain, as might be said of our whole body, is probably the creation of a blind watchmaker who created it after a great many trial-and-error attempts in the figure of mutations and selections that brought the product to its final status.

Science philosopher Richard Dawkins described in his book *The Selfish Gene* an idea according to which the entity that fights a survival battle in the killing fields of natural selection is the genes rather than the animal or plant that carries them. In fact, the latter serves as a tool and a springboard toward the future.[4] In this case, Dawkins stepped up in the resolution hierarchy from the whole-animal level to the level of the genes it carries. If we consider the opposite direction—stepping down in the resolution hierarchy—it seems that the entity that fights the survival battle in the inner part of the skull is a group of neurons or a column of neurons rather than a sole neuron.

The neural columns constitute the material dimension of thoughts, and they are actually the main operational units in the brain. Neural Darwinism takes place primarily at their level; thus, in the neural Darwinism process, the basic unit that undergoes the selection process is not the sole neuron, but a group of neurons that represents a piece of information, a certain perceptual recording, etc.

The Serengeti Plains Between Our Ears

As in the savannas of the Serengeti reserve in Tanzania, there are the Serengeti plains of the developing brain—unused neuron columns deleted as cannon fodder of the brain's evolution, exactly like a wildebeest that cannot run fast enough.

There are two types of evolutionary selectivity. The first is developmental and the second is acquired (fruit-of-life experience). Developmental selectivity takes place prior to birth; thanks to it, all creatures, including identical twins, are born with a different brain. The genes and their products dictate a general structure whose borders are binding. In spite of the similar general pattern of brain areas, however, each person has a unique brain. The prenatal differentiation level organizes the brain and creates a type of primary repertoire. The acquired change stage starts after birth—our experiences either weaken or reinforce, or alternatively, create or destroy the primary structural repertoire, though they do not cause prominent changes in the "big" anatomy—the one that can be seen with the naked eye.

Size Doesn't Matter—the Brain Version

We belong to the subspecies—Homo sapiens sapiens (modern intelligent man, subspecies of Homo sapiens). Our species, the Homo sapiens, derived from the Homo erectus (the erect man) some two hundred thousand years ago along with the Homo neanderthalensis species, which was extinct some thirty thousand years ago. Neanderthals had bigger brain volume (1500 cubic centimeters), but the prefrontal area of their brain (where superior brain functions are formed) was significantly smaller than ours—a strategic disadvantage that tipped the scales in the neural survival battle.

Main Characteristics of Information Processing in the Brain

At the arena of assumptions regarding the manner of brain function, two polar approaches have been competing for a very long time. One is the "locating approach," according to which each of the soul components is located in a specific brain area in which its neural correlate. According to the other approach, which is a polar alternative, thinking and memory functions are the result of the function of the whole brain, and the level of damage they suffer reflect the level of damage suffered by the whole brain.

There is truth in both approaches, but, separately, their explanatory power about the function of the brain is only partial.

Many brain researchers believe the human brain is more comparable to a Swiss pocketknife, which integrates various brain tools into various registers according to the "need of the hour" rather than a central, hierarchical processing system. Similarly, a functional pattern that has a greater explanatory power is a "functional systems pattern."

This third pattern of function merges the two former approaches and polishes their rawness. It is based on the assumption that the brain is a collection of functional systems. To clarify this idea, we can take the digestion system as an example. The stomach itself does not produce the digestion process, nor does the intestine, the pancreas, or the gall bladder as individual organs, but their mutual function and the interaction between these organs create digestion. In a similar way, our breathing relies on a combined operation of the breathing regulation system in the brainstem, the diaphragm, the intercostal muscles, and the lungs. The main complex functions of the soul are produce in a similar way- by the combined action of functional systems in the brain . Most mental functions, such as remembering, production of emotions, and dreaming, are products of complicated systems decentralized at various areas of the brain, and their reciprocal relations bring about the functions of the soul. Each of the components contributes its share to the big production. The functions of the soul do not reside within one of the structures that participate in their production; they reside in between them. Thoughts, like many other brain functions, are not born in specific brain areas but, rather, in between these areas.

The various functional systems integrate with each other modularly in order to perform designated tasks. The ensemble of neurons recruited for a specific task creates a designated register for the designated task in which the brain deals. The functional-units approach combines the two preceding approaches, and the synthesis creates a more reliable understanding of the reality.

The brain is like a musician with multiple instruments, and a mental task is like an accord—a sound composed of several subsounds. One can see brain function as a whole when all sounds merge into a single melody, or, when listening more carefully, one can distinguish between the sound of the harp and the sound of the cello and between the sound of the sitar and the sound of guitar—in other words, between the different functional systems that contribute to the specific brain function. When we perform a thinking task that requires concentration, accuracy, and timing, we act as a musical editor who can carefully select the musical soundtrack played in the halls of our skull.

Committed to Saving Energy

Adjacent brain areas tend to be parts of the same functional system. The mutual activity relies on the geographic proximity, so the information passes through shorter channels, which decreases the amount of energy required for information processing.

Due to the proximity, there is less depreciation in the quality of information, and the information is transferred, in a shorter period of time, to its target by means of bioelectrical action potentials. The depreciation in information quality and the time needed for the information to reach its target are usually in direct correlation with the length of the connection path and in inverse correlation with the thickness of the axon that passes the information, and are also affected by other causes, such as the level of the axon's "insulation" by the myelin coating.

Thus, one may say that the proximity between brain areas that process similar types of information derives from the devotion of our brain to the energy-saving principle.

The Brain—a Dance Floor, Boxing Arena, and Theater Stage

The brain is an ecosystem of functional systems. Sometimes they dance cheek to cheek, and sometimes they fight one another as rivals in a boxing arena in an attempt to dictate the brain's agenda. The brain may be compared to a confederation of autonomous districts that every so often challenge the central regime.

If we compare our brain to a theater, we may also say that the stage of our brain is filled with main actors and co-stars who constantly exchange roles. Their casting is ever changing (our brain is the stage, and all the neurons are the actors), and the actors move from the stage of awareness to the backstage of the brain and back. Alongside the central stage, there are also fringe theater stages at the dim corners of the brain.

Our Brain as a Montage Artist—Creating Fascinating Combinations

The conceptualization of reality, which derives from combining the outputs of different brain areas, shares similarities with the art of montage (from the French, "*montage*," assembling handicraft). In a similar way to a film editor who integrates individual photos into one filmic continuum, our brain integrates individual experience impressions into a sense of a whole experience.

There is a philosophical approach according to which conceptualization, in terms of time and space, is merely the creation of our brain and does not exist objectively.

The reality-compatible pattern, which is loyal to the directionality of the arrow of time, is imprinted on our perception records (it is done in the frontal lobes at the same time information is retrieved from our memory). Prior to this process, perception records in the brain act as anarchists. They do not obey the rules of time and place and appear in a nonchronological, nonsequential continuum, in a sort of seamless unity. At the same time, a "timeless" component also exists within the space of our skull. Events from different periods of time "take place" in our brain simultaneously.

Some might claim that the imprinting of pattern-of-time directionality and causality is a Houdini-like illusion produced by the brain.

The Domino Effect in the Brain of the One Who Looks at You From the Mirror

The brain of the one who looks at you through the mirror is composed of about a hundred billion neurons that communicate by means of about a hundred trillion inner connections. The interactions between the functional systems of the brain often make the activity in a certain area of the functional system create a sort of a domino effect that echoes throughout the system. The interaction between the different parts of the brain sometime presents nonlinear correlations. Occasionally, recruitment of a small number of additional neuron columns reinforces the level of overall processing in the brain significantly.

The Brain as a Pattern-Processing Apparatus

Our brain tends to recognize patterns of order within the input that reaches it, which is sometime chaotic. These patterns of order are actually patterns that offer the perception mechanism of the brain some tools that will enable it to organize and process the input. The pattern-recognizing approach, adapted by our brain, makes the brain recognize similarities between a current situation and a past situation. Using these similarities, our brain is able to choose an action plan that is based on proven past action plans. If the action plan fails, the brain relies on the insights and lessons learned as a result of the failure. The coping approach, which is based on pattern recognition, in spite of overgeneralization or oversimplification related to it sometimes, is essential so that we do not meet a new, unfamiliar world every morning. This approach sometimes requires "selective blindness" with regard to certain aspects of the new situation.

The Familiar and the New

According to a popular view adapted by many brain researchers, in cases when the perception impression mediated by the sensory organs finds a link to an existing perception pattern, an echoing loop that is based on a "stimulus that contains familiar components" is created. When the perception impression does not find a link to one of the patterns that are already encoded in our brain, it requires the allocation of more mental resources in order to create a totally new echoing loop—of a stimulus with unfamiliar characteristics. The human brain is a creature of habit. Built into it is a preference for old, familiar information over new, seasonal information.

Pattern Recognition Formed as a Cocktail of Genetics and Environment

Not all recognizable patterns are the product of former, repetitive life experiences acquired throughout the years. Some of these patterns are built in to our perceptional mechanism from the moment we are born. Different animals have different mixes of innate patterns and acquired ones. Even recognizable patterns at early stages of life require critical exposure to an appropriate environmental stimulus in order to carry out the potential of recognition. In other words, the pattern recognition ability, as with most of our mental skills, is a mix of genetic potential and appropriate environmental stimulus.

Generalizing patterns are like drainage basins. Most global patterns are created out of extracting various aspects of the world of phenomenon into one coherent insight, which is the drainage basin that serves numerous streams of perception impressions.

Structural Code Programming and Open Code Programming

All brain areas speak the same language—the language of the bioelectrical evoked potentials, which speaks itself in sentences of the intensity of the evoked potentials and their distribution over time and space. While using the same language for transferring information, however, the various brain areas are different with regard to their ability to organize and encode the information. Certain brain areas that use an "open code" method create an independent programming language that gradually improves itself. It seems that the most recently acquired brain areas, which appeared in a later period in the evolution of the human brain, have a better ability of self-programming. This type of programming is performed by means of creating complex information clusters designed to encode patterns of deeper layers of the phenomenon world. More senior brain areas have a more limited cognitive maneuvering ability, and they mostly rely on built-in wiring, which is similar to built-in code programming. Thus, for instance, the associative, multisensorial cortex, which merges the various impressions of the senses into a coherent reality perception, includes a relatively small number of prewired solutions in advance and is flexible enough to deal with the input. On the other hand, the subcortical, more senior brain areas in the arena of the brain development are mostly based on information processing patterns that have been there from an early stage, and their ability to change their coping approach is somewhat limited. It seems one of the factors that brought about considerable improvement in the abilities of the human brain is the development of areas that operate according to an open code programming approach as opposed to an older approach that is similar to built-in code programming.

Upper and Lower Routes

The Upper and Lower Routes—Warning note: Conceptualization of the different activity pattern of the neural paths in terms of upper and lower routes may sin in oversimplification.

A common definition of the lower route is neural networking that creates behavioral motivation, which sometimes relies on "volcanic eruptions" of raw urges. The lower route is composed of a changing mix of unconscious thinking processes, mainly emotion derived and relatively rapid. These rapid activity patterns are similar to heuristics (behavioral rules based on previous experiences). From this point of view, a human being is, in many senses, a creature of instincts.

On the other hand, behavioral motivation, which is based on the neural paths of the upper routes, is created by voluntary thinking processes, which are based on a conscious intention to make sense and are slower in nature.

The upper route operates at the conscious level. It often requires strenuous intentions and is slower than the lower route. This is probably what lies behind the advice to count to ten before reacting, which, practically, means to stall for time until the upper route starts working in an attempt to control the messages of the lower route.

There is still controversy over the structures included in each of the systems. The lower route includes central structures in the functional system in the brain called the limbic system, which includes the brain structure called the amygdala. The upper route includes areas in the prefrontal cortex, an area called the anterior cingulate gyrus, etc.

It seems that in the moment of truth, while we are "under fire" in life's battle, our brain ("brain" in the sense of a slow, rationalistic action pattern that mostly relies on the prefrontal cortex) tends to take shelter behind the heart ("heart" in the sense of an emotion-driven action pattern that activates a heuristic action pattern, which is executed rapidly and mostly relies on the limbic system). The researcher Joseph Ledoux, explained the survival advantage of this kind of 'taking shelter' by saying that when we allowing evolution to do the thinking for us at first, through the quicker route of the emotional brain, we gain the necessary time to think about the situation and act wisely.

Longitudinal and Latitudinal Roads

In each of the hemispheres, there are "longitudinal roads"—neural paths paved along the longitudinal axis of the hemisphere—and "latitudinal roads," which are paved along the latitudinal axis of the hemisphere and serve as connecting channels between the two hemispheres. Most of the paths that connect the amygdala, which motivates instinctive, emotion-driven behavior, to the frontal lobes, which motivate calculated and restrained behavior, are longitudinal paths.

The anterior cingulate gyrus and the frontal cortex are key areas with regard to emotion regulation and moderation of the stormy waves of the amygdala. The amygdala has particularly "touchy" buttons, and it is, in fact, the center of the emotion wheel, whose spokes reach various brain areas.

When there is not enough mental energy to create "behavior filters" by the virtual filter of the frontal lobes, "raw" behavioral patterns appear. These patterns teach us about more basic layers of our consciousness and the processes creating our behavior. In such situations, the upper route does not manage to enforce its rule.

It seems that almost all our actions are based on a mix of brain processing derived from both systems. Most of our actions derive from mutual activity of both systems, and, in any case, it seems that when one system rules exclusively, it is the lower path system. In other words, it seems that certain actions rely exclusively on the lower path, while there is doubt if any of them rely exclusively on the upper one.

In most cases, our actions are based on a mix of unconscious thinking processes that are relatively rapid and mostly emotion driven. Along with these types of thinking processes, there are also voluntary, slower thinking processes that derive from a conscious intention to let logic rule.

Clear limitations are built-in to the attempt to investigate the action pattern of the lower route, which is unconscious, by exploring the conscious upper route (for example, by giving questionnaires to subjects).

Dwarfs and Knights

During routine brain activity, the paths of the lower route are bustling. In their dark paths pace the agile dwarfs of the unconscious kingdom, who perform most tasks in the brain. They are the proletariat that enables brain civilization. The knights of the upper routes are not driven to action that often. They need time to adjust their armor and prepare their horses, so their action is slower.

The complex tasks performed by our brain cast the dwarfs' proletariat of the lower route and the knights of the upper route in different mixes. At certain times, in a pattern of topological juggling, we manage to combine the upper and the lower routes into a single route of action. For example, the field of behavioral economics, which explores patterns of behavior in various economics-related situations, demonstrates the great impact of the lower route, that is mostly derived from emotions, even in territories where processing is allegedly rationalistic and emotion-free.

The Esperanto of Neurons—the Universal Language of the Brain

The standard communication interface between neurons is the language of bioelectrical evoked potentials, which was defined as the keystone of understanding the information-handling processes in the brain. The uniformity with regard to the language spoken by the different brain areas follows a repetitive pattern in nature, known as the "saving artist"—saving while performing various tasks by means of using known common routes—as opposed to the approach of "finding a thousand different ways to solve a thousand different problems."

The "globalization" strategy, in which global processes are combined with local processes, is at the heart of the communication interface of the brain. The local language in different brain areas exists as a global pattern of the whole brain.

The various areas of the cortex use unified data processing approaches—a kind of algorithms of information processing—that are valid in different brain areas.

Various types of sensory input—eyesight, hearing, and smell—although they are mediated to the sensory organs in various manners: a barrage of photons, molecular vibration, and volatile molecules, respectively, mediate to the brain and act within it through the same language. The differences in the experience manifestation, formed by the various inputs of the different senses, derive from the different connection patterns of the various brain areas responsible for the absorption and processing of the sensorial input with other brain areas—not from the variation in the communication interface, which is unified throughout the brain kingdom (the language of bioelectrical potentials).

In other words, from the variance in the wiring pattern between various brain areas is the source of the ability of designated brain areas in which a certain input from a specific sense produces an experience manifestation, which has different qualities from the qualities of other manifestations that derive from inputs from other senses that mediate in the same language.

Information processing in the brain takes place, at least partially, through a fractal pattern (a similar pattern related to ascent or descent in the level of resolution). This language of bioelectrical potentials exists from the sole neuron level through the neurons columns—the whole hemisphere—to the whole brain.

The ability to activate a unified language algorithm throughout the brain kingdom derives from the fact that the letters of this language are universal throughout the brain, and those letters are bioelectrical action spurts. These are currents of electrical charge that are mostly mediated in an electrochemical manner and move among the neurons. The brain does not know a different language. When we look at a certain object, information from the various senses is transferred to the brain as patterns of unique action spurts that turn into a unique letter according to the manner in which they are distributed through time and space. The distribution pattern in space derives from the number of axons involved in the process of routing and placing the action spurts, or, in other words, from the three-dimensional neural networking pattern that participates in transferring the specific input.

The time-distribution pattern derives from the duration of the action spurts and the timing relations between them. All the inputs we perceive from both the external and internal (our own body) environments are eventually translated into the same language of action-spurt patterns over the time and space axes. This is the only language our brain speaks.

The words of the entire perception—the sensory input from within our body and from outside our body—are spoken only in this language. Without this unified communication pattern, our brain would resemble the Tower of Babel, with numerous communication protocols. Although the brain is unilingual in terms of the communication pattern, its products enable it to become multilingual. It "understands" the language of speech; the language of music; body language; and the language of math, and the unified communication pattern is the universal language that is responsible for its ability to speak various languages. The versatile function ability of the brain, which acts like a jack-of-all-trades, derives from all neurons speaking the same language.

The brain processes information and translates it into a single language—the pattern of action spurts across the time and space axes. numerous brain areas are capable of conducting a variety of tasks and even to change their "expertise," from specializing in a specific processing task to specializing in a different type of task. It does this mostly through changing the connection pattern to various brain areas, a phenomenon that is reflected in the term "brain plasticity." Familiarity with the language of the brain enables different brain areas to perform the processing, for both internal and external sources of input, which is mediated by different sensory organs. The brain, which always tends to be efficient and save space, takes advantage of all available spaces, as would a skilled real-estate agent. For example, it is known that adult blind people use areas in the occipital lobes in order to read braille letters. People who are not blind use these areas for processing visual input. Since the braille alphabet is mediated through touch, one would expect that brain areas that are related to such an input would become activated, but the unemployment forced upon the occipital lobes of the blind turns, with time, into full-time employment decoding the braille alphabet; thus, there is no such a thing as unemployment in the brain kingdom.

Helen Keller—Sees and Hears

The skull cavity is like a container of sensory deprivation; the brain does not experience the world directly. Rays of light cannot penetrate—it is dark in there. No sounds are heard—it is quiet in there. There are no fibers transporting pain within the brain, and direct damage to the brain does not ring the alarm bells of pain. Our brain, exactly like Helen Keller's, is like a blind and deaf person. It creates eyesight and hearing out of the pattern of waves that float across time and space and break on its shores.

Brain areas function as versatile information processors that are capable of processing a variety of input data. The various brain areas have the potential to be multisensory, which means they can perceive and process information originating in more than one sense. The visual cortex, for example, which normally resides within the occipital lobes, is able to handle input and processing of touch-originated information in blind people. The various types of energy are rendered into the universal brain language of bioelectrical potentials (action potentials). In most areas of the brain, there are no irreversible internships enabling these potentials to perceive and process information only from a single sensory channel. On the contrary, a flexible ability to perceive and process information originated in different sensory channels, according to the circumstances, is a prominent feature of the brain.

"Sensory conversion" is one of the terms describing brain flexibility, and it includes a promise for the future: super-senses that will enrich human experiences in a manner that is hard to predict.

The studies of brain researcher Paul Bach-y-Rita revealed a portion of this ability that is mostly hidden from researchers. He developed a rugged plate that was put on the subjects' tongue, and, after they practiced, the touch-related information produced by their tongue found its way to the visual cortex. There it was processed and translated into a visual experience. So, the licking of a certain rugged plate invoked a virtual image of a cube in the subjects' brain[5].

The Pattern of Processing and Maintaining Information in the Brain

Aphorism

Aphorism is a concise encoding of information—our brain's favorite information encoding method.

Our brain aspires to realize substantial aspects of the experience and preserve them in memory. Only through this concise pattern is the world, as we know it, compressed in the space between our ears. In spite of the different manifestations of entities in the world, there is a certain "preserved pattern" that preserves the core of the information, even when it is enveloped in an ever-changing cover.

Our brain uses an aphoristic information processing method in terms of quantity and quality of the information. Only a small amount of the sensory input reaches the brain areas designated to process it.

There is prudent estimation related to depreciation in the amount of information from the visual input while it is transferred through the nozzle of information processing in the brain that constantly narrows its diameter. According to this estimation, out of the almost unlimited information mine around us, an estimated amount of ten billion bits per second meets our retina. Our eyes "mine" only a sample of the input—only six million bits per second "leave" the retina. It may be connected to the limit in information traffic, which is caused by a wiring limit: each optic nerve contains a million axons, which are used for transferring information signals from the retina. Only about ten thousand bits per second reach the visual-processing area in the occipital lobes. After further processing, the information is fed into the brain areas that "build" the world model in terms of visual perception. Surprisingly, the amount of information that reaches the brain areas responsible for producing conscious visual perception is less than a hundred bits per second. This thin flow of information is supposed to represent the enormous body of information out there.

The representative aphorism is an effective processing method, but information depreciation and sample representation are built-in features of this method.

Preserving the Core of Experience—Encoded Aphorism

The aphoristic information encoding method enables summarization of experience-perception impressions. The ability to identify and encode the key components, which constitute the essential core, is at the basis of our capability to encode memories. It derives from our ability to identify repetitive sequences in the sequence of signals thrown at our brain.

Names are given to these "repetitive sequences." The names lie at the core of the summarization process—and this is what brain aphorism is about.

Brain Aphorism—the Nesting Nature of Information Preservation in the Brain

Perceptional recording might be compared to a colorful stone in a mosaic of perception memory that is composed of numerous colorful stones. Cells that focus on perceiving and processing of perceptual recording work together, and we move one step upward in the processing hierarchy toward a more comprehensive image. From now on, other areas are responsible for the more comprehensive image. These areas are like "gathering areas for single impressions." There is a sort of brain aphorism—a summarization of perception impressions in a certain ladder whose top rung is a brief conceptualization. In the case of a verbal conceptualization, the brain-aphorism mechanism is able to summarize it into a single word. This conceptualization is the preserved representation of the whole experience. A lower level of information encoding is contained and nests within a higher level. Thus, it can be viewed as a nesting structure.

Information learned in detail is represented at a lower cortical level of processing, on top of which a higher information processing layer is built for processing more complex information.

Cognitive Hierarchy
Hierarchy of Information Representations

Information representation is built in a hierarchical pattern. High areas of the representation ladder in the cortex store the representation of the overall essence, and lower areas store the subgroups of which the overall essence is made. The high areas contain the overall picture, which does not change rapidly. The low areas of the representation ladder store the smaller details, which change more frequently. Thus, for example, a high-representation area stores the knowledge that we are residents of planet Earth, and lower areas store the knowledge related to our location in terms of country, county, city, borough, etc. It continues this way until we reach the cortical, lowest level of representation, which preserves our real-time location—the surface under our feet—whose recordings change constantly with each step we take.

Expert Brain, Novice Brain

The high levels of information processing in the cortex owe their processing ability to lower levels of processing which processing for them the more basic and raw input patterns. When a new task is first learned, a large number of neurons participate in the learning process. After a while, when the task is already familiar and encoded in lower levels of processing, the higher levels of processing are "set free" for the sake of a more complex processing of the information. This is the difference between an expert brain and a novice brain with regard to performing a certain cognitive task.

The learning of thinking or movement skills becomes semi-automatic following some practice and memorization. Then, the skill being learned "goes down" as being encoded at a lower neural level.

Reading written materials is an example of the processing hierarchy function. While we are reading, we focus our attention on understanding the content of the paragraphs—the peak of our reading skills at this moment. Lower levels of processing, which operate on the unconscious levels, constantly identify the letters that sail on streams of sentences in front of our eyes. Once they take this task upon themselves, the lower levels release the higher processing levels to use their "brain power" for processing at the paragraph level.

The low-processing levels are responsible for the morphological aspect (shape and identification of letters). Higher processing levels in the cortex are responsible for the aspect of the content.

The odrer of the lretets in a wrod is not inportmat as lnog as the frsit and lsatlrtteesramieninthe corecrt place—our brain reads the words as a pattern.

A young brain, which has not had a chance to build the "upper floors" of world insights, enables us to understand abstract entities and processes the information at the more concrete levels—at a lower cortical processing level. When it reaches sufficient mastery of the information, a higher processing layer will be created that will process the information at a higher level of complexity and sophistication. Thus the brain climbs the ladder of information storing, toward the apple of the Tree of Knowledge.

In the brain of an expert in a specific discipline, there is a "concise encoding," which is also a preserved representation—a result of a prolonged, aphorismic learning process. Identification of the code word enables the expert to choose appropriate response options quickly. The brain of an intern is still at a stage where it deals with creating the hierarchic information-representation pyramid (which is similar to the terraced pyramid of the Maya people), and, unlike the expert, he is still far from the pyramid's top, a point that allows a bird's-eye view.

Insufficiency, which is built-in to the situation, characterizes the state of our brain. This is required to deal with a huge amount of information nowadays and to process it in shorter periods of time. Sometimes, certain brain functions fail to keep pace. It seems that information processing according to the aphorism approach might be useful when it comes to the contemporary task of swimming across the ocean of information whose level, though it does not depend on global warming, is expected only to rise.

A novice brain is mostly slower when it comes to complex tasks, such as the work of an artist, as a result of a lower level of exposure to manners of performing the task. In addition, a novice brain has yet to create complex sets at the top of the neural encoding ladder for the necessary skills for a smooth, flawless performance as in an artist's performance.

Thus, the novice is slower and cumbersome compared to an experienced brain, which is considered an expert with regard to performing the complex task.

In the brain of an expert, conscious focus of the perception regarding raw information, which is processed at lower levels of processing, such as single letters for experienced readers, requires significant cognitive effort, since it is routinely done at the unconscious level. It is possible that, with respect to the more basic information processing levels of complex skills, the novice brain will have some advantage.

Experience is the Name of the Game

When human beings are exposed to a new "discipline" of knowledge or to phenomena to which they have never been exposed, brain conceptualization tends to create an "overgeneralization," which means putting emphasis on the common characteristics and blurring the differences. The similar is perceived as identical. This state reflects the initial learning stage. On the other hand, the brain of an expert in a certain discipline identifies the nuances and tiny differences between similar phenomena and situations.

The learning curve of most skills is characterized by a sharp slope of the performance curve, which becomes more moderate as more experience is gained in the secrets of the learned skill.

Time changes in our brain sometimes serve as guides. As we grow older, we tend to deal with problems in a different way and rely more on identification of patterns. Instead of a long and tiresome ascent of the information mountain toward finding solutions, we find a path that elegantly passes the mountain of data—a path for the experienced, who are familiar with the topography of the problem.

Acquiring Skills
Sets of Sets of Sets

The cortex creates information representations for impressions that are perceived in a repetitive manner. These information representations go down the processing ladder from time to time and set free those areas at the top of the processing ladder for learning more complex and sophisticated relations—this is how an expert is made. Thus, the attempt to categorize and regulate the details of the output by means of sets of sets at the time of learning might hasten the process of learning the skill necessary for the creation of an expert. This way, sets are made through encoding, acronyms, generalization, nesting patterns (patterns that nest within more general patterns), etc. Such a way enables the creation of shortened encoding, which brings down simple memory representations to a lower level and sets free the higher areas of the cortex so they are able to absorb new materials. Words, as a means to create sets, enable conceptualization at a lower level of the cortex.

The Magic of Preserved Representation
Information-Preserving Rule

Old insights acquired by our brain during its stepping on life's paths are mostly here to stay. Insights that were encoded as patterns, whose acquisition was bought by resources of time and energy, and sometimes blood and tears as well, and which were not refuted in various experiences, become resistant to new information that contradicts them. This is how we avoid shaking our core insights as a result of the ever-changing world we live in.

This approach proves itself as an approach that provides a survival advantage in many cases in which the same essence is reflected in various manners.

A Loved One Remains a Loved One—The Magic of Preserved Representation

A central pattern of memory encoding in the cortex is as "preserved representations." The brain representation of a certain essence remains stable despite the presence of an ever-changing input—for example, when a person is moving away from us or looks sideways rather than forward. This capacity of the brain to preserve perceptual representation, although the related input is ever changing, is called "preserved representation."

Thus, for example, the entire visual representation of a familiar person might change miraculously due to factors such as distance, angle, and light intensity if the figure is apparent or partially hidden. Despite the different images absorbed by the retinal carpet, however, we conceptualize the familiar figure and gather the various inputs into a single coherent entity and identify it this way. Our ability to find meaning in the flood of ever-changing information that meets our brain depends on creating a preserved structure in the squirt of information.

The brain representation of the essential core of a phenomenon might change if contradicting input representations appear and "convince" the truth-determination mechanism in our brain that the new representation is preferable to the previous ones.

Memory storage and retrieval are done through relying on the preserved representations that allow us to create an internal representation model for phenomena that are external to us or internal (inside our body) to us.

The neural netting that encodes information has a flexible topology. The nets of neurons often retain their basic pattern, even when they are stretched, widened or shrunk, and this explains how the essential information core is retained.

The saying "Changed and yet the same, I rise again," which appears on the gravestone of the Swiss mathematician Jacob Bernoulli (1655–1705), was meant to reflect the nature of the logarithmic spiral. This saying also reflects a basic principle in brain information processing and in the pattern of our human existence.

The Modular Use of the Building Blocks of Thinking
The Cycle of Information

Information that was learned and assimilated wanders from its initial input areas, in which numerous neurons dealt with its processing, to higher encoding areas where it is assimilated and where fewer neurons are in charge of it. The neurons responsible for processing the initial input are now available for processing new input.

More familiar information goes down the processing ladder in the brain and is assimilated at the neuronal net at a lower level of processing. This descent is in accordance with experience. Information streams downward and upward in the cortex. The brain representations assemble and spread out.

The hierarchy of patterns with nesting features enables a modular assembly of basic patterns into a fabric of a more complex information pattern. For example, phonemes and musical notes are interwoven into a more complex work: the melody. The same thing is true for our language as we go up from the level of letters to the level of words and then to the level of sentences, which are formed into a complex tale.

An ensemble of neurons, recruited for a certain task, is combined in a unique way in a modular form for the sake of performing a designated task.

Common-experience components also become modular and can be used for forming memories out of new experiences.

Generalization, Overgeneralization, and Separation

Conceptualization of reality in a generalizing pattern is typical for the brains of infants, who are novices in our world. So, at the time of their initial acquaintance with the world, they see every winged animal as a bird. Later on, however, they develop the ability to distinguish between different winged animals and to identify a certain such animal as a stork and another as a raven.

The brain of an expert is characterized by the ability to represent subcategories (subgroups) of similar situations. In other words, it withdraws from unification (low resolution) to the differentiation (high resolution). Thus, it has the ability to react more "accurately," in a manner appropriate for the uniqueness of a situation. On the other hand, atavism (backward evolution) of the information encoding pattern characterizes the cognitive-decline pattern of Alzheimer's patients.

Alzheimer's patients at progressive stages of the disease tend to rely on an individual trait of an individual who belongs to a group and to attach the same trait to each individual they relate to this group. For example, if their daughter's name is Sophie, they will call all women "Sophie." A plausible explanation for this is the induction mechanism (from the individual to the general) transfer to over-generalized.

In fact, most Alzheimer's patients process perception impressions at a low level of processing, at the basic layers of the pyramid, while its higher layers are disintegrating and collapsing.

The process of dissolution and loss of information is the opposite of the learning process of children. First, the most detailed representation groups collapse, and the representation pendulum moves to the side of unification and raw representation of the perception impressions, which are blind to nuances.

The assembling nature of insights from a low level to a high level is a mirror image of the spreading nature of dealing with a task that requires lower processing, in which case the information "pours downwards."

It seems that, when it comes to the brains of Alzheimer's patients, nuances and subcategories disappear as time goes by and all perceptual representations converge into a general, raw representation that is blind to the nuances that distinguish it from similar, but not identical, perception representation. There is a type of unification of the representation of various perception impressions that share similar components. The distinction ability is lost.

It seems that it can be related to the fact that, with respect to Alzheimer's patients, the first area in the brain that is damaged is the dentate gyrus at the hippocampus, to which the ability to distinguish between pieces of information carried by the sensorial input is ascribed. The dentate gyrus labels shades and nuances before the information undergoes unification and generalization as it is processed at other areas of the hippocampus; when it fails, the distinction ability is damaged.

A General Nesting Pattern

The ability to generalize by means of categories in a nesting pattern (like matryoshka dolls that contain each other) is at the basis of information coding in the brain. Information-processing and

-preserving in the brain are performed in a nesting pattern in the sense that the information is assimilated by means of gradual categories. A more general category contains pieces of information from the cortical level beneath it through an aphorismic pattern. It is done throughout the processing ladder—brain processing is a multilevel process.

The taxonomic hierarchy of the kingdoms of wildlife and vegetation (species assembled into types, which are assembled into series, which are assembled into systems, etc.) is similar, in a sense, to the information encoding process in our brain, which spreads out downward and converges upward. Perhaps this method of categorization seems satisfactory to us, since it reflects the central information processing pattern in our brain.

Maya Pyramid for the Input, Inverse Maya Pyramid for the Output

A reaction output whose common manifestation is the movement of various body parts spreads out from the cortex downward and constitutes a mirror image of the input pyramid. The decision to activate the action plan encoded as "kissing my spouse" is interpreted into a sequence of instructions given to our facial muscles (and perhaps tongue muscles, also...). Here, as well, the more experienced ones have contingency plans that are appropriate for various sub-situations related to this action plan.

The cortex stores information in a nesting manner—the information spreads out "downwards," to a lower level of processing and is contained—and gets a shortened code name (in the spirit of brain aphorism) as it goes up the ladder of information processing in the brain.

As a result of the nesting pattern of information storing in our brain, at any given moment a superposition of information resolution is taking place.

We have an overall holographic insight, which consciously stores information at different levels of resolution and allows maneuvering between the changing resolution levels—from the contained structures to the unique details, and vice versa. For instance, when we watch a scene in a play, we simultaneously store the content of the scene—at a certain resolution level, the fact the scene is part of the play—at a more general level of resolution, and that the play is part of a festival is at even more general levels of resolution.

The Imperialistic Urge of the Representation Maps in Our Brain

Use-Dependent Volume of Representation

Due to the competitive nature of the brain's flexibility, cognitive skills, which are not commonly used, get a representation area and become smaller in accordance with decreasing usage. On the other hand, brain areas (representation areas) on which they are based are "conquered" by other skills that are used more frequently.

An "unemployed" brain area raises the appetite of representation maps around it, and these tend to adopt the deserted brain real-estate asset.

In the Brain areas that constitute structural infrastructure for information encoded maps - there is a subdivision: between the neurons at the center of the map whose commitment to the processing of the map-specific skill is high, and, on the other hand, the marginal neuron groups, which are similar to peripheral areas, whose level of commitment to the skill-specific processing is lower. The "peripheral areas" of a brain map are the cortical columns, which have weaker relations with their partners. They have higher potential to serve as "double agents," meaning to assist in more than one skill, and, alternatively, to desert their position and start serving another functional map that maps the processing of another function.

A painting called *A Hand Painting a Hand*, inspired by Maurits Cornelis Escher's painting, shows one hand holding a pencil and drawing another hand while the painted hand is holding an eraser and erasing the hand that is drawing it. This is similar to a situation in which one brain map expands "on account of" another brain map and recruits the neurons that served as infrastructure to the former map.

When a skill is required and the frequency of its use is high, it is more likely that peripheral areas of other brain maps will be attached to the map encoding the commonly used function.

The borders of brain maps might change within short periods of time as a result of the challenges the brain is facing (as chairs in a musical chairs game exchange ownership from one bottom to another).

When it comes to blind people, the representation of the finger used for reading braille letters is larger than the representations of other fingers. Several brain studies have found evidence for the assumption that a significant plastic change takes place in the brain of people who become blind in the course of their life and were not born blind. The occipital lobe, which is normally in charge of processing visual input, converts to processing the input of the sense of touch, related to processing of braille inputs, among the blind. As aforesaid, the brain does not like deserted real-estate assets, and active brain maps confiscate the neural infrastructure of inactive brain maps.

A voluntary study revealed that temporary blindness, induced by covering the eye of sighted participants and by blocking visual input, causes the visual cortex to start processing input through the sense of touch, similar to the process that takes place among the blind who read braille letters. After five days of total blindness, the participants whose eyes were covered were better at reading braille letters than those whose eyes were not covered. Conducting Functional Magnetic Resonance Imaging (fMRI) brain imaging among the former group of participants while stimulating the tip of their fingers showed activity in brain areas that usually process only visual input. In other words, the map that processes touch stimulations invaded the map that is in charge of the visual input and started to enslave the neurons that were previously citizens of the visual-processing kingdom. Another interesting finding is that twenty hours after the eyes were uncovered, on the sixth day of the experiment, no activity was shown in the visual-processing areas as a result of stimulation to the fingers, and the visual-processing map retrieved the stolen territory and its new/old citizens—the neurons.

Similarly, it was also found that reversible eyesight deprivation caused the auditory-input map to expand its borders and invade the areas in charge of visual input, which was deprived. This further proves that the mutual obligation between input related to a certain sense and a certain brain area that processes it is not tied down necessarily in a monogamous relationship.

Phantom Pains

Phantom pains are common among 95 percent of limb amputation cases, in most cases throughout the course of their life. The person who suffers these pains, strangely, ascribes them to the limb that is no longer part of their body.

The British admiral Nelson, who became an amputee after losing his right hand at the Tenerife battle in 1797, described that he sensed his vanished hand as an invisible entity. He claimed it could be seen as direct proof of the existence of the soul, since the hand that no longer existed in the flesh-and-blood sense continued to exist invisibly. According to his belief, the senses related to his "ghost hand" reflected what happened within his soul following the end of its physical existence.

Phantom pains are ghost pains. These pains are related to the brain, their addressee, which does not accept the fact that no one lives at the address from which the pain is supposedly sent. What seems like a cartographic fixation of the brain is probably an involuntary product of the flexibility of the brain map's outline. The representation areas of the "ghost limb" are conquered and turn into sensorial representatives of other areas in the body. Stimulation of a nearby area revives a ghost sense at the nonexistent limb.

Brain Maps Merger

When a new interface is created between two existing brain maps as a result of a certain experience—for example, an interface between the map that processes input related to touch and the one that processes visual input—a new supermap is created in which takes place the rule the whole is more than the sum of its parts. This is a new neural network in which the two existing brain maps are interwoven. Every new interface between two brain maps brings about changes in the original maps, since it echoes within them, and the echo induces changes in the connection patterns.

Thoughts are matchmakers of brain maps, which bring together thought creatures who merge into a chimera unknown in the mental world. The poets weave innovative creatures and original metaphors through map-matching that conceptualizes the meaning of words. A scientific idea can also derive from an induced interweave of maps of ideas.

Daydreaming can serve as a fertile mental ground for the creation of mutations in brain maps by causing a certain map to rise up into another one and interweaving these maps randomly, more or less.

A map that ceases to function also changes the maps that are mutually connected to it. This is how the auditory sense of people who become blind in the course of life is intensified as a result of improving the processing quality of auditory input in the brain map that processes auditory input. Layers of resolution are added to the auditory map, while the map processing visual input remains unemployed.

Among the deaf, peripheral vision is improved, probably to make up for their inability to hear sounds of faraway objects getting closer to them.

This mechanism can also explain experiencing sexual pleasure in "exotic" ways. For instance, in the sensory brain map the sense representation related to the sexual organ is close to sense representation related to the legs. Perhaps the interweaving of two sensory maps and a partial fusion of these maps can explain foot fetishism—the ability to experience sexual stimulation that results from stimulating the feet.

Other examples from the province of sexual exoticism: Perhaps the neural networking that causes the abstract mental state of "a sense of control," is interwoven due to a repetitive experience into the neural networking that leads to a sense of sexual arousal. In that way the "sense of control" and the sexual arousal are attached to each other among those who experience sado-maso games on the dark side of the sexuality moon.

Even in maps encoded solely for motor tasks, a fusion of separate functions into one map that combines them is possible. For example, music players who maneuver two fingers together while playing music on a regular basis sometimes lose the ability to move each of the fingers separately. The fingers' representation maps fused into one, and its activation causes the activation of both fingers simultaneously.

Resolution of Brain Maps

The brain maps are layered on top of each other, and each of them reflects the information in an increasing or decreasing rating of resolution levels. When dealing with a certain task, we choose a certain level of resolution that, in our opinion, "matches the nature of the task."

The brain is a weariless cartographer. It maps maps of maps of maps in a recursive pattern in which one map is contained in another map which is contained in a more general map, and so on. Each containing map adds encoded information.

The Plasticity of the Maps

The brain is a plastic system. It experiences constant change, but it is not an elastic system in the sense that it never goes back to its previous status, even if it performs a repetitive task or experiences a similar experience. As with Heraclitus's river, we do not have the exact same brain twice. The brain is ever changing.

The choices we make throughout our life derive from our brain, and they, in turn, in a feedback mechanism, affect the brain's structure and its future function.

"Neuro-plasticity" is a term that describes, inter alia, the process in which the world of phenomena around us shapes the function, the number of neurons and the relationships between them. Our brain changes its connection patterns at any given moment of our life.

The brain maps are fluid and ever changing. The constant change of representation areas of cognitive, behavioral and motorial skills is based on the fluidity of the connections between the neurons whose networking pattern is at the base of the skill.

The commitment of a certain brain area to a functional neuro-map, which actually means to the processing of a specific function, compromises its ability to carry the burden of processing information related to another function.

Events that take place in the outer world serve as change agents in the brain. Contemporary media, for example, remaps our brain; TV, cinema and computer games often confront us with a rapid, rhythmical and sometimes even "unrealistic" pace of events. Under the circumstances, the "quicksibility" (a portmanteau of the words quickness and flexibility, which describes the ability to quickly adjust to changing circumstances) are challenged to the edge, and brain processing performance is subject to accelerated depreciation. Moreover, in light of these sputters of stimulations that are poured down on our senses, in "super-realistic" intensity our brain creates reaction networks that are sometimes inappropriate for realistic scenarios and have the potential to cause acquired attention deficit disorder, which is reflected, inter alia, in frequent skipping of attention, a high level of distraction and low mental endurance.

As a cartographer that never sleeps, the brain constantly deals with mental mapping of the world. The four-dimensional mental maps map the three planes of space and the time dimension. They are constantly updated, almost in real time, as a reaction to an input, and enable the brain to create a reliable forecast as much as possible with respect to time and space.

With every new challenge, the brain redraws the roadmap necessary to reach the relevant destination. Part of it determines where to go and the other part determines how to get there, as in "Tell me where and I will tell how."

Neuro-plasticity contributes both to rigidity and flexibility, which is reflected in human nature.

The forming and strengthening of brain maps derive from the use-dependent rule of existence, and, on the other hand, wear and evaporation of brain maps derive from lack of use. Often a brain map is not fully operated, or, alternately, certain parts of the map are more active than others. This changing mix grants the unique shade to a familiar skill we have activated numerous times. Like the changing lighting pattern of a Christmas tree, with different bulbs that are turned on and off at different times and levels of intensity, the activity pattern in different areas of a neural map is formed by activation of the various areas in the same manner. Due to this scattering in time and space, a unique pattern is formed for our activity, which, even in cases of repetitive activities, each time we perform it has a unique signature.

The Brain as a Weariless Cartographer

Learning and De-learning

It seems that numerous learning processes necessarily involve parallel processes of de-learning of previous behavior patterns. Sometimes, these patterns do not match the new behavior pattern or, even, contradict it. Thus, for example, an honest commitment to a spouse requires moderation or even cancelation of our tendency to prefer our wishes to those of the other. Parenthood requires certain altruistic behavior patterns when it comes to the offspring, so their needs become preferable to ours, contrary to the tendencies common before we became parents.

A brain wiring network that is activated frequently becomes more resistant and more difficult to abolish. If it encodes a certain behavior pattern we wish to avoid, the de-learning process becomes more difficult, since the neural, structural framework already exists and confiscates some energy resources from the energy reserve of our brain. It is more difficult to cancel and dismantle an existing structural neural framework, or to weaken the links between groups of neurons, than to lay a new structural framework that will serve as a bed for sprouting of a new pattern of behavior. It is easier to teach an old dog a new trick than to annihilate an old one.

Acquiring mastery of a behavioral or thinking skill is done in at least two stages. First, there is an expansion of the neural networking dimensions that constitutes the brain map that encodes the newly learned skill. The more we practice the skill and get better at it, fewer and fewer neurons are necessary for its performance, and the performance becomes more efficient in terms of accuracy and duration. Faster performance derives from faster transfer of signals through the network that encodes the skill. As we become more proficient, the task is performed more accurately as a result of neural networking that encodes the substantial aspects of the task and contains less and less "white noise."

When neural networks that serve as infrastructure for behavior patterns clash with each other, it might lead to a situation known as "cognitive dissonance." A conscious preference of one behavior pattern over another, for a long period of time, might lead to de-learning of the non-preferred behavior pattern.

A possible "childish metaphor" for such a situation is the building of two Lego blocks models, which are based on the same building blocks. Preference of one model over the other involves the dismantling of the other (if it is already built) and using the dismantled building blocks for the creation of the new model. Our brain resources are limited with respect to the neural infrastructure (despite the huge arsenal of nearly a hundred billion neurons) and with respect to the available energy resources. Therefore, new learning often involves cancelation of older learning and recycling of the old infrastructure for the sake of building a new infrastructure of neurons networking.

It seems that we are able to upgrade the performance of tasks that are important to us by initiating "conscious cartography" in order to unstitch the fabric of neuron columns of the maps that encode the skills that are no longer needed. On the other hand, we should initiate weaving of cortical columns for new brain maps. A process of upgrading existing maps is also plausible by adding resolution to a certain skill by means of weaving additional layers and attaching them to the map that maps the specific function.

Brain Electricity

The Glowing Path—Electric Memories

Thought that brings its owner to enlightenment about a certain revelation is often described by comic book artists as a light bulb above the person's head. Indeed, at any given moment the electric activity in the brain can light up a 25-watt bulb. It is not yet considered an alternative energy source for our world, however. This bioelectric current is at the basis of our ability to preserve memories and create thoughts, and through its course from one neuron to another we can observe the glowing path of memory and thought.

Electrical activity of the brain takes place 24/7. In deep sleep, the activity is characterized by slow, low-frequency waves, which distinguish it from the fast outbursts that characterize the reaction of the neurons to stimulation while we are awake. Some compare the slow activity during sleep to the screen saver in our computers. Our brain cells are active around the clock, and even when experience impressions are not thrown on them intensively, they still retain a basic continuous level of activity.

Electroencephalography (EEG)—Tracing Brain Waves

Standard EEG is a graphic recording of electrical activity in the brain. Such a recording can be compared to eavesdropping to brain sounds through the skull walls; it records shallow activity of the brain rather than activity of deep structures at its core. Electricity is produced in the brain by means of activation of voltage-gated ion channels at neuron membranes.

The graphic manifestation of the electric activity in brain cells is wavy patterns that are called "brain waves." The popular measure for their characterization is frequency. The conventional measurement unit is the number of waves per second, which is measured in Hertz units. A popular classification of the electric activity in the brain, as it is measured on the skin of the scalp in an EEG test, is according to five main frequency ranges:

Gamma Waves Pattern—These describe brain activity waves at a frequency of 35 Hz and above (thirty-five waves and above per second). This pattern characterizes intensive brain activity.
Beta Pattern—These describe brain activity waves at a frequency of 12 Hz and above (twelve waves and above per second). It

characterizes a wakefulness status in most brain areas during everyday activity (standard working status).

Alpha Waves—Their frequency is 8–12 Hz (eight to twelve waves per second), and they characterize sleepy wakefulness and relaxation.

Theta Waves—These are waves at a frequency of 4–8 Hz (four to eight waves per second), typical of light sleep or deep meditation.

Delta Waves—These waves are typical of deep sleep. They are characterized by the lowest frequency: 0-¬4 Hz (up to four waves per second).

While recording brain activity at a given moment, it is common to find electric activities at different frequencies at different brain areas that appear simultaneously.

High-frequency brain waves (beta and gamma) are in correlation with fast, intense brain activity, with maximum concentration and higher levels of the neural mediators serotonin and dopamine, which contribute to serenity along with cognitive vigor. "Rapid thinking"—the status in which our brain produces thinking products at a fast pace—is known to be related to high spirits. (And, here, we face the question of the chicken and the egg, in other words—does the rapid thinking pace improve our mood, or is it the elevated mood that leads to rapid thinking pace?). On the other hand, low frequency brain activity is in correlation with a low level of serotonin and dopamine and is related to low mood and cognitive fatigue.

The EEG graph of a fully conscious, awake person reflects chaos within expected borders. It is difficult to foresee the exact activity-spurts patterns, but they exist within specific borders.

There is correlation between the modes of consciousness and the manifestation of brain waves: the alpha waves are expressed in a status of relaxation when we are awake and our eyes are closed. The theta waves are expressed when our attention beam scatters. The working assumption of most neurofeedback therapists is that the activity of the slow waves, especially the theta frequency, is in correlation

Selected Aspects of Brain Function | 63

with the scattering of attention and mental focus, while rapid-waves activity and, in particular, beta frequency, is in correlation with the focus of attention. When the main melody of our brain is played in an alpha rhythm, we are in a state of relaxed wakefulness.

Among children who suffer from attention deficit disorder (ADD), EEG tests often reveal an intensified activity of low-frequency brain waves (theta) and medium-frequency waves (alpha) in the frontal lobes and, on the other hand, scarcity of high-frequency beta waves.

Among other tasks, the EEG, like a seismograph, searches the brain for an "electric earthquake" of an epileptic seizure. Epileptic automatism, such as repetitive, involuntary movements, is the result of irregular electric discharges that sometimes interrupt with consciousness—our reality interface.

The Bull Experiment—Conditioning and Evaporation of Impulsiveness

In 1965 researcher Jose Delgado carried out experiments in which he inserted electric pacemakers into the brain of bulls. By means of the currents that were inducted into their brain by a remote control, the bulls were conditioned to attack and end the attacks abruptly, in accordance with the signals that were sent to them by the researcher. The bulls turned into a kind of marionettes activated by invisible threads that were actually radio waves that inducted electrical currents in their brain.[6] These studies later led to using chips and pacemakers in the brain in cases of diseases related to disorder of brain function.

The Distance Between Laughter and Crying–a Few Millimeters

Inserting a pacemaker that inducts electric current in specific brain areas is known as a treatment to relieve symptoms of advanced

Parkinson's disease. An experiment in which electrical pace making was performed in an area at the brain core—the subthalamic nucleus (STN)—among Parkinson patients in an advanced stage of the disease revealed that electric stimulation in areas in proximity to the STN by a few millimeters leads to a response of automatic laughter or, alternately, automatic crying. In other words, behavioral expressions that reflect opposite emotions (laughter and crying) were produced from electrical stimulation of brain areas that are only a few millimeters apart from each other.

Mystic Experiences and the Temporal Lobe

A study conducted by a group of researchers from the University Medical School in Geneva[19] in which electrical stimulation of the temporal lobes among healthy participants was performed showed that the stimulation led to induction of "out-of-body-experiences," such as a sense of separation from the physical body. Such experiences are sometimes reported by people who were on the verge of death due to physical illness. In addition, the electrical stimulation caused induction of visions of mystical content.

An Analog Brain in a Digital Era

The brain uses analog signal processing, which means the information processing in it is successive, unlike modern computers that digitally process exact but separated, non-continuous signals. The brain stores the information in a holographic pattern—i.e., in neural networks that exist in the three space dimensions. In addition, brain processing is also time-dependent (manifested as its 'chrono-architecture'), which adds another dimension to the three space dimensions.

The Meridians of the Mental Chi

As energy channels in ancient Chinese medicine, so neural paths in the brain channel the transmission of information necessary for various cognitive processes.

The neuron columns are the main scaffoldings in the building of thinking. The information flows up and down the columns. Brain processing is based on the neurons and the information dance that flows through them and between them.

The perception of phenomena in the outer world, within a uniform domain of time and space as a major component, derives from the perception of ourselves as the reference point. Inner body information impressions, originated in signals sent from our body, are mutually connected to outer impressions. The whole experience is a combination of inner and outer information impressions that are confronted with existing insights and forecasts related to them.

Daydreams

Constantly Active

Our brain never rests, and its energy consumption does not change significantly, whether it deals with a strenuous task or ponders leisurely.

The never-ending activity of our brain, which is like "a city that never sleeps," was reflected in the words of Hans Berger (the inventor of the EEG, which graphically describes the brain's electrical activity). In 1929 Berger said that, according to his findings, it is obligatory to deduce that intensive activity takes place in the brain not only during wakefulness.

In the past, the prevailing assumption was that when we are in deep thought, resting in our rocking chair, our brain is also at rest. However, contemporary studies show that even when we are at a vacation state of mind, our brain cells are full-time workers, and our brain is as active as a beehive. Various measurements, based primarily on quantifying the amount of oxygen and glucose consumption and blood streaming to various brain areas, from which we deduce the level of activity in different brain areas, showed that the "activity gap" between dreamy rest and intense activity is estimated to be less than 5 percent, and it can be said that the level of activity is quite similar in both conditions. Even when our thoughts carry a wandering stick, our brain is as active as when they are focused on an exciting computer game.

The brain is a tough employer; the neurons are employed 24/7. Brain cycles do not rest, and they are active throughout the day, seven days a week.

A main component of general brain activity, which is estimated at 60–80 percent of the entire activity, takes place in the brain cycles, which are unrelated to outer events. This invisible activity constitutes most of the activity in the brain, and the amounts of energy channeled into it are in proportion to it.

The Oasis of the Imagination

When reality is boring at the emotional level (a state of mind reflected in the expression, coined by author Salvoj Zizek, "Welcome to the desert of the real"), our brain craves a sip of escapism that will carry it away from the desert of its realness toward the oasis of the imagination. The "infomania" (tirelessly longing for new information) is also a type of escapism that puts our brain resources in the hands of the ever-changing world manifestations. The need for escapism causes many of us to wallow in the universe of news, which is created and destroyed momentarily—and this is the secret of newspapers and the news flashes on TV, radio, or Internet.

The frequent shifting of glances toward routine nuisances, which flash in our mental field of vision now and again, often prevents us from gazing inquiringly at the core manifestations of reality. The compulsive occupation with everyday nuisances, which consume our time and energy in an attempt to float in the whirlpools of the sea of life, prevents us from focusing on the fundamental issues. It seems that our tendency to focus on the essential issues is higher when we rest under the coconut trees in the oasis of our imagination.

The neural infrastructure that equips our thoughts with a walking stick and sends them to the lands of imagination is called "the neural network of the default system."

The Neural Network of the Default System

At times of rest and relaxation, the brain activates a neural activity pattern that is called "the default network." This network is active whenever the reality manifestations do not knock on the doors of our brain vehemently. This network is like the eternal flame; it is always active, and the height of its flames is in inverse correlation with the level of challenge in which the outer world forces our brain to deal. The more we are focused on an exterior task, the less active the default network is. When dealing with more demanding tasks that require more brain resources, the network's level of activity is low, since resources are needed elsewhere for a more urgent need. The correlation is fixed and nearly linear. The more difficult the task is, the less active the network is, and, on the other hand, whenever the brain deals with an easy task, the network's activity raises its stature.

The default network is mostly active when our attention beam is not focused on the outer world manifestations but, rather, when it is focused inward. The consequence is the inside vision (introspection). In addition, moreover, the default network is active whenever our brain is sailing in the river of life in a dreamy, reflective mood. The transition from directing our attention beam inward to directing it outward is formed in a continuous, persistent pattern, and our attention pendulum moves between the ends according to the circumstances.

The windows of our consciousness are similar to the windows in our life; sometimes the light that comes through the windows washes over our room, and sometimes we look from the inside at the view outside these windows.

The default system's level of activity lessens at once when there is a need to focus our attention beam on something else. It increases instantaneously when the need to focus our attention lessens.

When we "talk business," and are busy with a target-oriented activity, our attention focuses on the interface with the world manifestations, and introspection is abandoned. It might lead us to feel that we "lose ourselves" in the midst of the events with which we deal. Circumstances sometimes force us to be present in the "here and now" rather than in the promising zones of the future, which can be misleading.

A person whose default network is in a state of intense activity might look as if he is in a world of his own. In the provinces of default thinking, the only companion of a man is usually the man himself. The brain is as chatty as always, even when its owner is silent.

Assumptions Regarding the Role of the Neural Network System of Default

One common assumption is that the default system is involved in memory processing and in preparation of the brain for future scenarios. According to this assumption, the default system is a sort of a garden bed that enables the flowers of focused attention to grow in it, instantly, at the time of need, when a sudden stimulation requires a rapid response. In addition, the system is also believed to be a sort of super synchronizer that allocates frequencies and prioritizes activities and, by doing so, enables the various brain areas to talk to each other on their typical frequency ranges, preventing a chaotic situation in which one frequency is mixed with another.

A study carried out on mice that were navigating through a maze showed that the shooting pattern of bioelectrical potentials in their brain at times of rest was a sort of backward reading of the sequence of potentials while the mice were learning to navigate through the maze on their way to the cheese. This is the basis for the common assumption, according to which a rest period is also a "different type" of a learning period. The brain assimilates the information and expands its processing even when we are resting after the learning period. This might be the source of the expression "to sleep on it" as a manner of assimilating information.

Assumptions regarding the survival advantage of this neural network, which is also proved to exist in animals' brains, are related to simulations inspired by it and enables the preparation of the brain for "real situations." According to another assumption, the network is a type of a "field court-martial" of the perception impressions that perform a sort of selection (i.e., determines which of the perceptual impressions will be further processed and which will be abandoned and dissipate) according to the level of importance to the perception of the self.

The default mode in which we are in a weak interface with reality manifestations, in a type of daydreaming, is the comfort zone our thoughts tend to wander to, more and more, as we grow older.

At an old age, our brain is more prone to "attention leakage"; it is more easily distracted and becomes more "flooded." The difficulty of ignoring distracting information that constitutes "white noise" probably derives from the fact that the "enforcement" of attention that takes place in the frontal lobes is less rigid and enables "attention blinking" and division of attention. Then, the brain tends to get lost in the paths of the daydreaming kingdom. But some people believe that the loosening of attention enforcement is actually an advantage; the wandering of thoughts enables, at certain times, divergent thinking and a more comprehensive review of the situation.

Some claim the changing level of wakefulness of the neural network system of default inducts a tendency to use certain behavioral patterns that cause the variation in our response to different stimulations of the external world. At different times, we tend toward extrovert or less extrovert behaviors in accordance with the background voices of the default system. Its activity, which was once hidden, is actually the part of the puzzle that was missing in the past, and its absence often made our reactions seem chaotic. The fluctuations in the level of activity cause certain changes in our responses, which is one of the reasons our brain can be compared to the river in Heraclitus's fable.

Through active tracing of the function of the default system, researchers managed to predict when a certain individual was about to make a mistake in a computerized test more than thirty seconds before the mistake was actually made. The tendency to make the mistake was formed when the default system "took control," and, at the same time, the attention beam, which was supposed to remain focused in order to meet the challenge, was scattered.

It seems that we are dealing with a sort of a mix, and only at rare points our attention pendulum turns sharply and "magnetizes" to a single end—of internal focus or an external one. It seems that at every given moment, the pendulum is somewhere between these two observation posts.

The ability to choose the object of attention and remain focused on it often lessens with the ascent in the years' mountain as we grow older. The stimulations of the external world tend, then, to become more invasive and distracting.

The function of the default system constitutes a supportive garden bed for the germination of raw thinking seeds, and it seems that it has an important role in maintaining the quality of brain function. It might also shed a different light on the tendency of many who consider daydreaming a waste of time. It seems that sometimes it is better to be a bench player in the game of life, since the "bench insights" might come in handy when we go back to the noisy court.

The Mirror Reflecting the Image of the Self

Some ascribe the origin of the "self," the preserving of its identity, and its adjustment to the changing circumstances- to the activity of the default system. According to this approach, the default system is the structural and functional basis of the sense of self-identity. The supposition is that the system preserves a basic pattern of the "self," which is available to us so we do not have to reevaluate time and time again the contents of our self-identity: our memories, values, beliefs, expectations, etc. The eternal flame of the "self" is preserved within the function of the default system. It is like a mirror that constantly reflects the image of the self, and the sense of continuity is preserved due to the constant function of the system.

The Structural and Functional Basis of the Default System

The structural-functional level of the default system relies on areas that are close to the brain's midline. These are mainly the central areas of the parietal lobes, where massive activity takes place whenever a person reminisces about the past, and of the frontal lobes, whose role is, inter alia, to attempt reading other people's thoughts. Activity at these areas, which testifies to the function of the default system, was also observed in the brain of people under preoperative general anesthesia and, also, during light sleep.

The Functional Infrastructure—Superconductor of the Frequencies Band?

The bioelectrical, or electrochemical, signals system in the brain spreads across a wide continuum of frequencies in a range that goes from a low frequency of a single signal per ten seconds (a frequency of one-tenth hertz) to signals whose frequencies are more than one hundred signals per second (more than one hundred hertz).

The chaos in the transportation channels in the brain is prevented due to the priority certain rhythms of signals have over others, as in a hierarchical range. According to a popular assumption, the default system is at the top of the range, and it functions as a frequencies synchronizer. This system functions as the super-conductor of the frequencies band.

The oxygen consumption of the neural network of the default system swings slowly once in ten seconds, approximately (one-tenth hertz frequencies). Such a low frequency was also found in the electrochemical activity of the cortex. It is assumed that it reflects the basic rhythm of the default system. According to this, rhythm acts as the super-conductor's baton in the brain's frequencies band, and it serves as the reference oscillator, or Greenwich point, of frequencies of the brain.

When the Default System Fails

The function of the default system is disrupted among those who suffer from various neurological diseases. Such a disruption was observed in the diseases of Alzheimer's, schizophrenia, depression, autism, and post-traumatic stress disorders. The disruption patterns related to the function of this network vary across different neurological disorders. Thus, for instance, among people who suffer from attention disorder, the attention pendulum tends toward the end of over-focus when external stimulations are present, and, accordingly, the activity of the default system in their brain is lower.

It was found that in the brain of people who suffer from schizophrenia, on the average, the connections between brain areas that constitute the default system do not function properly and the entire system is overactive. As a result, the ability of those who suffer from schizophrenia to focus and react to external reality is compromised. The internal stream of consciousness in their brain is dazzled by inner lights that cast a shadow over the occurrences of external reality.

It was also found that the system is overactive among people who suffer from depression, and attention dedicated to a certain external task tends to slide back to the dominant activity of the default system. Another finding was a lower-than-average connectivity between the areas of the default system and areas in charge of motivation and reward-oriented behaviors. This might be the source of anhedonia—the lack of a sense of pleasure, which characterizes inter alia, a depressive condition. At the same time, the system is overly linked to emotion-stimulating areas. It might provide an explanation for oversensitivity among those who suffer from depression.

Among researchers who study the default system in relation to depression, there is controversy concerning the cause and effect. Are those who tend to daydream more prone to become depressed? Or is it the other way around: depression leads to scattering of thinking and daydreaming?

These findings support the possibility of observing the system's function as a diagnostic tool. A glance at the brain when it is "on vacation" might enable us to assess whether a disease such a depression is being formed in the brain and, if so, what the level of severity is. Remedial interference might help improve the condition of those who suffer from cerebral syndromes related to this neural system.

Memory in the Moonlight

A great part of the enigma related to the role of sleep is still in the dark, but there are multiple pieces of evidence that point to the importance of sleep with regard to information assimilation and memory preservation.

Numerous studies show that during sleep information that has recently been captured is being consolidated in our memory, and new skills are being formed into the neural infrastructure that encodes them. At critical developmental periods, sleep is essential for creating a web of neural maps that represent skills that have been learned recently. During an average night's sleep we experience four to five sleep cycles that last approximately ninety minutes each. In each sleep cycle, we move from superficial sleep (stage 1 and primarily stage 2) through deep sleep (stages 3 and 4) to REM (rapid eye movement) sleep.

Brain-wave frequency at the deep sleep stage (stages 3 and 4 of the sleep cycle) is mainly at the range of the delta waves, which are characterized by low frequency and high amplitude. In addition, the waves' activity at these stages is regulated and synchronized. These stages tend to take place at the first half of the night and often in proximity, in time, to falling asleep. Deep sleep, which is reflected in brain waves, whose frequencies are even lower than the frequencies characterizing superficial sleep, was found as essential for the activity of the hippocampus and for encoding the information our brain is exposed to during wakefulness. Decrease in the duration of deep sleep stages compromises our ability to assimilate new information. In the aging brain, deep sleep stages become shorter and shorter and sometimes even absent altogether. It seems that it is related to the decreased ability of the aging brain to assimilate new information.

In an experiment, a ring was heard at the points where the participants moved from superficial sleep to a deeper sleep, and, accordingly the frequency of their brain waves became lower. The ring woke them up and made them start a new sleep cycle over and over, and, in fact, prevented them from reaching the deep sleep stages. As a result, their memory performance was worse than before, and a functional MRI test showed decreased hippocampus activity. These findings supported the assumption that deep sleep is highly important to encoding new information.

We touch our memories with the fingertips of dreams. It seems that the REM stage, like deep sleep stages (slow wave sleep), contributes to turning short-term memories into a more durable neural web, which actually makes them long-term memories.

One challenging hypothesis that came up in the circles of sleep researchers was that the dreams are like "dustbins" of memories. According to this view, memories are consolidated during REM sleep, and those that are about to be deleted appear briefly in our dreams on their way to oblivion, and this is why we tend to forget our dreams. Nowadays, there are not many supporters for this view.

The Corridor of Experience—between Sleep and Wakefulness

At both ends of night, when we are half asleep, we experience unique modes of consciousness. When the influence of the arrows of Hypnos (god of sleep) has not yet fully diluted in our blood and has not yet fully taken control over our brain, we experience the hypnogogic state. At the other end of the night, when the effect of Hypnos's arrows starts to expire, we experience another unique mode of consciousness—the hypnopompic state. It seems that these are the times when we are at the edges of awareness. Upon awakening, the transition from the sleep-awareness state to the wakefulness-awareness state is gradual and provides a sort of "corridor" characterized by rare awareness states. When we wake up in the morning, the dream narrative is often cut off, and the self-system usually needs a few moments of grace to initialize itself. The self-model does not appear fully on the screen of consciousness, and we are at a stage that might be defined as "partial self." At this state, we feel a little confused and disoriented, and it lasts until the restart of the self-model is completed and it reappears, with all its glory, on the screen of consciousness.

The twilight zone between sleep and wakefulness presents a rare opportunity for us to glance through the porthole at the activity pattern of the consciousness.

Thoughts From the Borderland Between Sleep and Wakefulness

"Thought bewilderment" upon waking up (hypnopomp), before the self-model has had the chance to initialize itself and buy back the golden share of our brain and, at the other end of wakefulness, thought bewilderment prior to falling asleep (hypnagogia) are fertile sources of unique thinking.

The writer Robert Louis Stevenson said that many of his stories were based on hypnogogic visions, including his famous story of the divided self, *The Strange Case of Dr. Jekyll and Mr. Hyde* (1886), which was once described as a story whose basic plot thread was weaved at the junction between wakefulness and sleep.

Between Reality and Dream

Do dreams are a different reality? Or are dreams and reality pages in the same book? This question is ascribed to philosopher Arthur Schopenhauer.[7]

The thin line between dreams and reality is illustrated in a joke about a person who woke up startled from a dream in which he lost ten gold coins. A few minutes later, he came to his senses and decided to go back to sleep so he could search for the ten lost gold coins and find them.

There is also the familiar story from China about a person who dreamt that he was a butterfly. Upon waking, he started wondering whether he was a butterfly in a person's dream or whether what he thought of as a dream was actually the real reality and now he was actually a human being in a butterfly's dream.

In Lewis Carroll's book *Through the Looking-Glass*, Tweedledee and Tweedledum try to convince Alice that she is merely a character in the red king's dream and at the moment the king, who sleeps throughout the story, wakes up, her existence will end at once. Since Alice herself dreams the entire story, the red king is actually a character in her dream, as she is a character in his. This can be referred to as a dream within a dream within a dream—a recursive, multireflection picture.

At the "dream theater" that hosts a new performance behind our eyelids every night, we are the producers, directors, and stage workers. In this sense, studying the dreaming brain presents a keyhole through which we can glance at the backstage of the brain theater. On the other hand, we are also the viewers and the critics of the performance being performed at the hall of our skull.

Studies have shown that we remember approximately 5 percent of the dreams that visit us. Conscious dreaming is a state in which a person gains control over the plot of the dream—he is aware of the fact that he is asleep and dreaming. There are psychological methods, such as frequent reality testing during night sleep, that lead to an ability to be located in this unique consciousness mode. Conscious dreamers are called lucid dreamers, and, once their skill is improved, they are able to consciously, and in real time, write down part of the scenario of the plot of the dream theater that is unfolding in their brain.

According to a scientific article, when we dream about ourselves there is a certain age limit we never pass. It might be our ego that tries to upholster itself with pillows for prevention of shocks during sleep.

Dreamy Activity

Brain scans shows that while we dream intense activity takes place, primarily at brain areas in charge of processing raw instincts and emotions such as the libido and the aggression instinct.

At the same time, there is less activity in the cortex of the prefrontal lobe, which is in charge of regulating emotions and channeling raw instincts into more subtle, restricted paths. Thus, at time of dreaming, our brain is mainly motivated by raw emotions and instincts that are not given the polished touch by the prefrontal cortex. In this sense, dreams can serve as a window to the primeval basis of our being.

The Brain Under the Stars' Light, or Who is Not Sleeping While We are Asleep?

When the lullaby of Hypnos (god of sleep in Greek mythology) envelops our brain and makes us fall asleep, our brain enters a different functional mode. Even when we are asleep, however, certain areas remain active in our brain, which is like a "city that never sleeps."

When Morpheus, the god of dreams, visits us every night at the REM stage, our brain experiences a wakefulness-like mode in terms of the electrical rhythm, as reflected in electroencephalogram (EEG) test. While our eyelids are fluttering, our brain is dipping in the pool of neurotransmitters, which is characteristic of the dream stage of the "night swim." During the REM stage, the amount of dopamine in the brain is reduced and the amount of acetylcholine increases. At the same time, the muscle tone declines. This mode of sleep is characterized by high-frequency, low-amplitude beta waves similar to the state in intense wakefulness. The duration of the REM stages gradually increases as sleep progresses, and most of them take place during the second half of the night. The average duration of REM sleep changes in different periods of life. In the brain of a newborn, who has just emerged in our world, dream sleep constitutes about 80 percent of the overall sleep time, while in the brain of an adult it constitutes about one-quarter of the whole sleep time.

The Stuff that Dreams are Made of...

Some claim that dreams are riddles that our subconscious provides to our consciousness, as Sigmund Freud said in his book *The Interpretation of Dreams* on the verge of the twentieth century (1900). Others claim that dreams are like a dish of fragments of information, pieces of memories and chunks of fantasy, and the boiling of the mix is actually the accompanying soundtrack. Many believe that when we dream we continue to correspond with the events we experienced during wakefulness.

The parade of additional hypotheses regarding the nature of our dreams is a long one.

By the time we reach the age of seventy, we accumulate about fifty thousand hours of dreaming, which add up to five and a half years. A considerable part of our life takes place in this forgotten deposit of creativity.

Timeless Dream

Night dreams are characterized by a low level of activity in the prefrontal cortex.

Accordingly, one of the things that characterize dreams is a chaotic, zigzagging course of the arrow of time, since this particular brain area is in charge of assimilating the reality-matching time pattern in our experiences. This might explain the "mix of tenses" and the merge of past, present, and future into a single dreamy moment.

A Bad Dream

The sleep disorder known as RBD (REM sleep behavior disorder) takes place at the REM stage in which we dream about 90 percent of our dreams. The disorder is characterized by moving of the body, talking, or screaming while dreaming, as a reflection of active physical participation of the dreamer in the dream experience. Total flaccidity of skeletal muscles (atony), which is normal at this stage, is disrupted. Some might claim that the purpose of flaccidity is to prevent us from participating physically in the dream experience, which might lead to physical injury. Due to the disruption, the muscles in our body can move and may cause self-injury or injury to those who share our bed.

This sleep disorder was found to be in correlation with future development of degenerative cerebral diseases such as Parkinson's or Alzheimer's.

A faulty function of neural pathways, originating in the brain stem, is sometimes first expressed as RBD sleep disorder, which foresees degenerative cerebral disorder that is likely to appear within ten years or more from the appearance of RBD.

Post-Love Dreams

The cocktail of neural mediators (neurotransmitters) in the period after lovemaking often brings about a sense of physical fatigue along with mild emotional dullness. In my dreamy experience, on the second night "after," we experience "vivid dreams"—dreams that are perceived as more real than ordinary ones. These dreams are painted in intense colors, as the colors of the reality, and often leave a deeper imprint on our personal dream diary. It is probably related to the acute change in the cerebral mix of the neurotransmitters that is formed during love making and leads an acute change in the brain activity pattern afterward. The journey of the neurotransmitters swings back to the balance point.

Personal Perspective on Brain Function at Night

Personal observation: Many a time, the hour between three and four o'clock in the morning is considered the "dark hour of the soul."

At this time, my brain tends to experience gloomy hyper-realism. Perhaps the one that composes the emotional soundtrack played at this particular time is the right hemisphere—particularly the right frontal lobe, which has a tendency to interpret reality pessimistically. It might be the reason for the fact that people who suffer from depression and anxiety sometimes tend to wake up around this time and find it difficult to fall asleep again. Later on, toward awakening, our mood usually improves and our interpretation of the very same event is usually more optimistic. Perhaps the left frontal lobe, which tends to beautify reality and interpret it more optimistically, is the one that induces the emotional climate at this hour.

A sleeping pattern in which one-half of the brain is asleep while the other half is awake, alternatively, is called "unihemispheric slow-wave sleep" (USWS). This pattern is known in the animal kingdom and was observed among various animals: terrestrial animals, sea animals, and birds. This sleeping pattern was documented, inter alia, among dolphins whose hemispheres are asleep and awake alternatively during sleep. The teleological explanation to this phenomenon might be that dolphins, as cetaceans, must regularly bring themselves to the surface of the water in order to fill their lungs with air.

Perhaps a soft version of this "hemispheric dominance" pattern exists in the human brain during sleep, since, among human beings, there is no phase in which one hemisphere is fully awake while the other is fully asleep. It is possible, however, that during human sleep the two brain hemispheres take alternate shifts, and at every given moment one of them is more dominant with regard to setting the brain agenda. In other words, the mix of "hemispheric dominance" is relative; there is never a complete shutting down of one hemisphere while the other hemisphere is fully active but, rather, intense activity of one hemisphere at a time, like a seesaw that tilts to one of the ends alternatively. This supposition regarding relative "hemispheric dominance" during night sleep has not been satisfactorily proven yet.

The Assumption Regarding the Nightly Journey of the Emotion Pendulum

According to this assumption, the nightly tour in the land of emotions starts at the anxiety and melancholy regions and, later on, toward awakening, reaches the districts of positive emotions. Although it is not an "organized tour" that ends in the districts of happiness, and sometime the course is reversed, it seems that most of the time there is a basic tendency to visit brain areas that are more prone to optimistic interpretation of reality toward awakening as preparation for another day in the battlefront of our life.

The poet David Avidan once wrote, "A man wakes up in the morning, but the morning does not wake up within the man." Perhaps this line reflects a failure in the mechanism of reducing glumness toward awakening.

Emotional Sunrise

It seems that just before sunrise, a wave of neurotransmitters, comprised primarily of serotonin and dopamine, washes over our brain. The serotonin and dopamine are the main ingredients in the "pre-awakening" cocktail that promotes serenity and prepares us for the struggles of the new day that await us beyond the awakening.

Chapter 2
The Perception

Information Filters

A Tango of the Senses and Brain Interpretation—Forever

The philosopher Immanuel Kant claimed in his book *The Critique of Pure Reason* that "The mind can feel nothing; the senses can think nothing; only through their unity knowledge comes into existence." His insights contradicted the popular approach of the time, according to which the manifestations of phenomena march along the paths of senses toward our brain, in which they are perceived as is. Kant was among the first who believed that the sensorial input is routinely affected by the brain's interpretive attitude; thus, our understanding of the world does not represent the world phenomena as itself accurately, and various biases are built into the manner our brain perceives the world.

The Blessing of Forgetfulness

The spectrum of input patterns related to the world of phenomena is infinite. Our brain, which is finite in many senses, will meet just a fraction of these patterns; thus, it is important that, during its term on planet Earth, it conceptualizes and encodes the patterns most significant to it among the ones it comes across. This is why the information filters that are built-in within us are so important. At any given moment, showers of perception impressions fall onto our brain. They are perceived by the senses, which mediate between the world phenomena and our internal being.

The perception impressions mostly evaporate once they meet the brain cells as part of the strict filtering process that is meant to protect our pattern of useful knowledge. The reason for this is that most perception impressions are totally irrelevant or have "low relevance coefficient" for us. Information filters are like the doormen of consciousness, and they prevent useless impressions from inviting themselves into the lounge of our consciousness. The brain, in this sense, is the biological parallel of the coastal lands in the Netherlands, which are protected by high dams so they will not become flooded or immersed below sea level. When it comes to our brain, it is about a sea of perception impressions that are irrelevant for our function.

Malthus in the Era of Information Overload

We are swamped in information deriving from all over, and it seems that this information can be referred to according to the Malthusian principle. It was the economist Thomas Malthus who predicted that, while world population grows in geometrical progression, and, on the other hand, the food-production rate increases only in arithmetic progression, we are doomed to become victims of hunger's disaster. Information, as a whole, increases in geometrical progression, but the part that constitutes "useful information" seems to be increasing in arithmetic progression, and we keep on drowning in the sea of information that floods our consciousness. In this sense, it seems that it will be necessary to reduce the size of the filter's holes through which the brain selects the "useful information."

Reality Mining

The brain is a system that sorts and processes perception impressions in a selective pattern.

The encounter between information and neurons creates the patterns of neural networking that serve as the infrastructure for our unique worldview. In the pattern of creating reality within reality, our brain mines from the mine of impressions, which contains myriad of nature's phenomena, the selected perceptual impressions in order to achieve a reliable picture of reality. This is how our brain attempts to establish a reliable, linear and predictable story out of the zigzagging concoction of nature's phenomena, which is full of contradictions, oddities, and an abundance of surprises.

The pluralism of reality manifestations is reflected in the multiple combinations of the "languages of senses," which are used to describe it. That is due to the fact that our brain is "multilingual" in terms of sensorial experiences.

Our brain is capable of conceptualizing an experience with a changing, multisensory mix including visual, auditory, and textural representations and, sometimes, even through a monosensory representation pattern. On the other hand, the language of description (in the sense of the manner of sensory experiencing) suitable for a certain condition is often nontransferable into a different language of description (in the sense of a different manner of experiencing), since, as in poetry, it is "lost in translation" from language to language. Our awareness of conceptualization qualities, which cannot be transferred from one language of expression to another, leads to a more "accurate" mediation of reality.

We record selective impressions from the internal and external world of phenomena. At least three layers mediate and filter the world of phenomena on the way to our consciousness. One filter is built into the sampling limitations of our senses. The second filter is built into our perceptional conceptualization capabilities, and the third one is formed in focusing our attention on selective world phenomena and derives from our flexibility with respect to choosing the object of attention-focusing.

In other words, at least three filters stand between the external world of phenomena and our consciousness. We have only partial control over the first filter. The size of the sensory-input window can be increased, to a certain extent, by practicing using our senses. We also have partial control over the second filter, which organizes the perception of the sensory input, since part of it is determined genetically. We have more control over the third filter, however, which is formed by tuning the focus of attention, since this is partially a voluntary process.

As aforementioned, an interpretive framework of the world is assimilated in the raw sensory input. The input is sampled through a narrow sampling slot, whose size is determined by the limitations of our senses and through which only selective aspects of the world of phenomena are reflected. The sensory impressions themselves go through yet another layer of filtering of the brain's interpretive framework. Thus, the mediation of the "world" to our perception is biased, and, as such, there is always a risk of a "broken telephone syndrome," as in the game in which a sentence is whispered from mouth to ear in a human chain; when it meets the ear of the last participant, it is mostly different from the original message.

The ability to select the significant input out of the mixture of sensory input perceptions derives from an inherent capability that improves through practice and becomes a skill of directing the focus of attention along with the "sensory attention." It can be done voluntarily, semi-voluntarily, or involuntarily with regard to pieces of information that are "suspected" to be more relevant than others.

The sensory-information-processing brain areas react selectively to stimulations and enable us to separate the sheep from the goats—the words of conversation from the noise of passing vehicles and screams of sirens.

The focusing-of-attention stage is the stage in which our approach to life is manifest—for instance, a tendency to see the glass as half full (or, unfortunately, half-empty). As long as we do not compromise the truth and ignore significant facts, we have mental latitude to select the more positive interpretation of the situation and to focus on the positive aspects of reality.

Attention-focusing, when done involuntarily, is formed as a subconscious process.

Fences to Knowledge

The current borders of our knowledge do not allow us to go beyond the perception "shell" and to prove the existence of a certain entity in a manner that does not depend on our perceptual conceptualization ability. The organizing framework of perception creates borderlines that delimit the information pieces that can be perceived in our brain.

As aforementioned, we glance at the world through a narrow keyhole. The elements that appear before our brain pass through at least three filtering curtains. All of our world knowledge relies on these patterns in time and space, which penetrate through the keyholes, and the three filtering curtains.

There is constant mutuality between the three filters. They influence each other's filtering pattern. Thus, an irregular sensory input will capture our attention, and it will be able to recalibrate our "world perception" pattern.

The perceptual conceptualization system calibrates, to a certain extent, the diameter of the filter holes of the sensory input. One of the experiments conducted in this field demonstrated the mutual influence when an irregular input induces prolonged perceptual changes. The experiment showed that the visual worldview of people who wear prism glasses, through which one sees the world's image upside down (upper becomes lower, and vice versa) for a certain period of time changes the visual worldview of their brain, which learns to perform "inverted" processing. The new perception enables them to see the world as standing on its feet and not on its head. Once they remove the glasses, however, their brain perceives the world upside down once again, and only after a restorative learning period it returns to seeing the world as standing on its feet.

The level of selectivity of our perception is expressed as the diameter of the holes in a fishing net in the following fable.

A fisherman throws his fishing net into the sea. The diameter of the holes in the net is four centimeters. The fisherman notices that the measurements of the sea creatures that are caught in the net daily are always larger than four centimeters. He conducts multiple observations and concludes that the diameter of all sea creatures is always larger than four centimeters. The fisherman attributed his findings and the following erroneous conclusion to the sea creatures and did not even consider attributing them to the fishing net. Metaphorically speaking, there are built-in biases in the nets spread by our senses that are intended to fish the world's phenomena, and it makes some of the world manifestations drop from the grip of our mental perception.

Constant Gap

There will always be a gap between "all there is" in the world and our knowledge about "all there is"—a gap between the actual entity and the entity as it is reflected in our consciousness.

In the language of the philosopher, it might be said that an eternal gap exists between the epistemological (in the sense of nature as perceived by our consciousness) and the ontological (in the sense of nature as is).

Coping with our world's challenges should rely on the appreciation that our world perception and our understanding of the world will always be partial. We always choose selected aspects of the world (often unconsciously), and on the basis of these we compose a comprehensive picture of reality. The compatibility between this picture and reality is on a spectrum that ranges from high compatibility to total incompatibility (when the reality testing component is impaired).

The built-in lack of ability to have a complete knowledge derives from the fact that each point of view dictates sieving part of the information, and what we have is actually a selective observation, which is partial by nature. If we dive into the molecular level of the material, we will expel from our observation field its macro characteristics, such as color and texture. For example, if we focus on this level when we look at water, we will get two gases—hydrogen and oxygen—which lack the "watery" characteristics we are familiar with from our "natural-resolution" level in the world.

Reality Testing

Even when we try to document reality phenomena in an asymptotic accurate manner, the real reality seal can't be imprinted in our perception of the experience.

The intellectuals of ancient Greece searched for universal regularity with a great, almost religious, passion that yielded numerous practical rewards. This spiritual Golden Age is the basis of the outlook according to which our world is just a pale reflection of the real world of phenomena. The path toward the world of universal truths goes in the way of pure intelligence according to the heart-stirring fable by Plato, "Allegory of the Cave."

The fable describes a group of prisoners who were locked up in a cave when they were infants. They were tied in chains so that they could look only at an internal wall of the cave that was full of shadows. Outside voices were heard by them as echoes rolling inside the cave. The prisoners of the shadow cave can be seen as prisoners of their own consciousness, which was forced by circumstances to be based upon shadows and echoes rather than on reality as it is.

We must always walk along the cautious path which winds from healthy skepticism regarding the information that is not perceived by our senses to the knowledge that even the information that is perceived by our senses and processed by the organizing framework of our brain will always be incomplete. In this sense, we are all prisoners of the shadow cave of our skull, witnessing the pale shadow of world manifestations and using them to create our image of reality.

Yesterday's News

When we look at the sky's lantern that illuminates our nights—the moon—we actually see it as it was a little more than one second before we looked at it. (Its distance from planet Earth is about 384 thousand kilometers. The light speed that reaches our eyes is about 300 thousand kilometers per second). When we look through a telescopic lens at the Andromeda Galaxy, we actually witness it as it was 2.9 million years ago when our forefathers were still Hominidae and their brain was significantly smaller than our brain size. We experience the world of phenomena in the past tense. There will always be a delay (of different time gaps) between the time of actual occurrence and the point when our brain perceives the phenomenon and interprets it.

Relative Perception Eclipse—the Blind Mole Fable

The blind mole lives throughout its entire life in splendid isolation, in dark burrows, except for one instance that occurs once a year when it knocks on the walls of its burrow in order to find a mate, of the same species, for the annual lovemaking episode. We probably consider it an animal that suffers from severe perception eclipse, since it is not aware of other species of animals and plants, or the existence of the sun, for that matter. But, we are all "blind mole" with respect to certain aspects of reality.

The selective blindness pattern with respect to certain aspects of reality is, inter alia, age-dependent. In various seasons of our life we tend toward different interpretations of the same sensory input. Thus, most children looking at the drawing *Love Poem of the Dolphins* by Sandro Del-Prete will see only the dolphins. Their conceptual eyes do not notice the couple making love, although the visual pattern is visible to the "material eyes." On the other hand, "standard" grown-ups usually notice the outline of the man and woman in the midst of making love, and, most of the time, they will notice the dolphins only after they are instructed to look for other figures in the picture.

In the human existence and the multiabled cerebral universe, there is built-in ability to reflect only partial reality. Complete reflection and understanding of the world are beyond our grasp.

Thus, we must be humble and suspicious regarding concepts that claim absolute truths.

A partial understanding of the internal and external world of phenomena is built in within the function and structure of our brain. Since we cannot overcome the limitations of our senses, we will never be able to perceive all the layers of the external world of phenomena directly. On the other hand, the perception framework in our brain that relates to internal events is also limited, and it seems that some functions of our brain will remain hidden from the eye of our consciousness forever.

The concept of reality in our perception, as the concept of truth, exists on a spectrum—a changing mix of fictional and true reality. In the best-case scenario, it is a partial truth (with respect to its suitability to reality manifestations). We can say that "subreality," which is information-limited, is the concept that describes our worldview.

Context-Dependent Correction of Input Impressions

Since we are limited by the nature of our perception and can only perceive partial aspects of experiences, our brain completes information pieces that fit the spirit of the overall interpretation it gives to a certain event. Our brain initiates this addition and often does it unconsciously.

Information representation maps are adjustable whenever there are changes in the input for various reasons. These maps can compensate for the disruption of the input using a pattern of overall context-dependent correction.

For example, people who suffer from retinal injury might develop visual distortion called "fish eye," which is actually "rounding" of the peripheral visual input image. The distortion is clearly shown while focusing on single objects. When we look at a sight that is rich in details, such as multicolored scenery, however, the visual distortion gradually decreases and eventually vanishes as a result of our brain's ability to correct the input through "overall context-dependent correction" of the experience.

The above example supports the assumption that our brain enforces "reality sight" on the raw sensory input as it sees it and as it is expected by it. And, in this case, the cerebral-perception mechanism improves our reality perception and creates a better suitability to world manifestations as is. Our brain serves as perceptual glasses that correct the visual impairment of the eyes.

An example of the function of this mechanism in relation to the perception of audio input can be found at times when there is an impairment of the cochlea (the hearing organ within the inner ear), which converts sound vibrations into the language of neural potentials (bioelectrical signals). Following an exposure to intense noise, the function of the hair cells within the cochlea, the ability to experience sounds within the particular frequencies range of the intense noise the ear was exposed to, is damaged. The impairment is especially evident while listening to single sounds, but while listening to a familiar musical piece the impairment is much less evident. It is due to the brain's ability to "bridge the gaps" through "predicted but fabricated input by self-production."

"Filling the gap" with respect to audio input was also part of an experiment in which normally hearing subjects listened to a soundtrack of familiar music. Throughout the experiment, the music was stopped for short periods of time. Many subjects experienced the melody as continuous despite it being interrupted several times, since their familiarity with the music made their brain fill in the gaps of the missing sounds unconsciously.

The ever-changing world manifestations and their reflection as constant change in the sensory input pattern change the calibration of the maps of sensory input consecutively and enable the function of the context-dependent correction mechanism.

The mechanisms in charge of filling the perception gaps and the correction of the perceived perceptual impairment depend on another mechanism. Here we use compensating information from another sensory channel. For instance, people who suffer from retinal injury that causes the "fisheye" distortion might see the outline of a standard pencil as wavy. When the subjects also drag their finger along the pencil, however, the tactile sense enforces straightening of the lines, and, for a moment, the pencil "becomes straight." Nevertheless, the visual impairment reappears the moment the tactile input ends.

New Knowledge Changes our Interpretation of the Past and the Future

New knowledge acquired in the present constantly changes the way we look at the past and the future. A "heavenly" example of that is the acquired embarrassment felt by Adam and Eve with respect to their nudity right after they bit the apple from the Tree of Knowledge. A more "earthy" example is a child who has learned to read and will never be able to see the alphabet as enigmatic ink configurations again.

The Genetic Conditioning of Perception

Different sensory organs create different worldviews in the brain of different animals. Each animal lives in its unique ecological niche. In this sense, human beings, who are spread out over multiple climate zones, are an exception. Animals are required to have unique "world knowledge," which maximizes their survival ability within the specific ecological niche that is their natural territory. Unique sensory organs were created in accordance with the information required for survival. These organs mediate information transfer in a selective manner. Thus, the keyholes through which different animals sample the world are characterized by different diameter and coverage range by which their owners sample the all-embracing panorama of the world's phenomena. The genetically built-in memories also vary among different species.

Each species has its own basic framework of patterning the world. The species are different from each other with respect to the levels of freedom to change the patterning encoded in their DNA.

Our worldview is partially a "genetic dictation." As the toad sees only moving objects and static objects missing from its world's image, We are also driven by certain genetic dictations that provide guidelines to our "worldview." Our brain, which was shaped in the course of "natural selection," makes us organize reasons and processes in regulated frameworks of cause and effect.

The representations of world manifestations in our brain mainly derive from our biology and evolution. A fortifications line of *a priori* suppositions stands between our perception of the world and the world as it really is.

Our inherent perception habits enforce preconditions. We are used to thinking that nouns are usually selected randomly (Juliet tells Romeo "A rose by any other name would smell as sweet"). It seems, however, that at least part of the connections between audio input and visual input rely on a universal platform of a tendency toward certain linkage—a wired, structural linkage that is partially not culture dependent. It was illustrated by the "Bouba-Kiki effect," which was first demonstrated by psychologist Wolfgang Kohler in 1929. Kohler introduced two shapes to his subjects. The outlines of the first shape were round, and the outlines of the second shape were sharp. The subjects were also introduced to two meaningless words: "Bouba" and "Kiki." The subjects were asked to match the "words" to the shapes. The absolute majority of subjects chose to match the "word" Bouba to the shape with the round, subtle outlines, and the "word" Kiki to the hedgehog-like shape.[8] This experiment showed that sounds tend to create in our mind configurational outlines that "match" these sounds more than others. This is a sort of synthesis that is built into our perception. A study conducted in 2001 supports the findings of the previous study and showed that this tendency to link between soft sounds and a shape with a "soft" outline and, on the other hand, to link guttural, more difficult-to-pronounce sounds to sharp outlines is cross-cultural. A 95-percent correlation was found in the responses of English-speaking subjects from the United States, and among Tamil speakers in India. The absolute majority related the sound "Bouba" to the soft shape and the sound "Kiki" to the sharp shape.

A possible assumption deriving from these experiments is that we have an "inherent" perceptual platform for a human language that includes a built-in preference for certain sounds as nouns representing certain objects. It might be seen as a "soft onomatopoeia" (onomatopoeia is a word that describes an object, and it is the sound that the object makes). The findings found in the Bouba-Kiki experiment also reflect a natural tendency to prefer a certain sound-shape linkage to another.

The same inherent tendency to prefer certain sounds as representatives of certain objects, determined by their outline and overall shape, challenges the hypothesis related to the totally random correlation between sounds and objects in human languages—although it seems that this hypothesis is justified most of the time.

A Difficult-to-Pronounce Danger

A study found that nutritional supplements whose names were difficult to pronounce were perceived as more risky to one's health compared to those that had easy-to-pronounce names. It was true also for attractions at amusement parks. The difficult-to-pronounce attractions were perceived as more dangerous, yet more desirable. In the case of nutritional supplements, the danger was perceived as negative; when it came to amusement parks, it also had an attractive component.

The Perfect Picture: Water in Blue

Some claim that we all have, regardless of our cultural background, an inherent tendency to prefer certain artistic features. So, for example, most people from various cultures tend to prefer scenery pictures that include water over other themes, and there is a clear preference for various shades of blue. Thus, even when it comes to visual aesthetics, we share some kind of universal tendency that is reflected in preferring certain color and shape features over others.

Brain Interpretative Bias
Admissible Evidence at the Brain Court

The Evolution sets "reality testing" with regard to the authenticity of facts, which might be defined as "a fact is not a fact until the brain determines that it is a fact." In order to determine the admissibility of a certain fact in the brain court, expert witnesses are summoned. These experts are, in fact, a collection of insights that are etched in our brain as patterns that derive from past experiences of the brain's owner. The accused are the impressions of the senses and the interpretation offered for them by the organizing system of our perception.

An interpretation of perceptual record will be accepted as a "fact" and admissible evidence at the brain court according to its compatibility to the testimonies of the "expert witnesses," and, alternatively, according to the "strength of the evidence." In case of a contradiction between the witnesses' testimony and the "new evidence," the burden of proof will be on the shoulders of the "new evidence," and it will be quite a heavy burden. The brain court is not impartial—it is biased, and its policy is always to prefer evidence that matches experts' testimonies, which are actually encoded patterns derived from past experiences. The reason for this is that many of the past patterns were acquired through sweat and tears, and sometimes through blood as well, and they have accumulated "seniority" with respect to the interface with reality. It can be connected to the spirit of an old saying: "If you worked hard and found something, believe in what you found." Acceptance of evidence that contradicts the witnesses' testimonies will create a perceptual shaking and will require a perceptual change that involves changing, updating, or upgrading the core patterns that no longer match reality.

The human brain is doomed to slip over cognitive banana peels, some of which have universal characteristics. The multilayered reality enables fastidious gatherers to select only the aspects that match their initial assumptions. Thus, as in an illusive magic cycle, truths seem as if they are supported by an infrastructure of evidence. For example, when we think about past events, we tend, in retrospect, to undermine the importance of factors that do not match the final result and overestimate the importance of the factors that match it. This biased thinking is called "creeping determinism."

Our brain often acts according to an approach of preferring the plausible over the certain. The brain's interpretation (in our brain inasmuch as in animals' brains) of the impressions of the senses tends to prefer proximity over accuracy. A statistical proximity suffices to win support. A plausible, somewhat teleological, explanation for that preference might be that a quick decision based on probability is preferred to a slow decision based on complete certainty. The time gap between the plausible diagnosis and the certain diagnosis might turn out to be the difference between life and death. For instance, in an attempt to determine the specific type of a snake we come across, certain identification might take too much time and become our last experience of applied taxonomy. In this case, and in so many other cases, a rapid selection of the preferred reality interpretation is highly crucial. The amygdala—the structure in the brain that induces the sense of panic in light of reality manifestation that might be interpreted as threatening, restarts a very fast response pattern. The operational philosophy structured in the amygdala is based on the assumption that only the paranoids survive.

In stressful situations, when our heart misses a beat, our brain usually does not miss a beat of thinking. On the contrary, the dwarfs of thinking are hurrying in the burrows of our unconsciousness and activate a rapid response.

All is Relative

Interpretive bias that derives from a known truth is that human beings are more sensitive to relative levels of change than to absolute levels of change. There is a famous experiment in which the subjects are asked to dip their hands in iced water and immediately after in cold water. The result is that they perceive the cold water as warm. This experiment shows that past experiences directly affect the perception of the present (a situation that can be called "the recalled present").

In negotiations, we encounter bias derived from the "anchoring effect." The information fed into our brain becomes an anchor to be used in the future. The minute that a certain piece of information acquires a foothold in our brain, it becomes a reference point. For example, during a business negotiation between two parties, the party that makes the first offer has an advantage, since that offer is the basis for discussion.

An example of intended misleading of the cerebral interpretation mechanism is reflected in an experiment, conducted in the Netherlands, intended to improve road safety. As part of the experiment, the researchers creatively used trees that were planted at intervals and became shorter and shorter on both sides of a highway. The intervals between the trees became shorter and shorter as one approached a dangerous junction, and this created in our "tree-counting" consciousness an illusive perception that speed was gradually increasing. As a reaction to this illusive perception, many drivers slowed down as they approached the junction.

This is an example of taking advantage of biased perception for the sake of promoting road safety.

Ego-Sympathizer Bias is Built into Our Perception

We tend toward ego-padding interpretation of phenomena. Built-in bias of self-sympathizer interpretation is typical of our brain interpretation. The brain often tends to choose the most complimenting angle of our profile when we look, supposedly unbiased, at the mirror of reality.

Casinos cast their lots on the fact that people tend to overestimate their self-control and believe that their chances are above average. In this sense, the casinos deliberately defy the commandment "Do not put an obstacle in front of the blind," although, in this case, blindness derives from "free choice."

While we are driving our car, during a routine ride in the metropolis of reality, we all use our built-in shock absorbers. Ego-padding defensive shields mediate between our perception and the potholes of reality. These defensive shields are intended to soften the sharp curves on the reality road and prevent "overshaking" of our perception. We sacrifice the sense of reality road as-is for the sake of easier driving. Psychological convenience is preferred to a more accurate sense of reality potholes.

Ideological Coercion on Reality Manifestations

The chronicle of astronomy is full of fascinating examples of enforcing a brain's worldview, originated in idealism and perception of reality as perfect and flawless, on the factual reality of the world of phenomena. In this spirit, the preference was given to thinking creatures that were validated only in the "lab" inside the skull and were not confronted with reality manifestations and with impartial examination of facts. It was true with respect to the ultimate "truth" that prevailed for two thousand years: that of Aristotle's geocentric model, which puts planet Earth at the navel of the universe. Nicolaus Copernicus disagreed with this theory on behalf of the facts and developed the heliocentric model (which puts the sun in the center) and was forced to keep silent during his lifetime.

Another example is Plato's idea that the planets were moving in perfect circles—until the voices of the facts spoke from the throat of Johannes Kepler and refuted the dogma. Kepler "ellipsed" the circle by proving that the course of the planets is elliptical rather than round.

Optical illusions often rely on enforcing the organizing framework of perception on the raw information perceived by the eye. In other words, the right to have the last word or, in this case, the right to have the last sight is kept for the brain. Optical illusions serve as an additional example to the assumption that perceiving the pure objective status of reality is a utopian vision that cannot be attained by our brain.

The optical illusion called the "Rubin Vase" presents the conflict of whether it is a vase or two profiles—a single visual input that has two possible interpretations. Each of the interpretations is typical of a different action pattern in the brain, as is reflected in functional MRI imaging. The visual input does not change, but its brain interpretation changes and sheds different light on the information. The functional parallel of the interpretive difference is, as aforementioned, a different pattern of cerebral function when different brain areas function in different functional intensity.

Both interpretations are appropriate, but only one of them is possible for the brain at any given moment.

At any given moment, our brain "chooses a world" and validates a certain worldview out of the various options.

Senior insights also control our expectations from the sensory input. During a famous experiment that was carried out in France, tasteless and odorless red liquid was poured into a glass of white wine that was served to professional wine tasters. The tasters described the wine's quality in various terms related to red wine, although the red liquid did not change any of the white wine's characteristics except for its color. It seems that we evaluate taste also through sight, and even project our expectations on the sensory information, so our perception might be biased and less reliable.

The Framing Effect

Thinking bias, which is related to the framing effect, derives from the fact that the introduction of information, which means the "frame" in which it is framed, has a great effect on how we refer to the information.

An elegant experiment, conducted by researchers Amos Tversky and Daniel Kahneman,[9] assessed the framing effect among medical interns in hospitals in the United States. The interns were presented with real data regarding mortality, within five years from the day of the diagnosis of cancer, with relation to the treatment given to the patients: operation versus radiation therapy. Both groups were presented with identical data that were phrased differently. One of the presentations referred to the mortality rate, while the other referred to the survival rate. One of the groups was told that the survival rate of the operation is 60 percent, while the other group was told that the mortality rate as a result of the operation was 40 percent. The manner in which the data were presented led to a distinct bias in brain processing and, as a result, also affected treatment recommendations.

A similar phenomenon was also found on the other side of the fence among patients. It was found that the tendency of patients to agree to undergo an operation is two times higher when the survival chances are referred to in terms of an 80-percent success rate as opposed to describing the same data in terms of a 20-percent mortality rate.

A possible explanation for this cross-cultural human thinking pattern is that presenting data in negative terms activates a mix of brain structures in which the amygdalae get greater weight, and when data are presented using positive terms the activated cerebral pattern is different, and greater weight is put on the frontal lobes.

Another type of thinking bias derived from the framing effect is reflected in the fact that, in our day-to-day life, our brain tends to ignore the rule "Don't judge a book by its cover." Subjects in an experiment were given samples of identical wine, which was poured from bottles that had two different price tags. It was found that the tasters consistently preferred the wine samples poured from the bottles that had the higher price tag, as in the old saying "Life is too short to be drinking cheap wine." A similar experiment was conducted with medications. The subjects were patients who had recently undergone a medical procedure accompanied by pain. The patients were offered painkillers. One package of painkillers had a price tag of two and a half dollars; another package, containing the exact same medication, had a price tag of ten dollars. Sixty percent of the patients who were offered the "cheap" medication reported relief, while 85 percent of the patients who were offered the "expensive" medication reported relief. Our expectations have a direct impact on our experience.

Eyes Right!

Inherent spatial preferences exist in us. People with right-hand dominance tend to turn right at unfamiliar buildings, parking lots, or junctions. Thus, if you wish to choose the shorter queue at the bank or the supermarket, you better go for the one to the left, since fewer people tend to choose it.

Aspects of Processing Perception Impressions

The question how single sensory impressions are merged into a mosaic of unified experience is a riddle that has not yet solved sufficiently.

We project our expectations onto the world all the time. These expectations represent the patterns created in our brain as a result of past experiences. The perception of the present is at least partially determined according to past experience and our forecast for the future.

The input skips from one sense to another and from one area of information integration to another. When our dog barks in steadily growing volume, our brain predicts a "visual prediction" regarding its figure approaching our doorstep, and we are in readiness for its image to fall on our retinas. The prediction of a future input relies on patterns of past memories etched in our brain. These patterns were created due to the consistency of the sequences of signals conceptualizing them. The repetitive correlation patterns that compose the sequence of features of the phenomena are at the basis of the interphenomenal compatibility and, in this case, at the basis of the correlation between the barking sound and the sight of the dog.

Watching a film, like listening to music, is not conducted through a passive pattern, since our brain is constantly busy, consciously and unconsciously, projecting its expectations on the screen of consciousness. We rely on external clues with which we confront the creatures of our brain—memories and experiences from similar past events. We predict future scenes from this synthesis.

"Necessary and Sufficient" Sequence of Signals

Mostly, at a certain point in time, the momentary, specific input does not contain enough information to identify the overall pattern. In order to enable the brain to identify a "pattern," which means identifying a selected piece of information as an object or an entity, we need information that was perceived at a critical period of time. For example, the first syllable of a word does not identify a word, and additional information is required, which takes more time for mental processing, which will then enable the identification of the word. In order to perceive the meaning of the information, we must document a sequence of input signals for a unique and sufficient period of time. For example, we cannot identify a melody according to a single sound. A number of sounds may constitute a sufficient and essential unit of information. Information processing relies on pillars, similar to a proof in geometry. We need "necessary and sufficient information."

The Relevance Coefficient

Our brain is programmed to act according to a key of "relevance coefficient." It constantly prioritizes information pertaining to the immediate present and the interest map perceived as best promoting survival. What is less relevant, in this sense, gets fewer processing resources and less "brain time." In this sense, the brain is like a tough banker of mental energy.

For instance, when we experience physical distress, our freedom to concentrate on mental contemplation is taken away. The resources are expropriated from abstract thinking to the "here and now" in an attempt to improve our immediate situation. The brain is a trade-off expert that tends to constantly prioritize the maximal rewards in terms of survival.

Determinism and Randomality Under One Roof
Reality Has a Multifaced Narrative

The Chaos Theory and the Fractal Geometry Theory demonstrate the complexity of "both this and that," which is typical of the world of phenomena. In certain aspects of its being, this world relies on determinism, and, in other aspects, it relies on randomality. Both aspects live under the same roof. It seems that reality resides in between determinism and randomality.

As part of reality rewriting, intended to find the cause and the effect, our brain often acts as "a representative on behalf of" that is characterized by built-in bias related to impressions it perceives from reality and is not acting as an impartial detective for whom the truth is a guiding principle and who is determined to shed light on a mystery.

It seems that the system that organizes our perception guides us to perceive the world as less random than it actually is. We tend to view world events through a deterministic mirror in order to apply cause-and-effect rules.

Our Brain Faces Challenges of Probability

Our brain sometimes tends toward a bichromatic simplistic life view ("black and white") and finds it hard to deal with shades of probability. The language of statistics is not its mother tongue; when it speaks this language, it often makes grammatical mistakes and its foreign accent is noticeable. Studies show that our thinking tends to overestimate scenarios of low probability and, on the other hand, underestimate scenarios of high probability. This tendency is intensified in emotional circumstances.

In many cases, we do not act as successful intuitive statisticians. On a day-to-day basis, we often tend to ascribe probabilities to anticipated events mostly according to the emotional effect they have on us.

Our risk management philosophy is largely determined by the emotional system in the brain. The voice of reason (a product of brain areas that are in charge of rational analysis of reality) is often a second voice, although the desirable ratio is often opposite. The vibration of the emotional cords often surpasses the sounds of reason, which are played according to statistical inference and risks analysis.

In various gambling games we tend toward an extremely optimistic estimation, which is unjustified and unproven. In most common games of chance, our desire to win is usually statistically groundless, but our brain prefers wishful thinking to statistically proven thinking.

When we wear glasses with a probabilistic prism, we are less prone to be caught in the trap of a perceptional illusion. Watching the world through the probability prism enables prediction and recognition of the limitation of our ability to know.

Since the entire circumstances of our life, except for the start and ending points, involve various levels of uncertainties, existence "in the shadow of uncertainty" is to be expected throughout our life.

The Brain Versus Uncertainty, Randomality, and Determinism

Randomality at the Core of Reality

Are coincidences the hidden footprints of the wisdom of nature—which, as we know, likes to hide? Are they the tricks of God, who wishes to remain anonymous? Or are they the representatives of randomality, whose threads are woven daily within the fabric of our lives?

Divine dice: Determinism, derived from Newton's physics, greatly chilled the warm belief in free choice, but quantum mechanics, according to which determinism does not control the world of particles somewhat restored the blush to the cheeks of this belief due to the possibility of free choice, although it is challenged by uncertainty and randomality inherent in the world of phenomena.

Einstein, who disagreed with the spirit of randomality built into the world of phenomena that blow from the bellows of quantum mechanics, said then, "God does not play dice." However, it seems at least once in a while God does play dice. The conceptual revolution derived from quantum mechanics made it clear that the lack of ability to predict the future accurately does not depend on the person who makes the prediction and his knowledge gaps, but it is an inherent part of the system that is being predicted. Philosophers might put it this way: uncertainty is ontological—inherent in the situation itself—rather than epistemological—dependent on human perception.

Quantum mechanics is known for making crystal balls turbid. An applied definition of quantum mechanics is a scientific branch whose results are completely opposite to the ones we predict. In the world, where it exists, randomality has a place of honor. The capricious nature of circumstances seems like "natural law" in light of it.

Our brain is only partially prepared for the inherent uncertainty in our life, as reflected in the known saying: "Life is what happens to you while you're busy making other plans."

The Randomality Within

Randomality is also built into our "randomality scattering machine"—the brain. Our brain is intended to scatter the vagueness and randomality in the world of phenomena around us, but randomality also resides within our brain. While facing the randomality outside, there is the randomality inside—within our thoughts and reactions. Which component of our personality will take control at a certain moment? How predictable are we? We all know the expression "I surprised myself."

Sometimes, we are impervious to ourselves. The "inner mirror" does not reflect us, and we do not understand the motives behind our own behavior.

The saying—which sounds like an oxymoron—"All is foreseen but freedom of choice is given" means there is certain flexibility within determinism. In other words, although our life has defined boundaries, there is a certain amount of freedom within these boundaries. The complexity of our brain makes the prediction of our behavior difficult, even for the conscious self. Often, there is no one captain holding the wheel on the ship of our brain, but there are many captains who hold the wheel alternately in unpredictable patterns and for unpredictable periods of times.

The potion of neurotransmitters that will intoxicate our brain in unusual circumstances, and the responses it will create, might come as a surprise, even to us.

A possible, somewhat gloomy, inference is that we were not born with pure free will, but, rather, in the best-case scenario, we are blessed with a touch of free will merged in gallons of dictations based on genes and environment.

However, in order to protect our mental health, even at the unconscious level, it seems that we obey the ironic claim of the writer Isaac Bashevis Singer, who said, "We must believe in free will—we have no other choice."

In his old age, Einstein tilted at windmills by trying to refute the inferences of quantum mechanics. This theory of particles promotes the idea that randomality is built into nature and objectivity is nothing but a mirage when dealing with material reality. Randomality is also the core process of evolution, which enables, by the formation of mutations (random changes in the genome and the proteins that derive from it), the change in various species. But, despite the fact we cannot reach complete objectivity, and despite the omnipresence of randomality, we manage to land spaceships on Mars with a high level of accuracy, according to Newton's deterministic rules. Thus, we might say that the distortion in our prediction ability, which derives from the omnipresent randomality, does not necessarily mean failure as the final outcome. Often a deterministic, approximate estimation can serve as satisfactory bed for successful prediction about the future.

The generic patterns in our brain enable the brain to devise future scenarios, and, in this sense, it is rather a successful prediction machine, despite the uncertainty surrounding us.

Many opportunities in our life are not promoted by aggressive marketing. Sometimes, the front door to the rainbow is painted in gray.

Multiple discoveries and inventions have a shared parenting of an alert brain and blessed randomality. As scientist Louis Pasteur once said, "Chance favors the prepared mind." Chance is a hidden partner, more or less, of many spurts of thought derived from "happy accidents." Serendipitous discoveries (discovered by chance, unintentionally) are like a muse in the carriage of randomality. Serendipitous revelations constitute an important component in science. (Discoveries such as penicillin or the effect of Viagra on male erection.)

Some see a lucky event as a loan from the bank of destiny, while others see in it the fingerprints of Fortuna, the patron of luck. Thus, for example, one can observe Fortuna's dance with each arm of Paul the Octopus, which enabled it to guess the winner of each football match in the 2010 Mondial—out of the eight matches that were presented to it. In practice, it was presented with boxes that contained food and were covered by the state's flags of the competing teams, and its choice of specific boxes were interpreted as predictions of the winner.

Fortuna's touch is the magic touch of good fortune, which can sometimes present a huge opportunity without us paying the price for it. Such a lucky event is part of many biographies of exceptional, successful figures, and it is often the key to understanding their exceptional success.

Sometimes, the so-called randomality can be explained. An example of this is "Littlewood's Law." John Edensor Littlewood was a British mathematician whose insights led to the "miracle law," according to which any person can expect to experience a miracle at the rate of about one per month. The law was intended to refute the existence of phenomena considered supernatural and is related to the law of truly large numbers; according to this, when there is a large enough sampling size, highly unlikely events will occur in high frequency. The assumptions underlying Littlewood's Law are explained as follows. Let us define a miracle as an event that occurs in a frequency of one in a million. Let's assume that during the hours in which a person is awake and conscious, he experiences one "event" per second. Assuming a person is awake and conscious eight hours a day, he will experience one million and eight thousand events in thirty-five days. In accordance with the definition of the word "miracle," a person can expect the occurrence of a miracle once in thirty-five days.[10] Alternatively, there is a version that ascribes ten seconds to an event; thus, the frequency of miraculous events is reduced, and the average frequency of miracles becomes one miracle per year.

In reality, most decisions have to be taken based on partial knowledge, and the percentage of the uncertainty component varies since "certain uncertainty" is an integral part of our life. Uncertainty is a basic quality in the world of phenomena, as it also derives from the uncertainty principle in quantum mechanics.

While our brain predicts the future as an astrologer, it is worthwhile remembering the flaws in our prediction ability. The black swan is a concept that wandered from the lands of philosophy to the lands of social studies and economy and became a common term for describing an irregular event that is unpredictable in the horizon of our consciousness. Our prediction ability ("crystal ball skills") mostly relies on our memory (the ability to look back to the past) and the momentary perception (the ability to see the present as it is). However, since the skills related to the documentation of past and present are only partially reliable, foggy clouds obscure the future visions in the crystal core between our temples. When the past casts too long a shadow over our expectations, we tend to look at the mirror that reflects what is behind us, mistakenly thinking we are watching in a glass panel that reflects what is in front of us. Our imagination sometimes finds it difficult to come up with reliable previews for our future feelings.

"Absolute determinism" is a term that acquired foothold in the claims of the nineteenth-century French mathematician Pierre-Simon Laplace, who said that when the location of all objects in nature is known, as well as the natural forces that operate among them, an intelligent entity to whom this information is disclosed might be able to predict all future events. Nowadays, however, the common belief is that predications will never be perfect due to the randomality derived from the quantum mechanics theory.

The present projects itself onto our foresight. Most people find it hard to imagine a tomorrow that is very different from today. When we are satiated, it is difficult for us to imagine ourselves gulping food at fast-food restaurant. It is hard to imagine the taste of strawberry sorbet while chewing a salty bagel.

Our insights about the world do not contain full knowledge but, rather, "sufficient knowledge" to perform the various tasks of existence. As a result, our brain often operates in a pattern of a heuristic machine. Translating complex aspects of reality into simpler, heuristic mental schemes is, in a sense, a flattening of the depths of reality. But, in practice, this flattening enables the performance of tasks for a low energetic price (in mental energy coins) and with accuracy sufficient for survival.

The brain has its own version of the Cassandra syndrome. Apollo bestowed upon Cassandra the ability to see the future. Once she rejected his courtship, however, he cursed her and told her that despite the truthfulness of her predictions, no one will listen to them. This is how it went when she predicted the fall of Troy. Underestimating the prediction ability of the brain that results from overestimating its inherent limitations is reflected in the sense of missed opportunity we sometimes feel as we say "I knew it"—the feeling we have in light of an event we predicted but did not act toward practical implementation of it.

Partiality of Knowing

The term "fallibilism" refers to an approach saying we have an inherent inability to achieve knowledge at the complete certainty level. Or, as philosopher Karl Popper once said, "Our knowledge can only be finite, while our ignorance must necessarily be infinite."[11]

It seems the best understanding we can aspire to will be found at a distance of the "endless millimeter" from complete understanding; although it seems reachable, as it is an asymptote we will never be able to reach it.

Knowing there is no such thing as full knowledge is a sobering insight. Although the light puddle of the candle of science expands its boundaries consistently, the darkness of not knowing will forever be part of our life.

The origin of the logic discipline is found in Aristotle's rules of reason from the fourth century BCE. Logic is deductive by nature (it applies inference from the general to the particular) and deals with arguments that are based upon premises affected by derivation and inference rules. This is even though most of Aristotle's science was not supported by evidence but, rather, on "belief of the heart." As opposed to it, Galileo Galilei's science was based on experiments and was the basis for the empirical approach of modern science.

Belief, which is sometimes referred to as "knowing of the heart," is characterized by a lower level of certainty compared to evidence-based knowledge that was validated through an experimental (empirical) pattern.

Many of our brain's reality conceptualizations are formed according to the logical spirit of the scientific, empiric method, which means generalizing inference based on systemic accumulation of private cases. But the Achilles heel of this "reality clarifying" approach was pointed out by the philosopher David Hume, as early as the eighteenth century. Hume claimed that generalizations derived from private cases contain more information than the information that exists in the observations on which they are based. Does an observation of a thousand elephants, which are all gray, lead to the conclusion that all elephants are gray? Could it be that a group of albino elephants was hiding at a thick forest at the time of observation?

In empirical science, there is an inherent difficulty in reaching undoubted conclusions regarding the universal validity of the phenomena. According to philosopher Karl Popper, there is no scientific theory that cannot be refuted and theories are valid until proven otherwise. We all know the example of the hen that was devotedly groomed by its owner for many years until, one morning, it was sent to the slaughterer—a development it could not possibly predict based on its past life experience.

It seems an essential condition required from a scientific hypothesis is that it will be refutable, but shall remain valid until proven otherwise. Science, in this sense, is a succession of new hypotheses' building, which was built on the ruins of refuted hypotheses.

Jewish resources reflect the complexity of knowing. On the one hand, it is said, "Wisdom excels folly as light excels darkness" (Ecclesiastes 2:13); in the same book we find the quote "For in much wisdom is much grief" (Ecclesiastes 1:18).

Galileo Galilei added atypical restrictions for himself to the pursuit of knowledge when he said, "We – the human beings, as the owners of finite minds, should not attempt to deal with the infinite."

An ancient picturesque metaphor originated in China describes, in a simple, graphic manner, the ever-changing relations between the questions that were answered by human knowledge and the unsolved questions: the answered questions are presented as the perimeter of the circle, while the unanswered questions are represented by the area of the circle. Such representations demonstrate that, as the perimeter of the circle expands as a reflection of expanding human knowledge, the representation of the unsolved questions expands as well, since knowledge increases the number of questions regarding the unknown.

Aspects of Dealing with Randomality

The brain is constantly looking for patterns of order in the world of phenomena. Our brain is like Galileo's; he threw objects from towers, shook pendulums and rolled balls downhill in order to identify repetitive patterns, in the world of phenomena, that can be phrased as natural laws in an attempt to weave threads of law and order in the bulb of capricious manifestations of the world of phenomena.

The edge points of belief are reflected in the words of physicist Stephen Hawking, who said, "I have noticed that even those who assert that everything is predestined and that we can change nothing about it still look to the right and to the left before crossing the street."[12]

Many people who claim to have deserted their belief about the existence of the soul often look back, as did Lot's wife, at the so-called deserted belief. This is well described by the saying "There are no atheists on a falling plane." People usually navigate their belief in a pragmatic manner, as required by the complex circumstances of life and the way they are reflected in their inner world.

Loyalty to a certain view "until the end," without adjusting it to current circumstances in a manner of constant "reality testing," actually reflects an "adjusting disorder" regarding life.

Disciplines of knowledge, such as engineering, aspire to minimize the randomality component while relying on pillars of solid, factual information. When such disciplines interact with the world of phenomena, with its built-in randomality, the expectation is for an effect that will reduce the "average world randomality." It is so we add a "low-randomality-rate" component, reflected in the engineering products, to the randomality component of reality. Nevertheless, even if randomality might be reduced, it will never cease to exist.

Our brain has a hard time dealing with nonlinearity. On the other hand, we feel comfortable with linear causality. The "noise" that is the offspring of randomality tends to converge to the average over a long period of time.

The waves of randomality approach the shores of our life routinely. In order to reduce the risk of a tsunami hitting the shores of our life, flooding randomly, we must be extremely cautious when we approach the undisputed territories of the kingdom of randomality (such as the stock exchange). The demon of randomality might not be tamed, but it seems it can sometimes be locked in a bottle.

When our brain comes across life circumstances that pass a certain threshold of randomality (an intense randomality ingredient in the cocktail of our life circumstances), it might produce a chaotic thinking output. Our emotions tend to take the reins, and rationalism is tossed aside. One way or the other, the waves of randomality will wash up on the shores of our life. If we are cautious enough, however, and avoid reaching climate zones where the waves of randomality create a tsunami, we will statistically reduce the risk they will unexpectedly wash over the shores of our life.

Looking at the "whites of the eyes" of randomality from a short distance is a situation that brings about rapid emotional burnout.

We lay the fingers of our attention on life appearances in order to check the "world's pulse," but, due to the variance among these appearances, when this "pulse checking" occurs too often, we are exposed to an overdose of randomality and white noise, which shocks our consciousness. The more we sample life phenomena, the more we observe variance and the less we observe the core of the phenomena. Overexposure to randomality wears out the armor of emotions, and anxiety penetrates between the cracks, which is revealed and becomes a frequent visitor in our emotional climate living room.

Our mental wellbeing will benefit from taking some time off from constantly dealing with an environment rich in wear-out randomality. It seems that our brain is more intended to act as a firefighter, who puts out fires whenever they break and enjoys rest periods in between, than as a stock exchange broker, who is supposed to respond to the constant changing of stocks graphs in real time and at any given moment. We need our share of escapism, which enables our brain to react better in the next confrontation it will be forced to deal with. An attempt to reach constant optimization does not come naturally to our brain and is often doomed to failure accompanied by "cerebral burnout."

The Brain as a Machine Intended for Prediction and Dealing with Unfamiliar Input

Predicting the future is at the basis of our intelligence. Predicting the possible outcomes prior to acting is a survival tool that, as philosopher Karl Popper phrased it, "let your assumptions die instead of you."

Our brain serves as a crystal ball that tries to predict the future, as the famous science fiction author, Isaac Asimov, attempted to become a historian of the future.

The operation pattern of prediction in our brain is reflected in its producing possible future scenarios that attempt to foresee the future. In this spirit, the philosopher Baruch Spinoza once said that "by using imagination and reason we turn experience into foresight."

Past Memories and Prediction of the Future

There is a fascinating claim according to which the cortex can be defined as a system whose products are shaped by two main forces: memory and future prediction.

Our consciousness is like a virtual time machine that sends us to tour the land of the past and the land of the future while our body exists in the present.

One can conceptualize the brain's thinking operations as a memory system (that stores information about past experiences) that is the basis for extrapolation and predicting the future.

These two abilities mutually project on each other. Relying on past memories for the sake of predicting the future is a central processing pattern. Prediction enables the brain to place the predicted scenario in the field of the possible situations by combining memories of past experiences with updated pieces of information.

The course of sensory input passes through the sense organs upward to the areas of sensory input processing in the brain. At the same time, our expectations, derived from "world patterns" encoded in our brain, are thrown downward, toward the sense organs, and affect both the incoming input and the processing of the input in our perception.

Between the upward flow and downward flow of perception, there are feedback loops that feed the input with output that predicts the expected input. At the same time, the feeding of updated input constantly updates the output that predicts the future input. The creation of a future memory, which is an expected-future scenario, mutually affects the past memory in a pattern of backward bias.

Brain prediction is often based on probabilistic assumptions. Whenever there is an occurrence that does not match the prediction, the focus of attention turns to it immediately in an attempt to clarify the reason of deviation.

To Know Yesterday What is Known Today

"Life can only be understood backwards; but it must be lived forwards," said Soren Kierkegaard.

The concept of eternal recurrence is expressed in the book *Thus Spoke Zarathustra* by Friedrich Nietzsche. In this book he proposed an experiment in which a person relives his life over and over in perfect accuracy.[13] Was it an unperceivable torture given the fact that he was not able to apply improved insights, derived by past experience, in order to change his actions for the best?

The pleasant (sometimes frustrating) conceptual exercise that aims to reveal how acquired insights can improve our performance when we are faced again with the same type of challenge is described in the film *Groundhog Day*. In this film, the hero is caught in a time net and is forced to relive a specific day of his life over and over again until he improves his conduct, after which the arms in his life's clock will circle again. The memory of the experience—reliving the same day over and over again—and the acquired insights help the hero improve his coping, and he goes on to win, picking the fruits of experience and memory. Finally, an escape route from the vicious magic cycle of endless repetition is revealed. The ability to apply acquired insights is at the basis of turning one's brain into a "life's expert brain."

Prediction of Inputs

Our momentary perception is created from the actual input and our predictions regarding the input. When our input is partial and lacking, our brain "fills in" the gaps with predictions based on the probability of events, which is based on patterns of past memories.

Our expectations affect our interpretation of the input in real time. The prediction spreads out to the various modes of input; the predictions take place in relation to all input patterns, such as visual input and audio input—all at the same time. At the single-cell level, prediction is expressed by the start of activity within the neurons before they are activated in practice by the input. The tracks of perceptual input are two-way tracks. Expectations regarding the next input pattern that goes downward are constantly fed at the same time that input signals from the sense receptors move upward to the brain. This is a process of prediction according to input patterns that were perceived almost at the same time, combined with familiarity with similar input patterns that were encoded in our memory. The prediction takes place in various input paths: prediction regarding the next sound we will hear, the next sight we will see, etc. The neurons in charge of mediation of the sensory inputs are activated before the pattern of input signals actually arrives; when it finally arrives, a comparison is made. If the input matches the expectation, it is processed in the brain at the level in charge of routine events. When the input does not match the expectation, however, it is transferred to another brain area that serves as "exceptions committee," which attempts to comprehend it and at the same time find similarities with rarer memory patterns known from past experience. When there is no compatibility with a certain past pattern, the new input is classified as the "new kid on the block," and its sequences are learned anew. It seems that the prediction ability, which attempts to predict the next input pattern in advance, is an essential ability in the repertoire of our intelligence qualifications.

The process of expectations projection can be seen as the opposite of a flashback; it's a type of flash-forward as an extrapolation of knowing the present. Most predictions take place out of the consciousness's area of jurisdiction. The predictions are often based on approximate statistic estimates. The preeminence of man, with respect to intelligence, largely derives from man's ability to make predictions regarding types of patterns that are more abstract and those that have a longer series of patterns. When the action is successful, and the neurons predict the pattern of incoming input correctly and repeatedly, the sequence of the input patterns is fixated as a series and becomes an ongoing representation (which is, in fact, memory in its constant figure—the engram). Learning the sequences of signals (the patterns) and their inclusion as series is an essential factor in the creation of preserved representations.

Circumstances—Dependent Matching as a Basis for Changing an Action Plan

Welcome to Realistan. When we are involved in daily "survival tasks," our predictions aspire to grip the grounds of reality. Cycles of feedback constantly recalibrate them in an attempt to reach maximal compatibility between them and the outline of reality ground. Our inner oracle constantly attempts to predict the secrets of the future and answers the eternal question "What happens next?" with changing answers according to the direction of the wind vane of circumstances.

The brain attempts to predict the next reality appearance it is about to meet at any given moment. At any given split second, a number of possible courses of action are proposed by our brain. The top level of processing chooses one of them, and the action starts, according to the course of the winning proposal, down the slopes of the neuron column that planned the policy of the "winning proposal." At any given moment, all neuron columns undergo adjustment in order to improve their level of attractiveness in accordance with data from the external world and inner feelings. For instance, when a task involves catching a ball, several proposals are made for activating body muscles. The premotor cortex chooses one of these proposals and prepares a "selected action plan." At this split second, the motor behavior starts to roll according to the outline of that proposal. When this split second is over, the plan might change, and preference might be given to some other neuron column that proposes a different motor plan that is better adjusted to reality in real time.

We do not always choose the best plan, and sometimes we repent our faulty choices regarding the selected column, as when it turns out that another column had a better plan. At any given moment, several gates lead to several paths down the motor-behavior path. Some of the paths are partially overlapping. When our prediction turns out to be erroneous, our decision goes back upward in order to reach a different decision about a behavior path that will start, again, from a higher level. Changing the selected course of action is done both horizontally and vertically, which enables us to change the course at any given time.

Mental flexibility means changing courses at interchanges of the train of thought according to needs that are sometimes urgent and sometimes rare.

"Transposition" is a term taken from the world of music; it describes moving between different musical scales. While we are performing a task, our brain constantly performs mental transposition of the motor and behavioral action plan so our performance will match the circumstances as well as possible. Our brain is like a professional musician who easily performs transposition from one scale to another. Our brain is a skilled "snakes and ladders" player.

The brain often acts as a jazz pianist who improvises the sequence of fingering—the encounter between the suitable finger to the suitable key—almost in real time, while constantly projecting his expectations regarding the future direction of the music that is being played.

One of the main roles of the frontal lobes is to set up a project outline for future moves. This process is carried out, while we are skipping between various motor and behavioral action plans, in order to match them with the changing circumstances as well as we can in real time. Maximal compatibility betters the outcome. For instance, during a basketball game, there is constant matching of the angle of shooting according to our changing location on the parquet. We constantly move between alternative action plans in order to find maximal compatibility between our responses and the world of phenomena around us. The flexibility of brain processing is at the basis of this ability.

Basal Ganglia as Action-Plan Routers

At the depth of the core of our brain, there are clusters of neurons called basal ganglia. This is where action intentions materialize and become sequences of operation signals sent downward toward the muscles. The basic function of the basal ganglia is similar to super routers that select the course of the railway on which the train of action is about to ride. The prefrontal cortex outlines action plans, and the basal ganglia channel the operation signals, like train wagons that are sent to the most suitable railway; they have a crucial effect on materializing the action plan. By virtue of their function as routers, they contribute to the skipping between "alternative" action plans in accordance with the changing circumstances.

Input Prediction and Core Preservation—in Action

Despite feeling that the "world image" is experienced as coherent, and matching events in real time, the truth is that our eyes constantly move and sample the world from different angles. The brain patches together the pieces of pictures into a single, coherent picture. Our eyes examine the surrounding in jumpy movements called "saccades"— eye movements at a frequency of three times per second. Thus, a different reflection of the world is projected on our retina three times per second, but we still experience the visual experience as continuous, smooth, and devoid of pitfalls.

Another example for the brain as an entity that bridges missing information and fills in gaps is the blinking of our eyes, during which momentary darkness takes place for splits seconds—the duration of blinking. Mostly, we are not aware of the momentary darkness formed every time we blink, since our brain preserves a visual, perceptional continuity and attaches the sights that were perceived prior to the blinking to the sights perceived afterward, and they all become a seamless continuum.

The perceptual continuum is preserved, mainly, due to compensation mechanisms of the prediction that complements the voids in the input and due to preservation of a substantial core in the presence of various appearances of the observed objects (such as different angles of vision with respect to the same object). Prediction is formed through a combination of the memory of past experiences and updated pieces of information.

The Virtual Time Machine

A mostly unconscious skill is allocating the plausible duration of time for the occurrence of a certain event. We base duration predictions both on external world events and on our brain conduct and performance of actions.

When a certain event exceeds the predicted time limits (as an unpredicted fold in the fabric of predicted chrono-architecture), we become alert and perform an "automatic check-out procedure" in order to assess the deviation.

Essence Extraction at the Service of Input Decoding

Familiarity with the sequences of signals characterizing the core of the essence of the encoded world appearance is possible through the aphorismic extraction of the input. Such an acquaintance sometimes allows for the retrieval of the information encoding layout regarding the same world manifestation, even when the information input is partial or disrupted. In other words, it makes it possible to infer the whole from the partial; the right from the disrupted.

The Role of Prediction in Processing an Unfamiliar Input

A vague or partial input, which is difficult to interpret, might sometimes form a "blind spot" in the perceptual field of vision due to its vagueness. Such a challenging input sometimes requires prediction to complete the picture. The partial contribution of the input and the complementary contribution of the prediction combined enable the production of a "comprehensive conceptual picture," which is not necessarily reality-compatible.

Cryptanalysis of the Enigma

When we come across a situation in which all familiar rules are broken, and the phenomena seem chaotic and unpredictable, our brain switches to "intensified-operation" mode. This is the action pattern that is reflected in brain imaging (functional MRI) as multiple light spots that represent the multiple activated areas in the brain, like a mass convention of fireflies that, in an attempt to find clues that will help decipher the unfamiliar phenomenon.

The Course of a Challenging Input

A "challenging" input that does not fit in with familiar patterns is transferred to higher levels of cortical processing until it is deciphered or, alternatively, until it is classified as an "abnormal," "breakthrough," "game-changing" input or, alternatively, until it is deciphered as an erroneous input derived from some sensory distortion or built-in vagueness. According to common belief, such an input is the preferred input, to be processed by the right hemisphere.

The greatest mountain ranges in the world—the Pamir mountains, Karakoram mountains, and the Himalayas—formed as a result of a geological collision of forces. The tectonic plate, called the Indian Plate, moved northward and tumultuously met the tectonic plate, called the Euro-Asian Plate. The huge mountain ranges were created at the meeting point of the tectonic plates. In these areas, there are signs declaring that "continents collide here." Metaphorically speaking, brain areas in which "continents collide," in the sense of substantial ideas that contain conflicting contents, are the areas in charge of settling cognitive dissonances; as in the mountain ranges, here it is also a "high" area with respect to functionality, which is in charge of complex functions. At these brain areas, complex processing is able to decide between colliding pieces of information that takes place.

Environmental Stimulation as Perception-Shaper

In an experiment conducted on kittens (and we shall not refer here to the problematic moral aspect of such an experiment), the kittens were deprived of an exposure to lines or horizontal objects within the narrow window of time of the first few months of their life; their living environment was shaped in such a way that it did not include horizontal objects. This lack of exposure caused the occipital cortex, which is in charge of processing visual input, to omit from its world image, which is actually the cat's visual world, any signs for the existence of horizontal objects. The lack of exposure to horizontal objects leads to perceiving a visual world that is vertical and diagonal, with no horizon. The mature cats that survived this experiment of selective deprivation of a sensory input as kittens later suffered from permanent impairment related to their ability to notice horizontal objects in space and respond to them. They repeatedly bumped into these objects and suffered injuries as a result. Their visual perception was disrupted in a manner that was partially irreversible.

Similar results were observed in a conditioning experiment conducted on fleas. The conditioning was related to acquired perceptual failure with respect to the space planes. Fleas are capable of jumping to a point that is sixteen times higher than the length of their body. If human beings had this ability, they would be able to reach a height of twenty-five meters by jumping. Fleas that were put into a box of matches jumped as usual and repeatedly bumped into the box roof. After a while, following multiple attempts, there were no more knockings that indicated they were hitting the roof of the box. The fleas adapted their jump to the height of the box of matches. When the fleas were released from the box prison, the conditioning did not disappear, and they kept jumping to the height of the box even when they were free to jump as high as they wished. The boundaries of their vertical world were permanently conditioned, with no return, to an artificial height that did not match their natural skills.

The following is another example of conditioning that results in behavioral fixation. In an experiment conducted in Germany, the researchers created a field of artificial blue and yellow flowers. They added honeydew only to the yellow flowers. Following repetitive wandering among the flowers, young bees that were brought to the field discovered the correlation between the color and the honeydew and, after a while, totally abandoned their visits to the blue flowers, focusing solely on the yellow ones. At this point, the researchers inverted the correlation and this time added honeydew to the blue flowers only. Contrary to the expectation that a renewed learning process will enable the bees to track the changes in the correlation, the bees continued to visit only the yellow flowers, despite recurrent disappointments. The uncompromising persistence never ended until the bees' extinction out of hunger and exhaustion.

This acquired perceptual failure can also be found in the world of humans who undergo types of cruel conditioning. Acknowledging the processes might enable us to reduce the impact of this conditioning.

"R" and "L" Are Not in Japanese—Time Windows for Acquiring World Understanding

Exposure to "core stimulations" that create the world's image in our brain is essential. Phonological differences between sounds of various languages create a "sound distinction," which is unique to the speakers of a certain language. Thus, for example, native speakers of Japanese have a hard time distinguishing between the "R" sound and the "L" sound, since in the Japanese phonological environment, to which they are exposed and according to which their brain creates sound patterns, there is no need for such a distinction. When they are later exposed to a different lingual environment with a different sound lexicon, they will still find it difficult to distinguish between these two sounds.

Today we are no longer debating the question of "nature versus environment" but, rather, focusing on nature through environment. It seems that the environment is a sort of filter that enables the parts of nature that penetrate it to be expressed.

A common assumption among brain development researchers is that the time window for the best acquisition of a second language is approximately up to the age of eight. Later, it is more difficult to acquire the phonemes that characterize the language that is not the mother tongue and internalize the different pronunciation of syllables. It seems that our mother tongue imprints rhythmic patterns in our brain that serve as a super-pattern for also processing vocal input that is not lingual.

An example of such an effect on shaping our audio perception is found in an interesting correlation between the mother tongue and absolute pitch. Among musicians who are native speakers of Mandarin Chinese, which is considered a tonal language, there is probably a much higher percentage of owners of absolute pitch in comparison to musicians who are native speakers of non-tonal languages (such as English).

A different example of the effect of environment as perception-shaper derives from the fact that we sometime sense emotional arousal but often mistake its origin.

A scenario taken from the telenovela of our life: When young men experience a rush of adrenaline during a bungee jump over a stormy river and soon after meet an attractive woman, they tend to initiate contact with her more than their peers who have not had the same bungee-jump experience. A plausible explanation for this might be that the emotional arousal triggered by the jump is ascribed, at least partially, to an arousal derived from the presence of an attractive woman near them. Due to the ascribed effect, there is more willingness on their part to initiate contact. Life events are mingled together; on the other hand, the "refreshing time" of the screen of emotional consciousness is often slower than the rhythm according to which new experiences knock on the doors of our consciousness. In such circumstances, the emotional impression of the recent past has not yet been omitted when, in fact, we are already in the midst of experiencing a present experience. The resolution of the "source of emotion spouting" is often too low, which makes us more prone to interpret the origin of a certain emotion erroneously. Some of the reasons are hidden from the eye of our consciousness and exist in the unconscious layers of our soul.

Environmental causes also create behavioral bias. Studies show that our "shadow characters," which means behavioral tendencies we try to conceal from public eye, come to life when the light is dimmed. It seems dim illumination tends to illuminate our "shadowy personality"; those aspects of our personality we consider unattractive step, with the diminution of light, to the front of the stage. When the illumination is dimmed, the moral level of our behavior tends to go down. Participants of a study who were observed in dim illumination, or those who were equipped with sunglasses in broad daylight, were less fair toward their peers compared to subjects who were observed in intense illumination with harsh light. A dark environment makes us feel hidden, and as a result we tend to reveal darker parts of our personality, like young children who believe that when they close their eyes, others cannot see them.

Perception Seasons of the Sensory Organs

Calibration of the perception skills of our sensory organs changes along various time axes. In general, it is possible to mark a cycle of change along a course of a whole day, a month, and the different periods of life. Perception undergoes constant change, and this variance is present throughout life.

At the "infancy season," the perception skills are calibrated as a result of tuning in to and focusing on stimulation to which exposure is frequent. For example, with respect to hearing, the core of the hearing organ—the spiral-like part of the inner ear called the cochlea—monitors in a differential pattern low sounds at its basis and high sounds in its sharpening top end.

A normally functioning human ear is capable of distinguishing sounds at a range of thirty to twelve thousand vibrations per second—from the whisper of a lover in the ear of her companion to the noise made by jet engines on our way (with the lover?) to a tropical island.

Along the time axis of twenty-four hours, the emphasis of perception changes. When it gets dark, the visual input is reduced and the perception skills based on other senses, mostly the audio sense, become sharper. Along the monthly time axis, it has been found that sensitivity to smells changes in circular patterns among women at the time of the monthly period.

Babies' ears are attentive to the whole spectrum of rhythms during the first year of life. Afterward, their hearing becomes tuned to the rhythms they were exposed to, and their sensitivity to other rhythms fades away. Thus, we can claim that there is a developmental window for rhythms that is open during the first year of life. A "rhythmic preference" is imprinted during this window of time, and it accompanies us throughout life. In this sense as well, the brain is the creature of the culture to which its owner was born into.

At the other end of the course of life, when we are old, there is a rise in the threshold of stimulation that manages to pass through sensory sampling slots, and the range of stimulation which is experienced becomes narrower.

The Brain Hegemony Over the Senses

The light of reality penetrates our brain through the prism of the perceptual system, which naturally unstitches the light and changes its manifestations, as prisms do. Reliable monitoring of reality derives, inter alia, from the ability to distinguish between sensory stimulations originating in the external world of phenomena and contents originating in our inner world. In a sense, this is the ability to distinguish between reality and fantasy. Our perception is formed in between.

The interpretive framework of the brain enforces premises on the reality perceived by the senses. Often, we enforce beliefs and stigmas on the perceived world image. In fact, this is *a priori* coercion, which precedes the perceived experience.

The term "mimesis," which derives from the Greek word meaning representation or imitation, refers, as a metaphor, to the representation of reality. Our brain actually attempts to produce its own version of mimesis.

Any perceived piece of information is interpreted according to our perception insights.

The philosopher and Noble prize laureate Henri Bergson once said, "The eye sees only what the brain is prepared to understand." He indicated the two phases that are essential for the process of vision.

The framework of perception also challenges the assumption that seeing something with our own eyes is the best proof. During an experiment, strangers who wandered around a university campus started to talk with people who happened to walk near them. During the conversation, two people carrying a door blocked their line of vision between them so they could not see each other. The interruption lasted for one second, during which one of the carriers replaced the stranger who initiated the conversation. When the door carriers moved on, the subjects of the experiment were faced by a different person, who continued the conversation from the point at which it was interrupted. Only seven out of the fifteen subjects reported the disappearance of the first person they conversed with and his replacement by another person. It seems that among the other subjects, "mental inertia" concealed the change from their consciousness.

Objective, impartial knowledge is actually wishful thinking. We all have an interpretive framework in light of which we interpret the input from the external world.

The version of reality produced by our brain is biased by nature. In a sense, our brain is an information manipulator that directs the "show of reality," which is partially correlated with the reality as-is. This correlation is mainly intended to enable our survival.

Our interpretation of the world partially derives from the inherent interpretive framework in our brain, which makes us more prone to choose and prefer certain interpretations. Subjectivity is built into all of our brain's interpretations of reality's occurrences.

Perception is a hermeneutic skill (hermeneutics is the theory of interpretation, mostly of written texts, but it also refers metaphorically to other disciplines).

Our interpretation of reality derives, to a great extent, from conditionings of our consciousness, which are sometimes Pavlovian conditionings.

Perceive Reality as is—the "Polonius Test" and "Bee-Like" Vision

The Roman author Polonius claimed that the ultimate test for a painter is his ability to draw a flower so vividly that bees will mistake it for a real flower. In other words, to conceptualize reality exactly as it is.

In this context, there is a story about King Solomon, who was assisted by a little bee when he tried to solve the riddle of the Queen of Sheba and distinguish between a real flower and an artificial one.

Some might claim that creating a photograph-like drawing is considered low imaginative art, which does not produce anything new, and that art should aspire to create "new worlds" that do not exist in our world. This approach sees the work of art as an expression of a unique interpretation of reality that is the unique brainchild of the artist. Any brain perception of reality, however, derives from interpretation of the brain, according to different "levels of interpretive freedom," as an expression of different levels of proximity to reality as it is. The bee, by the way, sees colors of flowers within the ultraviolet range. A yellow buttercup is perceived by the bee as purple; thus, the bee also has a unique, "bee-like" interpretation for the colors of flowers.

Perception as an Adjustable Procrustean Bed

The perception-organizing system is built-in in us like a procrustean bed with definite boundaries. It is possible, however, to tune and adjust this bed, and we can take advantage of this flexibility in order to better our ability to perceive and process raw information coming from the senses.

Our basic worldview, which is subjective by nature, often distorts the perception of external inputs. An applicable insight related to this is circumstances that require strict adherence to the outline of external reality; it is recommended to censor non-objective influences in an attempt to draw near to the line of reality up to an asymptotic proximity.

The Resolution Dial

We need to deal with different phenomena in our world according to different levels of conceptual resolution. Thus, for example, a telescope is suitable for watching the stars but not for reading the newspaper. Judging an input regardless of the resolution is similar to an attempt to navigate with a map with no scale.

There is a built-in resolution dial in our brain. We are used to viewing many things in "daily-life resolution," which might be referred to as the direct "natural level" or the mesoscopic level. For this level, we do not need to use "perception-expanding" devices such as a microscope or a telescope.

The measurement horizon of the human consciousness ruler is mostly limited to routine life resolution. Our brain is tuned for dealing with time and space occurrences of daily activities. It is more difficult for us to deal with time and space dimensions that do not fit the resolution level of the routine dealing of our brain with reality occurrences.

We tend to categorize the world according to distinct perception categories that restrict the boundaries of our thought. We are entities that exist at the mesoscopic level, which is in between the macroscopic world and the microscopic one. We are used to defining concepts within the boundaries of the existence plane, which is essential for our survival, and it is hard for us to stretch our conceptual boundaries beyond the borders of the mesoscopic space and the time resolution of human life.

The subjective decision regarding the target value toward which to turn the resolution dial determines the quality of information we are about to receive and, as a result, very likely our behavior with respect to this information.

Each level of resolution, by its nature, leads to an inherent loss of certain information regarding the entity we focus on. The resolution is applied in a task-compatible pattern. Different tasks require different levels of resolution. In a mostly "automatic" pattern, we tune our senses to selected levels of resolution in order to dig up the information that seems important at the specific time. The information-digging-up strategy is dynamic.

Our commonsense promotes practicable insights of daily-life resolution to a level of axioms. These axioms are valid at the mesoscopic-resolution level, but their validity grows weak at other levels of resolution. For instance, let us consider the firmness quality of a stone. At a microscopic resolution, we can see that the stone we kick contains, like most solid objects in our world, mostly a vacant space. But this resolution is not suitable for everyday life in which, when we kick a stone, our toes bump into clear, solid firmness.

The intuitions that serve us faithfully in our daily-life routine lose their validity when we wander, in our mind, up and down the scale of resolution toward the subatomic world or, alternately, the wilderness of galaxies.

The Multi-Resolution Present

The ability to hold in our consciousness representations of several levels of resolution simultaneously results in an elaborate, reliable, and clear representation of reality.

Fractal patterns enable us to infer from a certain level of resolution to the level that is one step higher or one step lower from it. The recurrent self-similarity feature, which is embodied in the fractals, enables us to infer the complete visual image from each of its parts.

When More is Less…

The border of "no information" appears when we maneuver between different levels of resolution. In the material sense, at each level of resolution we use to assess information, pieces of information drop from our jurisdiction and become invisible to our eyes. For example, when we look at planet Earth from space, we cannot notice the outline of small volcanic islands spread out in the Pacific Ocean, but if we go down and reach the tops of the coconut palm trees growing on these islands, we will no longer have an ability to see the outline of the various continents. For each piece of information there is a typical "no information border." When we cross this border of limitations, it vanishes from our sensory screen but not, necessarily, from our perceptual screen.

The mesoscopic level, which represents a medium level of resolution—the level we most use for processing daily-life world occurrences, with no mediation of "sensory input intensifiers"—is supposed to "contain in its memory" the information from the two levels in between it.

Culture and science—which is an important component of the culture, enable us to expand our perception in the sense of creating a hierarchy of patterns related to the nature of the world. The ability to create patterns that are more and more comprehensive sometimes derives from focusing on increasing levels of resolution, which allows us to see the small details that were invisible before. Sometimes we go up to the resolution levels of the micro and nano world in order to conceptualize the world better than we are able to at the meso and macro levels.

Our less comprehensive perceptions become a subgroup within our more comprehensive ones. This is what happened to Newton's mechanics that became a private case in Einstein's general theory of relativity.

It makes sense to predict that the brain of the future human being will contain patterns with more expanded drainage basins. Many insights will become more comprehensive.

The Tango of Senses and Perception

Any conceptualization that is created in our brain is the combined result (gestalt) of the senses' impressions from the outside and organizing principles in the perceptual mechanism. Our response to an event is in the texture felt by the fingers of the one who touches, in the ears of the one who listens, and in the eyes of the one who looks—and in the brain's processing of these inputs. In other words, the response is in the senses' input and in the input's perceptual processing.

Conceptual understanding (intellectual vision) is essential in order to come up with a reality-compatible interpretation of sensory input.

The brain does not always see what the eyes see, and vice versa.

There is no complete overlap between the sensory field of vision and the perceptual field of vision. The sights reflected from our retina are not necessarily the ones reflected in our mind's eye. Looking at a certain object is completely different from seeing it.

We all suffer from blindness of thought of a certain type, in light of phenomena seen by our eyes, in the sensory sense. We are different from one another with respect to the location of our perceptual blind spot and its diameter, which means in the fields of our ignorance, in the sense of "interpretation failure," to the data of raw sensory input.

Scientific evidence shows that information flows between the brain and the sensory organs in a bidirectional pattern. The sensory organs feed the brain with information, but the brain also has the ability to feed the sensory organs with information, and this path is probably essential with respect to the production of various sensory hallucinations. When the brain is constantly fed with information, the flow of information in the opposite lane becomes moderated and restrained. When there is lack of feeding on the part of the senses, however, the brain feeds the sensory organs with raw materials, which might turn into sensory hallucinations.

Feedback relations between the sensory organs and the brain take place in an upward-downward pattern and in a downward-upward pattern. This is the type of relationship shared by the brain and the ear. The cochlea, the organ in our inner ear that is in charge of transforming air vibrations into bioelectric impulses, infuses audio input, and, at the other end, the brain calibrates and modulates the cochlea according to its momentary preferences.

Representation maps of visual, audio, and other types of input are organized like topographic maps so that they truthfully represent the gradient of sensory impressions.

With respect to the areas of sensory input, they are more similar to a "retinotopic map" (*retino*, as in retina), which maps the input by creating linkage between specific input areas in the retina and specific processing areas in the brain, which are in charge of the input of visual impressions from those specific areas in the retina. Similarly, the areas of initial audio input are organized in a pattern of a "tonotopic map" (*tono*, as in tone), in which there is linkage between specific areas of initial audio input in the brain and specific tones.

Pictures "speak" louder than words. The number of brain cells that process information originating in the sense of sight is larger than the number of cells that process information originating in other senses.

Memory researchers use terms such as "very-short-term memory" to describe what is done by external perception stimulations at the time of their occurrence in basic consciousness.

As for the input of the sense of sight, pictures in modern films, which appear at a frequency of twenty-five hertz, are perceived as a continuous visual input. On the other hand, old movies, whose pictures appear at a frequency that is lower than fifteen hertz, are perceived as fragmented—as a sequence of isolated pictures. We might conclude that the freshening time of the "perceptual screen" with regard to cinematic visual icons is one-twentieth of a second, between one-fifteenth and one-twenty-fifth of a second. Lighting conditions and the level of contrast between the observed object and its surroundings influence the life expectancy of the visual icon. For example, the reflection of lightning in a dark night will exist for a longer period of time compared to a visual manifestation of more moderated contrast.

With respect to our hearing sense, it seems that most of the times a sound echoes in our audio memory for a longer period of time—between one second to three seconds—before its impression vanishes (this type of memory is called "echoic memory").

And in association to that a bitter-funny saying can be cited: " the speed of sound waves is sometimes a strange thing, there are things that my parents told me when I was a kid and I hear them only now"…

Attentional Blink

The system that processes perception impressions needs minimal freshening time between one perception stimulation to the next. The time gap that is created from the end point of processing a certain sensory stimulation to the point when the sensory organ is ready for perceiving and processing a new stimulation is called "attentional blink," and it lasts for several thousandths of a second with respect to visual perception stimulations.[14]

The period of "attentional blink" restarts the lifetime window of the present-missing sensory stimulation (a stimulation that is no longer trapped in the net of our senses but whose impressions have not yet vanished from the sensory organ). Visual dragged images (like the image that is dragged on the retina after looking at a glaring object, which is a sort of sensory hangover) share similarities with earworms, which are melodies or songs that play in our ears even when, in practice, the audio stimulation no longer exists.

At the basis of these phenomena is the fact that our senses rely on recalibration, freshening time, which prevents them from reflecting world phenomena exactly in real time.

Perception and Sensory Illusions

In Buddhism is interwoven the idea that our senses bind our consciousness and only after our death will we be freed from the tyranny of our senses and experience reality as it is.

Some wander to the border zones of philosophy provinces and wonder whether the physical phenomenon at the basis of the sensory stimulation exists when there is no perception to perceive the stimulation. In philosophical wording, the question that can be asked is, do falling trees make a sound even when there are no ears around to listen to the sound?

The phenomena of sensory and perception illusions are part of our daily routine and an inexhaustible source of psychologists' tricks. Here are some of the exotic illusions:

Simultanagnosia

Simultanagnosia is a phenomenon in which there is a difficulty in perceiving, or a complete lack of ability to perceive, all of the impressions as a whole; they are perceived as isolated impressions out of context. This phenomenon might affect various inputs in the senses. The sound of a car horn or a cat's howling while we are walking down the street might be perceived as an isolated sound out of the context of "environmental sounds that are frequently heard on the street."

Visual simultanagnosia is common among people who use drugs that cause hallucinations. The observed objects are perceived as isolated entities that lack mutual relationships.

It is possible that a simultanagnosia-like mechanism is part of the world-experiencing mechanism of people who suffer from autism, and it makes it hard for them to tell the wood for the trees. On the other hand, it intensifies their ability to distinguish between isolated trees.

Synesthesia

Synesthesia is a situation in which stimulation in one sensory channel is translated into stimulation in another sensory channel, or is merged into a multiple-sensory experience.

Synesthesia is built into our sensory perception. It might also be defined as a phenomenon in which the outline that encircles the various senses becomes blurred. An input in a certain sense is perceived by the brain to be an input in a different sense. A certain sight is accompanied by a certain sound and a certain texture is accompanied by a certain smell, etc.

When we put something edible in our mouth, such as cherry ice cream, the sense of taste does not derive solely from the taste buds on our tongue but from a combination of the information originating from the taste buds and the texture and smell of the food. "Aroma" is the term used to define the essential contribution of the sense of smell to the sense of taste. Thus, the sense of taste inherently involves synesthesia.

The more esoteric synesthesia is the one that creates unconventional interfaces between senses, such as a synesthesia of sound and taste. Recently, some findings have shown that sounds might calibrate the sense of smell. This is a synesthetic pattern called SMOUND.

There are synesthesia of color-sound, color-smell, and other possible combinations. People who experience such synesthesia can talk about the taste of the wind or the color of a sound.

Babies' consciousness is a "beginner's consciousness." Some claim that a baby's perception is a synesthetic chaos, since babies' senses have yet to cross the threshold of differentiation that distinguishes between the various senses. Some relate it to overconnectivity between processing areas of various sensory inputs, which characterizes the brain of a baby and is retained, to some extent, in the brains of adults who are endowed with synesthesia. Studies that can support this hypothesis show that the frequency of synesthesia is high in childhood and low in adulthood.

A quoted estimate regarding incidence of synesthesia in adults is one per two thousand. There is an inherent difficulty in assessing this ability, however, since it relies on self-reporting, and the estimated incidence is controversial.

The process of thinking might be considered a synesthetic creation in which ideas originating in one brain are merged with ideas originating in other brains, creating conceptual hybrids.

The Hallucinations Generator

The brain is a "stimulation junkie," since it requires constant stimulation. In the absence of stimulations, it generates activity that simulates sensory stimulation, such as audio hallucinations that are common among the deaf at an advanced stage of their life, or visual hallucinations that are common among the blind at an advanced stage of their life.

A similar phenomenon was observed in situations of lasting external sensory deprivation. In an experiment in which the subject stays within an isolated container, not exposed to any external stimulation, it was found that the brain makes up consciousness impressions, which it pretends are real sensory impressions.

A syndrome called "Alice in Wonderland" is a type of perceptual distortion. People who suffer from it experience delusional distortions in the visual input, for example, with respect to perceiving the size or shape of objects or body parts. The distortions are mostly related to micropsia ("Lilliput" vision—objects are perceived as smaller than their real size) or macropsia ("Gulliver" vision—objects are perceived as bigger than their real size).

The syndrome results from a distortion in the function of neurons located at the temporal lobes, where an electric disorder takes place that results in visual distortion. This syndrome demonstrates how much sensory impressions perception relies on the processing processes of the perception-organizing system in the brain.

Color and Consciousness

Colors can trigger a certain behavior pattern based on innate conditioning. For example, the expertise of the golden silk spider is setting honey traps that attract insects to their death with sweet, vain promises. The spider weaves its webs in yellow, which is the common shade of flowers that produce honeydew. The yellow shade magnetizes insects in a pattern of inherent behavioral conditioning like the irresistible song of the sirens.

In the world of human beings, color also has explicit and implicit effect on our behavior. For instance, pink is known to induce universal relaxation and tranquility. This is why the walls of prison cells are often painted pink. "Prison-cell pink" reduces the level of anxiety and aggressiveness among those who find themselves there.

Another example is taken from a consumer survey that showed that when a yellow shade was added to the words printed on a "7-Up" soda can, subjects reported a marked taste of lemon, and when green shade was added, subjects reported a marked taste of lime. The color colored the taste as well.

The Journey of Marco Polo in the Land of Smell

The Distance Between the Smell of an Orange and the Smell of the Sea

The attempt to quantify with a qualitative-subjective dominant aspect comes with great difficulty. It is true for all sensory inputs. The study of the sense of smell is especially affected by this.

In the vision domain, the "color code" has been cracked, meaning the correlation between the wavelength of a beam of light and the accompanying sense of color is known. It is also true for the "code of sound"—we know the frequency and intensity of the sound waves that are typical of certain musical instruments. But it is much more difficult to draw a reference axis on which the location of the smell molecule will enable the prediction of the smelling experience.

With respect to visual input, we can claim that the "difference" in the length of the waves between purple and orange, which is about 35 nanometers, is almost half of the "difference" between green and orange, which is approximately 65 nanometers. As for audio input, we might say that the difference between Re and Mi is always the same as the difference between Re and Do, but what is the distance between the smell of an orange and the smell of the sea?

Smell Me a Sheep

A mental image of smell is rather difficult to convey, perhaps because of the relatively small cortex area devoted to the processing of smell in the human brain.

It seems that, even in those senses to which a "high subjectivity coefficient" is ascribed, however, there are more universal elements than was believed in the past. The assumption that the sense of smell is a sort of a bench player in comparison to other senses has become less popular. Despite the inherent difficulties in the research of smell, several fascinating studies have been conducted in this field.

Smell researchers assume there are about one thousand different types of smell receptors, some of which are not even expressed in practice. Each receptor can respond to several types of smell molecules, which are volatile molecules that carry a specific smell. On the other hand, each smell molecule can activate several types of receptors. The estimated assessment of the human smelling spectrum is approximately ten thousand smells. Thus, the boundaries of the perceptual universe of smell are rather extensive, considering the world of colors we perceive through the sense of sight is based on the receptors of the three primary colors: red, blue and green, which make up all colors and shades of our visual world.

The correlation between the chemical and physical qualities of the smell molecule and the sense of pleasure or displeasure it creates in the person who smells it used to be mostly unknown.

Researchers have created a map that places smell substances in a multidimensional space based on a number of features, such as the number of carbon atoms, weight, and level of solubility in water. This map enables us to assess the distance between the locations of two separate smell substances on the map in a measurable manner. It turns out that smell substances people perceive as similar are also located close to one another on the smell map. It seems that the differences between the smells of various substances are based on universal regularity, which depends more on the chemical and physical traits of the smell substances and less on personal, subjective preferences.[15] Now, it is much easier to understand the laws of the state of smell that were considered a mystery not so long ago.

Polecat or Lilac?

It was found that the smelling experience is based, inter alia, on "universal scents." The variation of the chemical and physical traits of the smell molecules is correlated with the sense of pleasure or displeasure caused by the smell, as perceived by people of various ethnic backgrounds. In other words, the level of pleasure or displeasure invoked by a smell is a universal trait shared by people of different ethnic backgrounds. The so-called subjective dimension of the sense of smell relies, to a great extent, on the objective dimension of the physical and chemical traits of the smell molecule. A possible (teleological) explanation for that derives from the fact that the sense of smell is designed to help us map the chemical aspects of the world of phenomena, so these aspects determine the subjective sense that rely on a universal codex of smells. This is how the road was paved for the ability to predict a sense of pleasure or displeasure felt by people prior to the actual smelling. Thus, for example, the smell of men's sweat was proven to induce a sense of relaxation among women (they love them making an effort).

The sense of pleasure or displeasure originating in a smell is also determined, to a great extent, by the context in which it is perceived. The smell of Gorgonzola cheese might be considered appetizing among fans of this cheese, when it is attached to the cheese itself, but even those fans will not wear Gorgonzola perfume under normal social circumstances.

Chapter 3
Attention, Consciousness, and Awareness at the Center and Margins of the Attention Beam

The Awareness Spectrum

The brain was developed as a means of creating adjustable behavior patterns of an organism to the world. In order to fulfill this task, the brain controls the constant division of attention between the inner world, which is defined by whatever exists within the boundaries of our skin, and the outer world phenomena outside our body. The proportion of attention divided between these two worlds changes continuously.

Our awareness changes along a spectrum. There are things we are fully aware of, and they are located at the center of our attention beam. On the other hand, there are things we are only remotely aware of, and they are located at the "margins of our attention beam."

The periphery of our attention cycle represents a background of consciousness. In the periphery of our attention cycle, there is the twilight zone of consciousness, a zone characterized by partial shadowing, in which the perception impressions are dimmed but are not fully concealed from the eye of consciousness.

The ability to focus on a narrow, selected aspect of sensory input derives from our ability to focus our attention beam on it.

We tend to focus our attention on "essential" components of an experience and the elements of it, which, when not perceived, might not cross the threshold of awareness or might, possibly, remain at the margins of consciousness.

An easy-to-apply experiment demonstrating the act of focusing attention is the attempt to remember the shape of digits on your watch, assuming it is not a digital watch, without "refreshing your memory" by checking the watch. Many people consider it a tough task, despite having looked at their watch numerous times before. The reason for this is that, when we look at our watch, the focus of our attention is given to the pattern that shows the time. We move through the information related to the configurational aspect of the digits without noticing them, aiming at the information related to the pattern of time. The impressions related to the configuration of the digits are mostly dull, and our attempt to invoke this information and revive it fails most of the times we are asked to do so.

An experiment that resulted in similar findings showed that the configurational features of a one-cent coin (President Lincoln's profile, the word "Liberty," etc.), which is a very common coin and is used on a day-to-day basis by Americans, were remembered by only one of the ten participants—and this person happened to be a coin collector.

Focusing Attention According to the Gambit Approach

"Gambit" is a term taken from the world of chess and refers to a deliberate sacrifice of a chessman in order to gain an advantage in the game. Our focus of attention also "sacrifices" aspects of reality and ignores them while focusing on chosen aspects of reality in a selective pattern in order to maximize the benefit of absorbing information.

Voluntary regulation grants us the partial ability to determine the diameter of our attention beam and its intensity so that it fits the relevant task.

Our attention system enables us to tune into the stimulations of the external world and our inner world. The stimulations from our inner world alongside stimulations from the different organs of our body include input of the world of phenomena within the brain, and it is also done through selective filtering.

Different brain structures are in charge of selective attention to inner brain input and of filtering various types of "inner white noise." Failure in the process of filtering and regulation of signals across various brain areas is typical of Parkinson's disease. Some of the involuntary movements that appear in some patients who suffer from the disease originate in the "white noise." This is the output of brain areas that, under normal circumstances, is deleted and censored from the brain's action plan. Among people who suffer from Parkinson's, this output is not deleted and reaches, in its raw shape, the movement centers in the frontal lobe.

The Smell of Fear

Perception impressions that are not exposed to the sun of awareness might unconsciously affect our thoughts and actions. Our reality perception is sometimes affected by subconscious sensory impressions. An example is an experiment we might call "the smell of fear."

The participants in the experiment were women (who, compared to men, have higher sensitivity to most types of smell molecules). Some of the women were asked to smell a pad that had been soaked in the perspiration of men who had watched a horror film. During the experiment, the women watched photos of human faces with a variety of expressions. The women who smelled the pad interpreted the facial expressions as more threatening and scary than the women who had not been exposed to the "smell bias." A plausible conclusion (which is only an assumption) might be that the sense of fear is accompanied by a typical smell component that is discharged in perspiration and has not been identified yet. This component is sensed by our brain and leads to a subconscious interpretive bias reflected in the tendency to interpret a certain facial expression as more threatening.

The Focus of the Attention Beam as Wiring our Habits

We often act according to alternate repetitive sequences of focusing our attention on information related to performing a task, its performance in practice, refocusing it on a different data, and so on, in a continuous manner.

The material expression of information assimilation in a learning process is the creation of a new networking pattern between the neurons. Such learning involves focusing our attention beam on the sequence of data related to the skill. Our attention is, to a great extent, the chief imprinter of the profile of proficiency at the mint of the brain's skills. In its absence, or at times when it is faint, the portrait of the skill becomes blurred and loses some of its characteristics within a short period of time.

With respect to the problematic aspect of acquiring addictive habits, often a precondition for a sustainable change in the brain's wiring, which causes the creation of addiction, is consciously focusing the attention on the object of addiction in a repetitive pattern. As a result, the three-dimensional wiring network that stores the information about the addictive behavior becomes strengthened. For instance, focusing attention on the contents of online porn, beyond the critical threshold of occurrences, creates the change in the wiring that is at the basis of the addiction.

The Wonder of Reading as an Example for the Maneuvering Ability of Focusing our Attention Beam

While we read a book, the attention beam produced by our brain focuses on understanding the contents of the paragraphs. This is the complex processing stage through which our brain grants meaning to the sequence of letters.

The configurational aspect of the font of the letters usually does not reach the conscious level but remains at the unconscious level. This fact will change in cases where the configurational aspect has an irregular form, which will shift our attention beam, or if we voluntarily focus our attention on this aspect in case we become interested in the shape of the letters. The maneuvering range with regard to the aspect on which we focus our attention enables us to skip from one resolution of information processing to the next, from the most basic level to the most complex one.

Attention Filters

It is impossible to encode and store all the impressions we experience. There are attention mechanisms that filter and delete a large amount of information at the early stage of the perceptual input. The encoding is done under the auspices of our attention and derives directly from our ability to focus our attention.

Attention-filtering mechanisms determine the fate of small information fishes that penetrate the holes of the sensory perception net. Some of them are sent from the sensory organs to the brain, and some of them are sent to the ocean of oblivion. Perception impressions that survived the sensory input stage and reached our brain but did not manage to cross the wall of brain imprinting are also sent to the ocean of oblivion. The attention filter is like a firewall that protects the contents of consciousness.

The attention skill is measured, to a great extent, by the ability to selectively seal our brain from most pieces of sensory information. The level of the sensory input river constantly threatens to flood our brain and disrupt its activity.

The level of attention at the time of the experience serves as a predicting element regarding the lastingness of the related memory. The level of our attention determines the extent to which we are "present" in experiencing the perception impressions. Twilight zones of full consciousness, but faint attention, are familiar to all of us, and the perception impressions from such instances are likely to be shallow or disappear altogether.

The saying "Yes, there were times when I forgot not only who I was but that I was, forgot to be" (Samuel Beckett, *Molloy*) can be interpreted as being present in a moment in the present without full consciousness or attention—as an absentee present, which leads to failure in encoding the events of a certain instance.

Given a fast pace of stimulations, which challenges our attention capacity, stimulations that do not make vibrations in the pool of emotions, and that are emotionally neutral, tend to drip beneath the threshold of perception. On the other hand, stimulations that "make waves" in the pool of emotions, and that are emotionally charged, tend to be caught.

A delay in the expected feedback of the input (as in a situation in which we expect to see figures on a computer screen promptly but they appear on the screen a few seconds afterward) might lead to a 'cognitive sweeping' and the wandering of attention to other provinces.

Initiated intensification of the level of attention is possible by means of charging the perception impressions with emotional contents that are relevant to us. Such charging leads to intensification of attention and increases the chance of intensifying the memory of specific information. In fact, it is an initiated stimulation of the amygdala and its enslavement as an instrument in the orchestra run by the frontal cortex.

By being aware of the different "intensity levels" we apply to a stimulation from the external or internal environment, and through channeling and focusing our attention, we can gain a certain maneuvering area of voluntary control. It can be achieved by intensifying or dimming the intensity of attention and by controlling the direction of the attention beam and its diameter. In such a way we determine which aspects of the world will benefit from more extensive processing. The aspects that are located at the center of our attention beam are granted a higher level of awareness. On the other hand, it is possible to determine which aspects of the world will undergo more limited processing in the brain. These will be pushed to the margins of our attention beam, and the level of awareness granted to them will be low.

The dimming of attention and the directing of its beam are the main tools used for applying levels of perceptional importance to various world aspects. Improving the perception skills and focusing of attention can be based on practice in voluntary omission of irrelevant information and perceiving relevant information. Controlling the aperture of our attention eye enables us the maneuver of a zoom-in or zoom-out type, which enables the eye of perception to see the forest that bursts out from a group of trees or, alternatively, a particular tree in the forest, according to our needs.

When we face a challenge with inherent stimulations that distract our attention, we can use our focusing attention ability. We might be assisted by the ability to create voluntary "sensory deprivation" for certain aspects of reality in order to focus our attention on selected aspects of reality. As with Odysseus' sailors sealing their ears with wax when approaching the sirens who sang their devilish songs, voluntary avoidance of certain aspects of an input enables us to focus on a particular task and decrease the level of distraction.

A Zen insight related to bow-shooting, according to which "the archer basically directs to himself," might be interpreted as saying that voluntary focus of attention on an external target requires focusing of internal attention, which must come first.

Skips of Attention

As a television camera brings to the screen events from changing angles and locations, so our attention beam focuses alternately on the changing objects of interest. The skipping between objects of attention is possible by means of shifting the projector of attention. Thus, sometimes a "marginal" stimulation, which once was at the margins of our attention focus, relocates to the center of focus. Division of attention means the focus of attention alternately shifts between different objects of attention.

Attention transitions can take place in a smooth pattern of a continuous transition between one object of attention to the next, or in a skipping pattern of shifting the focus of attention from one object of attention to another.

It seems the ability to consciously change the object of attention focus originates in a dialogue between the areas of sensory input in the cortex and a brain area called the thalamus. This ability enables us to alternately zigzag between the small letters and the titles and to focus our attention on pieces of information that usually slip from the eye of consciousness.

Impressions of the Senses Craving Consciousness Screen Time

Our control over the allocation of the cake of mental resources in our brain is only partial. In most cases, we pass the knife, which is in charge of dividing the cake, to the hands of the capricious reality circumstances that give generous pieces of the cake to random world events. Internal or external distractions "steal" slices from the screen of consciousness.

The confrontation arenas of daily life magnetize our attention and the thoughts at the front of consciousness and, most of the time, prevents channeling of mental resources to matters at the core of our existence.

The rhythm of life, which confronts us with reality manifestations with great intensity, prevents us from thinking deeply about basic existential questions and researching them. In an indirect manner, it enables us to repress the fact that we are not immortal and to avoid a thorough cogitation about our existence.

Focusing our perception on the input selected from all the components of information that compose the experience while we change the focus of attention is a process that heavily consumes mental energy. Our attention tends to be "snatched" by irregular or unexpected events, such as a loud, sudden noise.

An example is a clock's cuckoo that strikes thirteen times. Phenomena that do not match the routine order of the world usually catch our attention in an attempt to track down the source of irregularity. Our attention beam is shifted at once toward the source of irregularity that challenges our expectations regarding world phenomena.

This fact is taken advantage of at the theater. In order to retain the audience's attention, the performers often wear phantasmagoric costumes, full of heart-stirring tricks, since it is known that extraordinary elements tend to magnetize attention.

"Oh, Damn It"—Primeval Behavior Spurts

When something that contradicts our immediate expectations happens and, at the same time, we are challenged by a task requiring immediate focus of our attention, we are prone to slips of the tongue and spurts of raw behavior. The explanation for this is probably that these tasks swallow our mental energy; not enough of it is left for the system in charge of behavioral filtering and censorship, which is located mostly at the frontal lobes.

"Automatic behavior spurts," moments of loose association ("flight of ideas"), spontaneous, immediate reaction for stimulations and exceptional memory modes in which memories from different times are merged—all of these are situations that might serve as a keyhole for the worlds of unconsciousness and preconsciousness. In such situations, a hidden door, which is usually hidden from the gaze of the normal mode of consciousness, is opened, and forgotten plots from our past materialize.

Some see such forms of behavioral output as a draft that serves as the spokesperson of the stream of consciousness without mediation, proofreading, or editing at the frontal lobes.

The Zeigarnik Effect is named after psychologist Bluma Zeigarnik, whose studies exposed the tendency of "unfinished matters" to take control over inches of the screen of consciousness until they are taken care of.[16] Such a tendency is used in television series and in films that end at a dramatic moment before closure. It is done in order to maintain viewers' interest. This psychological "trick" is called the cliffhanger effect—like a person who clings to a cliff, and what the future holds for him troubles our mind until the next episode.

An example for the echoing intensity of unfinished matters, which is cross-generational, is the proof for the claim of mathematician Pierre de Fermat. Some consider him the patron of "unfinished matters." In 1637, Fermat scribbled a note at the margins of a book whose content echoed in the field of mathematics for hundreds of years. This note was known as the unfinished proof of Fermat and remained on the screen of collective consciousness of mathematicians for a long period of time. The proof was finally found in 1994 by a British mathematician, Andrew Wiles, who had searched for it for seven years, and it was explained and described in more than a hundred pages. The proof that has been "hanging on a cliff" for 350 years finally found solid ground.

Mental Juggling

The aim of mental juggling is to "keep all the balls in the air," which means to perform all tasks successfully at the same time.

Control areas in the brain act as a "semaphore" (a term taken from computer science that refers to the mechanism that synchronizes several processes taking place simultaneously) and try to channel attention resources according to the quantity and timing that match the tasks.

The ability to divide our attention and focus our attention beam on the information that washes over the screen of consciousness at a very rapid refreshing pace is one of the abilities that is most vulnerable to the ravages of time as we become older.

Over-Dispersion of the Focus of Attention

The over-dispersion risk of the beam of attention is reflected in the famous saying "the one who is everywhere, is nowhere to be found."

Dispersion of the attention beam leads to a split screen of consciousness on which several films of input experience are projected in parallel.

Sometimes, the attempt to spread the fan of consciousness to its full width in order to deal with a challenging task, physically and mentally, causes a failure of attention resources.

When the same brain areas are required to perform several tasks simultaneously, the transportation channels within them become crowded; performance is slowed down and might even become disrupted. Multitask division of attention is considered an admirable skill, which is glorified in our modern work environments. In a situation in which our cognitive resources skip from one task to another in real time, however, the safety margins, which are in charge of accurate performance, become narrower, and we are more prone to make mistakes or perform defectively. Some see multitasking as a vicious myth that puts us in danger of crashing into the wall of reality. Thus, for instance, a study conducted by the American Ministry of Transportation showed that approximately 80 percent (!) of crashes between motor vehicles took place while the drivers were busy with another task in addition to driving, such as dealing with their mobile phones.

Too many accessories in the driver's cabin, including a multitude of indicators, displays, and buttons, might also create an overload of information that might steal essential attention resources from the task of driving.

A correct management of the "workload," which is reflected in resource allocation that matches brain capacity and prevents over-dispersion of the attention beam, is an essential condition to high-quality task performance.

Underneath the Crust of Consciousness—the Land of Subconsciousness

Major parts of the thinking processes are formed in the shadowy zone of the subconscious, and some take place at the brightly illuminated areas, which are exposed to the sun of consciousness. Our mental life is a mix of conscious and unconscious thinking.

At each stage of our life, the relations between these thinking types change according to our situation. When we are busy with a task that follows external world phenomena, it seems that the mix includes a higher percentage of conscious, task-oriented thinking. While we are ponder in daydreaming provinces, the pendulum moves toward more unconscious thinking. It seems that, along the continuum of our mental life, however, the two types of thinking are present in different proportions except for at times of deep unconsciousness. In such cases, conscious thinking is probably totally absent from the stage of our mental life.

The term "unconscious" has different layers of meaning. On the cognitive layer, there is common agreement that most of the information processing processes probably take place at the unconscious layer, or, metaphorically speaking, behind the back of consciousness. Freud's version of the term "unconscious" refers to the estimated location of the repressed, the locked garden of the forbidden, and, as such, it is disputed with regard to terminology.

Similar to the light coming from a ray of sun, which meets our eyes eight minutes after its creation in the sun, it seems that many of our decisions are formed at the unconscious layer and our consciousness becomes aware of them after some delay, although it tends to see itself as the initial source.

Brain researcher Benjamin Libet[17] conducted a series of experiments in the early 1980s. The experiments showed that brain activity—which is related to movement and called the "readiness potential," during which the nature and timing of movement are determined—is formed half a second prior to the forming of a conscious sense of desire to perform the movement. In other words, a distinguished, separated neurological process is formed at the unconscious layer and initializes the action before the conscious layer makes its contribution, thus leading to the plausible interpretation that the decision originator in our brain is found at the unconscious level and our consciousness, which is the home of free will, is informed in retrospect.

A series of experiments conducted by Libet supposedly shows that we do not do what we want to do, but, rather, want what we do. Ascribing decisions to a conscious activity is a supposedly false perception. Some saw it as a scientific, empirical proof for the belief that decisions are not formed at the conscious layer; thus, they are not born from free will, which is only a misrepresentation. In the spirit of this interpretation, we might claim free will in the brain is a "slave" in its own house, forced to want whatever was previously decided.

The activity of the readiness potential precedes the conscious sense of desire to perform the action, which reinforces the wonderment regarding the status of consciousness in the decision-making hierarchy.

If we accept the interpretation, according to which the decision originator in the brain is actually present at the unconscious layer, and the conscious layer is informed about the decision only in retrospect, it might be connected to a supposedly strange approach called "epiphenomenalism." According to this approach, conscious thinking that derives from free will has no effect on actions, and it is a type of by-product of brain activities that does not correlate with our behavior in practical terms of causality. If we feel that conscious thinking is the "big brother," here is even a "bigger brother" in the brain who outlines the behavior channel through an unconscious pattern, and the "big brother" of conscious thinking is informed only in retrospect—although it attempts to ascribe the initiation of the action to itself in an illusionary pattern.

A strange possibility that arises from these controversial findings is a technology that will enable us to monitor the readiness activity in the brain in almost real time. Then, we will be able to know, before the person himself, what this person would like to do in a few seconds and respond to it—for example, uttering a sound that will characterize activity that prepares the lifting of the right hand before the person knows he wishes to lift his right hand. In such a way, the natural order of things will become disrupted, and the direction of the arrow of time will be inverted, so that the effect will come before the cause. In other words, the implications of a future act will be known before the person "voluntarily" decides to perform it.

The hierarchic direction of the chain of command is being questioned. Does consciousness determine the policy and the source of orders, or is it only a junior commander that receives reports in retrospect and falsely believes that it holds the general's baton? Is our free will only a pawn in the chessboard of our brain whose moves are dictated by processes concealed from our consciousness? Is the approach that sees man as a creature of free will just naive, delusional humanism?

When it comes to reflexive behaviors, such as withdrawing our hand upon touching a blazing object, there is no doubt the decision to withdraw our hand was not taken at a hierarchic scale topped by areas of consciousness. The same is true for the tapping on the kneecap, which makes our leg jump—a reaction due to activity at the spinal cord level that does not involve the brain. Is it possible, however, that the same process also takes place with respect to more fundamental decisions? Could it be that the royal gown of consciousness is only a version of the king's new clothes?

There are numerous methodological objections regarding Libet's experiments, so the validity of the findings is controversial.

A more comforting possibility is that the conscious layer that is led by free will has a power of veto with respect to decisions that are formed without it, which means its signature is required in order to execute an action.

The "Inner Boundaries" of Knowing

Subliminal thinking processes that take place below the threshold of consciousness are a central layer of our cognitive activity. "The knowing unconsciousness" is a kind of a twilight zone to which information is streaming and from which information is passed to the conscious areas, and it is an active, vibrant area. It often processes data at a higher speed than conscious processing. This is why, for example, at times of emergency, drivers press the brakes before they are aware of the need for an urgent stop.

The brain probably processes supraliminal information (which crosses the threshold of consciousness) and subliminal information (below the threshold of consciousness) in a similar way. Contents from these two lanes, which meet the brain cells, cause similar effects.

Our brain is proficient at hearing the unheard and seeing the unseen—in the sense of input processing at the unconscious layer.

Subliminal stimulations shape many of our decisions as an input whose impressions, or shades, do not appear on the canvas of consciousness. As a result, part of our behavior is not navigated by consciousness. The navigator is often our subconscious.

Thus, sometimes our brain is affected by a state of mind characterized by anxiety and concern whose cause is a mystery to our consciousness.

Brain Priming

Priming means exposing our brain to information that does not cross the threshold of consciousness. Subconscious priming creates an emotional charge that, in its turn, channels our consciousness to select a certain interpretation of reality or a certain behavioral pattern.

An experiment was conducted during which the participants saw words on a screen for a particularly brief period of time, so the content of the words would not cross the threshold of consciousness. Some of these words had a violent message, while others had a peaceful message. When the participants left the room after watching the words on the screen, they came across a mock situation. The participants who were exposed to the words with the violent messages tended to react according to a more violent pattern of behavior compared to those who were exposed to the words carrying the peaceful message. We are exposed to unconscious influences more than we would like.

In a similar experiment, students listened to a supposedly random list of words, most of which were related to old age: elderly, hoariness, lonely, wrinkles, etc. After the experiment ended, the students walked from the lab to the elevator more slowly than usual. They behaved like old people, although they were unaware of this change in their behavior. The words caused behavioral suggestion that was concealed from their consciousness.

The Role of the Unconscious with Respect to Thinking

"The dark matter" in the universe, which is concealed from our eyes, can be compared to the subconscious (which is concealed from the eyes of our consciousness). But, while we estimate the share of the dark matter as one-quarter of the mass of the universe, our subconscious constitutes the central processing layer in the brain.

Nowadays, most brain researchers believe that most brain processing functions take place in the unconscious layers. Studies that attempted to estimate the relative part of consciousness in determining our behavior showed that its part is rather low—some say as low as 5 percent of our actions, which means that 95 percent is determined in the unconscious layer. The lion's share of the plots in our brain is written anonymously in the subconscious layers.

Some underestimate the relative part of consciousness even more and claim that it can be compared to the area taken by a bonfire at the heart of Australia (metaphorically speaking, the entire area of the Australian continent around the bonfire of consciousness represents the unconscious layers of our soul life).

"Go to your bosom; Knock there, and ask your heart what it doth know." (William Shakespeare, *Measure for Measure*).

The volume of our knowledge is larger than the volume of our consciousness. We know more than we know we know. Some of the information in our brain is inaccessible to the conscious self.

Our secretive brain is constantly active. Most inferences inferred by our brain are unconscious inferences, concealed from the eye of consciousness. Our consciousness suffers from agnosia (lack of knowing) with respect to a great part of the information that exists in our brain.

Information processing in the brain partly takes place in a mode of lack of cognitive awareness, and it has served as the basis for numerous magnificent brainchildren in the history of human culture.

An "exotic" example of a behavioral outcome that is based on information processing at the unconscious layer is "blindsight." The term, which sounds like an oxymoron, refers to patients who suffer from damages in the occipital lobes of the cortex, which is in charge of processing visual input that is transferred from the retina through the paths of vision. A functional damage in this area leads to "cortical blindness"; people who suffer from it become blind, since this brain area, which is in charge of visual awareness, is damaged. People who suffer from this damage consciously claim they are blind and cannot see anything. When they are asked to guess the name of an object presented to them, however, their guess is correct in a way that cannot be attributed to random guessing. A plausible explanation for this phenomenon is that a certain visual input is led through the paths of vision, also, toward other brain areas that do not produce "sight awareness." This information, which is contained at the unconscious levels of processing, affects the behavior of the subjects and the way they read the world.

Visual input of a person who suffers from the "blindsight" syndrome is not exposed to the sun of consciousness, and it is active at the unconscious layers of visual input processing.

"Forced memory" is a phenomenon that resembles the blindsight syndrome to some extent. It can also be found among people whose memory was damaged as a result of certain brain injuries, and they find it hard to recall certain events. When they are asked, however, to "guess" the details of a certain past scenario out of several possible scenarios, they manage to guess the correct scenario at a ratio that exceeds random probability.

This could be called "the zombie within us"—it is a well-known fact that certain cognitive skills (such as implicit memory and blindsight) exist even without a conscious experience; they are, rather, in a mode of "zombie" information processing.

Numerous pieces of evidence show that the zombie component is a central one, and some even claim it is the main component of our human essence.

The Traits of Consciousness

The spectrum of hypotheses regarding the nature of consciousness is broad.

The creation story of human consciousness as we know it today, and the time that it first appeared in the skulls of our forefathers, is shrouded in mystery with regard to chronology. As far as the geographical location is concerned, however, it is common to relate it to the African savanna.

Some claim that consciousness, as a trait that emerges from the brain, is a by-product of brain complexity that was created "unexpectedly."

It is agreed that the brain is essential for the creation of consciousness. But is it also a sufficient condition? In other words, is it all about the material essence of the brain?

A person's thoughts accompany him wherever he goes. They are rooted in his physical body, but, on the structural level of the brain's galaxy, there is no sign of intelligent life.

Thomas Huxley said that the sudden emergence of consciousness out of the interaction between neurons is as surprising as the appearance of the genie following the rubbing of Aladdin's magic lamp.

Perception and consciousness impressions dance the tango within our brain. According to one of the hypotheses, we receive sensory impressions all the time, but we become conscious only once we refer what we receive to our inner categories of experience.

Some brain researchers relate consciousness to a permanent brain resonance of forty hertz, and once a rhythmic correlation between different brain areas that process input from the senses is created, a unification of various shreds of input is made and a moment of consciousness is born.

The mode of consciousness is determined by the activity of routes originating at the brain stem. The content of consciousness, which is mostly formed in the cortex, derives from the routes that channel the sensory input of the external world of phenomena and the internal physical input to the brain.

Representation maps of our internal physical condition, which represent various indexes—such as glucose level, level of hunger and fullness, muscle tonus, level of fatigue or wakefulness, and in parallel- the representation maps of the external world of phenomena—are all hidden in our brain. The interface points between the various maps create linkage that connects, at every given moment, the sum of internal indexes to the sum of outer world phenomena. The self-world couplings serve as the bedding on which our perception of experience sprouts.

It seems that, as the awareness of the outer world refers to objects that are external to our awareness and are represented in our perception, our awareness of the occurrences within our body is also nothing but a perceptual conceptualization product. The quality of the external world conceptualization depends on the mediation of the senses and the conceptualizing framework in the brain. The quality of our inner world conceptualization also depends on the information that is channeled in the inner senses—for instance, the sense of hunger or fatigue—and on the brain system that conceptualizes our inner world. Due to systemic limitations, both in the system that mediates the perception impressions and the system that conceptualizes the perceived information, we are prone to make mistakes with regard to the creation of world models that match external reality and the creation of world models that match our internal reality. We might, therefore, make mistakes regarding the motives behind our behavior or, even, regarding the perception of ourselves.

It seems that our perception of the kingdom called the "self" is fluid. This was illustrated in the "rubber-hand experiment," during which an artificial hand is assimilated in the body scheme and is perceived as part of our body once the senses fall into the trap of illusion.[18]

The Cocktail of Experience

How do different pieces of information merge into one coherent and unified experience, which is at the basis of our consciousness experience? According to common belief, it takes place as a result of signaling and reciprocal signaling between reality-representation maps, sensory-input maps and representation maps of the "internal weather" (our feelings, physiological functioning measurements, etc.).

Some think that consciousness is formed out of the confrontation of perception impressions with our inner categories of experience. The perception impressions (the received stimulations) are confronted with "the world image," which is retained in our long-term memory, and our consciousness is based on the insights created in this confrontation.

Narrow Hips—the Perimeter of "Momentary Awareness"

The different aspects of reality, around the time we are exposed to them, can be retained in our "working memory."

This type of memory, which is like the "screen of momentary awareness," is sometimes called "memory at the time of performance." Information is retained in it in quantities that change in accordance with the type of input. For example, we can represent, at any given moment, an average of four pieces of visual-spatial information, such as the order of touching bricks that are scattered around the room, and seven pieces of verbal information, such as the order of digits in a telephone number we wish to remember. Considering the extensive volume of data processing in our brain at any given moment, we might assume that most of the brain processing takes place in a layer that is not accessible to the conscious layer and thus is not represented on the screen of momentary awareness.

Attention, at its different levels, takes place continuously. In an attempt to present a formula conceptualizing cognition as dependent upon attention, scientist Francis Crick suggested that the formula of cognition is attention multiplied by memory at the time of performance. Memory at the time of performance is the border zone between past and future, and, as such, it is actually the "representative of the present" in our existence.

The Ethics of Consciousness

The decision that certain modes of consciousness are worth experiencing—the ones that honor our essence as human beings and enable us to fulfill the human potential of each and every one of us—and, on the other hand, that other modes of consciousness are undesirable for the individual or society as one (for example, the mental dullness or catatonia that is experienced by drug addicts) is a controversial issue.

This issue is mostly raised with respect to using psychoactive substances that affect our awareness.

My Body on the Outside

The "otoscopy" experience can be described as an experience during which a person feels as if he is watching his body from the outside, and that the observing entity he is identified with is floating above his material body. This is a description that is usually connected to near-death physical conditions. A similar description can also be attributed to people who suffer from an electrical brain disorder originating at the temporal lobes. In studies, such experiences were connected to abnormal changes in brain activity at the border areas between the temporal lobes and the parietal lobes.[1]

The Scope of Cognition Modes

The scope of cognition modes starts with deep unconsciousness and ends with full wakefulness (which, I hope, is the current mode of most of my readers). The pendulum of attention moves between these ends, and its location depends on various factors, including emotional motivation, physical vitality, etc.

These definitions are inherently problematic, since cognition and consciousness are subjective experiences; due to their essence, it is hard, sometimes, to estimate the level of their presence in another person.

In the Absence of Consciousness—the Vegetative State

Victims of brain injury might fall into a coma. In this state, their eyes are closed, and the movement of their limbs is a reflexive reaction. The lucky ones overcome this terrible condition; others pass away; and others come out of the coma but remain unconscious, which means they remain in a vegetative state. In this state, the differentiation between the two main components of consciousness is reflected. The two components are wakefulness, which exists among the patients in the vegetative state, and awareness, which is the component that is probably absent among patients in the vegetative state. It seems that they are in a state of unaware wakefulness as opposed the state of aware wakefulness that characterizes the brain of a healthy person.

A cycle of sleep–wakefulness characterizes the existence wakefulness in the function of the vegetative brain. The eyes of such patients are alternatively open and closed, and they perform body movements that are probably reflexive and not task-oriented. The chances of overcoming a vegetative state lessen as time goes by. Today, it is believed that if a patient does not recuperate from a vegetative state within a year after brain injury or within six months after suffering brain damage related to oxygen supply to the brain, the recuperation chances are slim. Such patients are defined as suffering from a permanent vegetative state.

An EEG examination finds slower electrical activity during sleep that is not REM sleep. The findings of EEG with regard to activity in the brain of a person in a vegetative state are not significant, however, and at this stage do not provide prognostic evidence that helps assess the chances of recuperation.

Functional brain imaging, performed through methods such as PET examination, reveals that the level of metabolism in the brain usually drops in a vegetative state to values that equal less than half of the values in a healthy brain. Significant variance was found among the findings, however. In certain brain areas, such as the associative cortex, mostly located at the frontal and parietal lobes, and which deals with complex processing of sensory input, it was found that brain metabolism was extremely low or even nonexistent among people suffering from a vegetative state. It seems that the processing of sensory input takes place in the brain of such patients at the basic, initial levels of processing. The synthesis with other sensory inputs for the purpose of creating a multisensory, combined fabric of information is reduced or even nonexistent, and thus it seems that the awareness of sensory inputs is also absent.

Is an unconscious person deprived of cognition? The answer to this question can be found on a spectrum and depends on the level of shallowness of cognition. It refers to threshold proportion rather than to binary dichotomy.

It is also true for the level of "absence of cognition" during anesthesia in an operation—it moves along a continuum, and it is not a matter of "all or nothing."

It seems that cognition is the basic brain activity that constitutes the essential condition, though not the only condition, for the existence of consciousness. Consciousness requires brain activity at a volume and intensity that cannot be performed by the brain of a person who suffers from a vegetative state.

Chapter 4
Aspects of Inner Motivation and Reward

Motivation to Act

Don't give up, Sisyphus; it is just the peak . . .

The gods decreed that Sisyphus would roll a heavy boulder up a hill. When the boulder reached the peak, it would roll back down, and Sisyphus had to start all over again. Sisyphus was made to roll the boulder up the hill over and over again in a vicious, exhausting cycle that was ultimately futile. The component of purpose and reward is an essential part of motivation. When we believe that a certain act is futile and will not be rewarded, we might be ruled by the Sisyphus syndrome, which extinguishes the inner motivation necessary for performing the task.

The work of thinking is tiresome. It is sometimes intentionally cut off in the middle as a result of frustration in light of what seems to be a purposeless march in the paths in the brain that do not lead to a desirable destination.

The motivation to continue the climb up the mountainous path, despite the trembling knees, is the "motivational component." In this sense, what brings mountain climbers to the top of the mountain is the head and not the feet. This component is mediated mostly by the neurotransmitter dopamine. The dosage of dopamine in the synapses of the neural network has a critical part in determining the distance we reach and the number of paths we tour before we give up and stop our pacing on a certain thinking or behavioral path.

Boredom is mostly reflected as under activity of the inner motivation system.

A common behavioral expression of dopamine deficiency at the junctions of neural networking is lack of initiation (apathy).

New desires are constantly born. We desire ever-changing objects that come one after the other. The life expectancy of satisfaction derived from fulfilling our desires is short and mostly cut off as soon as a new desire appears and demands its realization. A new North Pole is born time and time again for the compass of our desires. Our desire-related mechanism is characterized by inherent dynamic imbalance.

Our yearning for satisfaction is a natural desire. As usual, however, the dosage makes all the difference, and when this desire takes exclusive control over the steering wheel of our life ship, and totally pushes aside valued tasks related to the "meaning of life," we lose a great deal.

When Pavlov's bell rang and his dogs drooled, the entire world of psychology got wet. The studies of B. F. Skinner[20] had a similar effect. Skinner is the famous representative of behaviorism. In his studies, he used extreme, estranged reductionism whose substantial limitations are reflected in omitting the "intracranial" processes from his worldview. The behaviorists totally and devotedly avoided models that included the mental processes that take place in the space between our ears. They referred only to stimulations and the resulted behavioral effect. In his experiments, Skinner demonstrated the power of rewards and reinforcements with regard to shaping behavior.

One of Skinner's main findings was related to the difficulty in rooting behaviors that are irregularly rewarded. One version of such behavior is the groundless belief of persistent gamblers that the day will come when they will "beat the casino." In the long term, this pattern of behavior exists in a manner of momentum that feeds itself. The importance of positive reinforcement, which sometimes has a great role in shaping our behavior, is a finding that should serve as a guiding principle for parents and educators.

In his studies, Skinner exposed the important role of the basal ganglia, which are clusters of gray matter in the brain that probably have a central role in encoding behaviors that are defined as habits.

The behaviorists' vision was that, if we adapt a strict training regime, we will be able to reshape our behavior and add improved abilities to our basic behaviors' repertoire. Training may lead to behavioral betterment. Methods that were directly developed from Skinner's reservoir of operant conditionings (operant conditioning is a basic concept in behaviorism that refers to changes in behavior patterns caused by reinforcement) are used as an efficient tool in treating various types of anxiety disorders (phobias) and sometimes manage to eradicate them. An example of such methods is the systematic desensitization of the sense of anxiety through "flooding"—ongoing exposure to the anxiety-causing stimulation.

The Reward

Freud thought that human behavior might be explained on the basis of two main motives: the pursuit of pleasure and the avoidance of pain.

The race to happiness, which exists out there, beyond the current moment, is a common motive of long-distance runners, though it often remains an unfulfilled promise (a type of fata morgana).

The pleasure and its reward: Our brain composes a sense of pleasure and satisfaction out of neural activation patterns that involve activities that improve genes' survival (eating, drinking, sexual intercourse, etc.).

The system of rewards, a functional system that relies on a number of brain structures, grants a hedonistic value tag to results of various activities according to past experiences (hedonistic memory). The estimated hedonistic value of different acts (which is the estimation of "potential pleasure") is an important factor with respect to our basic motivation to perform these acts.

The dopamine nectar drips its intoxicating drops as early as the stage of anticipation. It might mean that a person who buys a lottery ticket pays "the tax of the foolish," but it is seen this way only from a rationalistic point of view, which calculates the slight chances of actually winning the prize. The anticipation period between purchasing the ticket and the announcement of the results is a period characterized by elevated dopamine activity, which induces an exciting sense of arousal in the brain of the person who bought the ticket. This sense is reflected, inter alia, in sweet dreams about using the future prize.

A tug-of-war competition constantly takes place between an action pattern that is expected to bear fruit immediately and an action pattern whose fruit are expected to ripen sometime in the future.

Functional brain imaging studies show that an action whose effect is expected to be manifested in the future, and which requires postponement of gratifications, mainly involves brain areas in charge of planning and logic, such as the prefrontal cortex. On the other hand, an action whose effect is expected immediately involves "short-fuse" brain areas, especially the nucleus accumbens area, which is intended to take actions whose rewards are expected "here and now." The result of this inner-brain confrontation is determined by the brain areas that perform the more intense activity.

While anticipating a reward, the nucleus accumbens brain area, in charge of expectations, is active and inducts a pleasant arousal. This feeling often exceeds the reward itself with regard to the hedonistic value (induction of the sense of pleasure). The mere anticipation floods the nucleus accumbens area with dopamine. It also turns out that the level of dopamine is highly sensitive to the value of anticipated reward and less affected by the chances of winning it. It is thus more likely that a higher number of people will participate in a lottery with a higher prize than in a lottery with a lower prize, regardless of the winning probability.

In most cases, the reward matches the effort; the size of the reward often correlates with the level of effort invested in it. An orgasm is thus a potential reward for long months of courtship and nail-biting.

When it comes to relationships, we are sometimes required to sweep the dust of routine off the path connecting the spouses. Novelty rewards and activates the cycle of rewards in a circular pattern.

The main center of pleasure in the brain is located in the cluster of cells of the nucleus accumbens, which is connected, inter alia, to the brain areas of the limbic system and the amygdala. It might be defined as the main temple of hedonism in the brain. Stimulation of this area, which is rich in receptors for the neurotransmitter dopamine, leads to the subjective sense of satisfaction.

The desire to activate this area and experience the accompanying sense of pleasure is a central component of the inner motivation for many of our actions. Some might claim, somewhat judgmentally, that when the basis of a certain thought or act is the desire to offer a sacrifice to the god at the temple of hedonism, the motive is morally inferior.

Many of our actions, or even most of them, however, are based on the pattern of this mechanism. The profile of the sense of pleasure is drawn by a three-dimensional network of neurons that mostly exist in proximity to the nucleus accumbens area. These neurons are activated through bioelectrical potentials that are mostly carried on the waves of neurotransmitters such as dopamine, endorphins, and probably noradrenaline as well.

The result of activating this neural network is the induction of the subjective sense of enjoyment and pleasure.

Attention Disorders and Hyperactivity—Unsatisfying Satisfaction?

When there is no sufficient arousal of bioelectrical potentials in the reward system or, alternatively, when the wave of arousal fades away too quickly, the emotional result is a rapid fading of the satisfaction experience.

It seems that this pattern, which is abnormal, is at the basis of various behaviors that are usually perceived as exceptional, such as the behavior related to attention disorder and hyperactivity. It seems that the sense of satisfaction experienced by people who suffer from this disorder has a shorter-than-usual life expectancy. The search for excitement that will arouse a longer-lasting pattern of bioelectrical potentials and induct a lasting sense of satisfaction is a potential cause of hyperactivity and the constant skip from one activity to another in search for satisfying stimulation.

The dosage of neurotransmitters and the distribution pattern of their matching receptors are the main measurements that affect the regulation of the intensity of bioelectrical potentials. A leading, evidence-supported hypothesis is that at the basis of attention disorder and hyperactivity (ADHD) there is a lasting disturbance caused mainly by the abnormal effects of dopamine and noradrenalin.

The Brain and Economic Decisions

The Wallet Between our Ears—Aspects of our Function as Homo Economicus

A model according to which man always acts rationalistically does not take into account the biases inherent in the human thinking pattern: our behavior as human beings is a mix of common sense and a lack of common sense.

An "algorithmic person" who always obeys the strict rules of common sense and who has served as a reference figure in many traditional economic theories is a figment of imagination and does not really exist.

Thus, in the sphere of behavioral economics, which is sometimes referred to as neuroeconomics, the top and bottom lanes are active at the same time, and rational and irrational decisions are taken in a mishmash manner.

When judged according to cold logic, Homo economicus might often be seen as an irrational figure, since considerations like life philosophy and emotions have a significant effect on his economic decisions.

Economic negotiation between people is, in many cases, a brain battle concerning neural superiority.

Our consumer brain is an easy prey for the sharks of the advertisement industry, who sometimes use "black magic" and influence people who are not aware of the fact that they are being influenced. Marketing wizards can sell heavy-metal CDs to participants of vipassana workshops or shaving cream to the Taliban people. Advertising and marketing methods use guerilla warfare against our brain, which is often captivated in their hands.

When people are making a consumer decision regarding purchasing a certain product, the nucleus accumbens wishing to purchase a product that might cause pleasure, and the insula, which is distressed and disgusted by the cost of the desirable product, often confront each other. The prefrontal cortex is like a spectator who just watches the confrontation without contributing much. The marketing wizards try to help the nucleus accumbens win by dimming the insula's function by means of offering discounts, installment payments, etc., and, on the other hand, intensifying the function of the nucleus accumbens by highlighting the product's desirability.

Happiness and Wealth

Studies show that wealth reinforces the daily experience of happiness when it lifts people from severe poverty to a position in which meeting the basic existence needs is relatively guaranteed. Beyond that, the contribution of wealth to happiness is small. The validity of the above conclusion, found in several studies, is a source of great controversy, as most people feel, at the intuitive level, that the correlation between wealth and happiness follows a linear pattern, at least until the high peaks of the cash mountain.

The mental toolbox we use when making economic decisions relies, inter alia, on heuristics, statistical predictions, and intuition. Heuristics are rules of thumb, or super-memes, that constitute the mental infrastructure of the bottom lane. They are at the basis of many of our decisions, since they are rapid and do not require much mental energy, but, on the other hand, they are rich in "white noise" and often soaked in mistaken inferences.

The predictions produced by our brain are prone to mistakes with respect to scenarios based on probabilistic assumptions. Sometimes intuition is also "highway to the mistake."

We tend to make a great part of our major economic decisions on the basis of emotion rather than on the basis of logic. For example, studies have pointed out that the emotional price tag attached to loss is higher than the emotional price tag attached to gain in the same rate. In other words, the bitterness we feel after an economic loss leaves us with a stronger impression than the sweet flavor of gain in the same rate. An assessment assessed the height of waves of negative emotions following an economic loss as two times higher than the height of waves of positive emotions following a gain, taking into account all necessary reservations with respect to the difficulty to quantify qualitative data and the unique circumstances of each case.

In terms of the main coin of pleasure in the brain—dopamine—a broken expectation creates a deep hole in the dopamine graph, the depth of which is twice the height of the hillock created by a fulfilled expectation.

A study in which a comparison was made between decisions made by bees and decisions made by humans had interesting findings. During the experiments, the human participants had to choose between two options: pressing the "risk" button that might lead to a gain of four units but only in 80 percent of the cases, or pressing the "safety" button that led to a certain gain of three units. Most of the participants choose the second option of guaranteed success, although the gain was lower than the gain in the first option. A similar experiment was conducted with bees. They were introduced to two types of sugar solutions of different concentrations. The first solution had a higher concentration of sugar, but the possibility to reach this solution was only at a ratio of 80 percent, while there was a 100-percent success rate with regard to meeting with the less-sweet solution. The bees, like the humans, distinctly preferred the second option, which carried the 100-percent success rate.

Allegedly, the action pattern of humans and bees leads to a lower expected profit, but these findings may be perceived as empirical verification of the validity of the insight according to which the bitterness resulted from a loss is twice as intense as the sweet taste (double meaning) of gain. Following this hypothesis, in the experiment with the human participants, which was composed of five experiments, the expected results were as follows: When choosing the "safety" button, the expected profit is fifteen units and, on the other hand, choosing the "risk" button consistently is expected to yield sixteen units (an expected success rate of 80 percent, which means four successful attempts out of five attempts, each of which produces four units of profit), but the fifth attempt, which statistically leads to a loss, will bring along a penalty, in the emotional sense, that subtracts eight units of profit (since the bitterness related to loss is twice as intense as the sweetness of gain). In such a case, the expected profit at the emotional level would be eight—almost half of the expected profit of the "safe" approach, which seems to be the more widespread approach in the animal kingdom.

Prediction of economic behavior, which is based on the hypothesis that humans make economic considerations by using only a rational point of view, has a limited efficacy. The stock exchange arena is ruled by the amygdala, as it is ruled by the frontal lobes, and sometimes the latter are even less prominent there.

Routine experiences bring with them an emotional suggestibility that is capable of conditioning our behavior. While we are driving, sometimes the song we listen to include implicit effects that prime our decisions. Such effects might also be concealed in a newspaper's subheading that caught our attention, or a sentence we heard from our colleague. All of these create an emotional charge that directs our tendency toward a more cautious or a more daring decision. In the random bush of priming factors, the incidence of logic-based paths is not high, which results in a difficulty in behavioral predictions in the economical context and other contexts. It also explains the difficulty of predicting stock exchange courses exclusively based on rationalistic models.

"The economic reasonable man" is an elusive entity that is a hard nut to crack.

The Fall Inside

According to a study, the attitude of stock exchange investors is more cautious in the fall, and they tend to buy fewer shares during this season. It seems that the chilly winds that carry the falling leaves chill the investors' enthusiasm as well. The season of the year creates a season of the brain, a sort of emotional climate that serves as a substrate that sprouts seasonal wishes, which might change as a new season comes. It seems that the weather creates priming, which leads us to prefer a different risk management policy painted in seasonal shades (or even daily shades, or shades of the hour).

Dim Monday, Sunny Friday

It was found that sunny days are more likely to end with appreciation of rates at the stock exchange, while cloudy days are more likely to end with depreciation of rates. It was also found that the days at the beginning of the week are more likely to be painted in gloomy shades and are more prone to depreciation of rates, while the days closer to the weekend are more prone to appreciation of rates.

The "fear index" (referred to as the VIX), which attempts to quantify the sense of anxiety among stock exchange investors, demonstrates, through its fluctuations, the waves of hope and despair that shake investors. The amygdala is the main generator of these fluctuations, which depend on the level of its activity.

"The ultimatum game" is called this because the owner of the property presents an ultimatum to his partner—"take it or leave it"—and does not leave any room for negotiation. Studies show that when the owner of a property offers a lower share of the property's value (usually a value lower than one-third of the value), the resentment toward the "unfair offer" wins, and it is rejected by the other party. This is an example of a nonrationalistic reaction, since getting any share of the property is better, in terms of economic common sense, than getting nothing. This tendency rules until the proposed share constitutes a "reasonable sum," and although it still constitutes only one-third of the prize, its absolute value is higher, since the basic sum of the prize is higher. In this case, due to the change of circumstances, the childish anger is put aside, and the insulted ego is replaced by the content ego and is willing to accept a third of the property. Such a scenario is a battlefield between emotional reactions and rationalistic reactions.

The insula is the source of the sense of insult when one of the participants is offered a sum he perceives as "inappropriate." The prefrontal cortex is the source of "economic rationalism," whose aim is to close the deal so we will "see the money." The frontal part of the cingulate tries to mediate between these two "behavioral recommendations," which often conflict, as the entity in charge of arbitration, at times, between inner-brain mental confrontations.

"The dictator game" is similar to the ultimatum game except for one significant aspect: the party to whom the offer is made must accept the offer. When there is no option of rejecting the offer, and thus no need to make a decision, there is no emotional-cognitive dissonance, which is common in the ultimatum game.

At the time of competitive sale, when more than one participant is involved, such as in a tender or an auction, a different cognitive dynamic takes place. The sense of fairness or lack of fairness with respect to the offers, as is commonly reflected in the ultimatum game, is usually replaced by "cold" economic considerations.

Different life situations involve significant information gaps. Our knowledge is never complete. Sometimes, as in a game of chess, the information on life's board is exposed but the plans of our opponent are concealed. On the other hand, in a game of poker very little is exposed, and the intentions of our opponent are concealed, as well.

Poker-like situations are like a jiggly dance with the unknown. In the game played by the masses, such as the game of the stock exchange, asymmetric information (reflecting an information gap—additional information held by some of the participants and concealed from others) sometimes grants a crucial, unfair advantage. Dealing with scenarios that include various levels of built-in uncertainty, and to which the reaction is time restricted, serves as the catalyst for decisions that do not necessarily rely on common sense.

Chapter 5
The Story of Dopamine

As an example of the importance of neurotransmitters that compose the biochemical cocktail in which our brain resides, this chapter focuses on one of the main neurotransmitters—dopamine.

The effect of dopamine was clarified in 1954 during an experiment carried out by two researchers, Peter Milner and James Olds. They put an electrode at the core of a rat's brain at a spot that later turned out to be the nucleus accumbens—which has a central role in producing the experience of pleasure in the brain. A constant electrical stimulation of this area made the rats wrap themselves up at the corner of their cage, overwhelmed with pleasure and ignoring all aspects of basic existence—food, drink, sleep, and a love life—until, at a certain point, they died of thirst. Afterward, it was found that as a result of the electrical stimulation, waves of dopamine flooded the area of nucleus accumbens, and its importance in creating a sense of supreme pleasure was made clear.[21]

The experiments of researcher Wolfram Schultz improved our understanding of the roles of dopamine and shed light on them. Shultz inserted a needle into single cells that communicated through dopamine in a monkey's brain. Shultz made a sound, after which he dripped a few drops of apple juice into the monkey's mouth. The cells responded to the apple juice with applause of dopamine spurts. Later, upon learning the correlation between the sound and the following drip of juice, the cells started to react with spurts of dopamine as soon as the sound was heard. Shultz called these cells "prediction cells," since they matched their activity to the reward scenario—the dripping of apple juice—and calibrated themselves, with fine-tuning, in a pattern of constant feedback to the level of reward and sometimes even to the lack of reward.[22]

Dopamine is responsible for navigating many of our behaviors, since it seems that it calibrates the dial of motivation to perform them. It is not an exclusive contributor to the calibration of the dial—other neurotransmitters also take part in this complex function—but the role of dopamine is central.

If we step up the resolution ladder toward one of Freud's main arguments regarding the central role of the libido in regulating our behavior, we might claim that it is not the libido, per se, but the dopamine, and the attempt to increase its level, which is at the basis of the libido, that constitutes this central layer in our behavioral motivation.

With respect to this claim, we might add that the intensification of dopamine levels caused by the image of the pretty Helen in the brain of Menelaus, King of Sparta (as a "mental, virtual icon," as she was not there physically), led him to send a thousand Greek ships to Troy.

Another "Greek illustration" of intense inner motivation guided by dopamine is the Sacred Band of Thebes, which was composed of 150 pairs of male lovers who formed a united group with the help of which the Thebes army managed to beat the tough Spartan soldiers. This elite troop was almost completely annihilated in the battle against the Macedonians. This is a cynical example of a cruel link between concepts that are "spiritual enemies": love and war (and a mirror image of the naive advice to "make love; not war"). In this case, the love and companionship among the troops, mediated by the dopamine and oxytocin neurotransmitters, were used to serve the war machine.

We might even claim, then, that it is not money that makes the world go around but, rather, the increase of dopamine levels in the brain that derives from the joys money can buy and in second-order conditioning—the financial profit itself.

Dopamine also has an important role in determining the "borders of the self." When the normal function of dopaminergic neurons (the neurons among which dopamine mediates information transfer) is disrupted, the borders of the self become blurred. People who suffer from schizophrenia (whose cause, according to a common hypothesis, is disruption in the tracks of dopamine and its receptors) thus often complain about hearing their inner thoughts as external voices. This is an example of the removal of partitions between "private territory" and "public territory," in the sense of the experiences they have.

Dopamine inducts a tendency for visual hallucinations and delusions. Its biochemical derivatives, noradrenaline and adrenaline, might also contribute to this. Dopamine is the main contender for the title "the main prediction substance in the brain." With the help of dopamine, the brain predicts the future and interprets reality.

In ADHD, there is reduction in the level of dopamine and noradrenalin derived from it. As a reflection of that, the key areas in the brain whose activity is reduced, in cases of ADHD, are the prefrontal cortex and the paths connecting it to the dorsal cingulate gyrus and the basal ganglia, located at the core of our brain. As a result, a difficulty might develop with respect to "executive functions," which are mostly the responsibility of prefrontal brain areas. These might include difficulty in focusing attention and preventing it from shifting toward distracting events at the margins of the attention beam, difficulty in mental resistance (the ability to preserve a lasting thinking effort), difficulty in refining reactions (sublimation) that might hurt other people, and so on.

A saying that briefly expresses the functional difficulty characterizing people who suffer from ADHD arises from the

"Berkeley theory," according to which "ADHD is not a problem of knowing what to do, but rather of doing what is known."

Using medications that reduce dopamine activity in the brain (referred to, in professional jargon, as "neuroleptic medications"), although sometimes necessary, especially during psychotic episodes, carry the side effect risk of scarecrow-like behavior: lacking initiation and motivation.

The claim that in many senses we are nothing but "biochemical marionettes" is painfully reinforced in light of such conditions.

Genetic variance with respect to the amount of dopamine receptors is a popular explanation for a behavior pattern that hurts its owner. Nowadays we know of five types of dopamine receptors that differ in terms of their reactions. A mutation of the gene that causes reduction in the density of dopamine receptor D2 was found, in several studies, as correlating with the tendency to suffer from various types of addictions such as obsessive gambling. It seems a plausible explanation for that is related to the finding that, among those who have a relatively small amount of D2 receptors, there is a reduction in neural reaction to negative feedback. As a result, the negative implications of the addiction leave a more subtle impression on these people compared to the impression left on those who have a normal amount of D2 receptors. As a result, it is more difficult for them to learn how to avoid the risks. On the other hand, it is possible that the impression left by the sense of satisfaction fades away more quickly than usual, which makes them more prone to taking risks, which leads to an increase in the level of dopamine.

Dopamine Dysregulation Disorder

Among people who take medications that increase dopamine's level in the brain or activate its receptors, a prominent behavioral change sometimes takes place. As part of the behavioral change, behaviors involving high risk, as well as behaviors that can potentially cause great excitement, in the sense of stimulating the reward and pleasure system in the brain, are intensified. Such behaviors include, inter alia, frequent gambling and hypersexuality. Alternatively, behaviors characterized by an obsessive nature are developed, such as compulsive, obsessive behavior that swallows the mental reservoir of the patient and steals a great part of his time resources. It seems that hyperactivity of the paths of dopaminergic neurons, especially at the functional system located at the core of our brain, called the "mesolimbic system," is at the basis of such behaviors. This hyperactivity intensifies the craving for a spurt of dopamine, which accompanies a moment of excitement such as the moment of winning a large sum of money or, alternatively, the moment of orgasm. In other words, dopamine intensifies the craving for more dopamine up to the "verge of fireworks," which leads to relaxation and a temporary pause in the race to increase the activity of dopaminergic neurons. After a while, the system is calibrated in such a way that only a stimulation of exceptional intensity—"supra-normal," whose intensity is higher than the intensity of routine stimulations—manages to increase the level of dopamine. As in a vicious, magical circle, this phenomenon reduces the level of pleasure caused by the familiar and routine—a concept termed "hedonistic erosion."

About 15 percent of Parkinson's patients who are treated with medications that increase the level of dopamine (called "dopaminergic medications") develop a phenomenon characterized by difficulty in controlling urges and a low level of inhibition, which is reflected, inter alia, by an obsessive gambling urge. Patients who have experienced this described a surge of pleasure that floods their brain at the time of winning. Such a wave derives from a spurt of dopamine at brain areas that network the "cycle of reward." The surge is formed as a result of the heightened availability of dopamine, carried on the waves of dopaminergic medications, and it makes patients experience "addictive happiness," which serves as a motive for continuing to gamble.

Dopamine has a great effect on our moods. Parkinson's patients who are at a state of severe decline in dopamine content in the brain (the professional term used to describe such a situation is "off") often feel melancholic and claim their "vital spark" has been extinguished. A low level of dopamine inducts a sort of "biochemical anhedonia" and a wide variety of low moods from dysphoria to major depression.

Supra-Normal Stimulation

"Supra-normal stimulation" is the term coined by the Dutch biologist and Noble Laureate Niko Tinbergen. It refers to a stimulation characterized by higher intensity compared to natural stimulation and formed through the tendency to intensify the reward and increase the intensity of dopamine spurts.

The current repertoire of stimulations in our era allows for a variety of supranatural ("supra-normal") stimulations, which turn the dial of excitement in our brain toward values it is not designed to reach naturally. Habituation to stimulation at a certain level of intensity makes it unsatisfactory down the road, and from one experience to another the unidirectional push of the stimulation threshold is formed upward in order to retain the same level of excitement.

In order to increase the level of dopamine, an ever-growing intensity of stimulation is needed, which is similar to the phenomenon of drug addicts needing larger quantities of a drug in order to retain the same effect.

An example of a contemporary, troublesome behavioral effect related to supranatural stimulations
  is the preference of online pornography to a flesh-and-blood partner.

Dopamine and Co.

It seems that a central ingredient in the divine nectar that nourishes our brain and wheels our life is dopamine.

Focusing on a single neurotransmitter, out of the many neurotransmitters that sail in the streams of our brain, is not intended to underestimate the essential role of other neurotransmitters. Thus for example Two other neurotransmitters that cooperate with dopamine to enhance the feeling of pleasure are Enkepalin (an opioid 'made in the brain') and Anandamide (the brain's version of marijuana).

Many neural messengers that constantly change in terms of location and concentration take part in complex brain functions. There is no single neurotransmitter that navigates complex brain activity exclusively, but dopamine, as aforementioned, has an important role in numerous key brain functions.

Dopamine at the Service of Genes

In a totally nonaccidental manner, situations that lead to "dopamine spurts" are closely related to improving gene survival. For instance, sexual orgasm involves a chance of creating a new creature that will carry some of the genes into the future.

On the day we are born we already have an overdraft in terms of genes: a "performance demand" of the libido bound with a fine from the ones who take their time. The fine is the sense of inner tension, which consumes energy resources, as a reminder that you still have to pay your periodic debt to the genes and fulfill your evolutionary role. The sense of pleasure is inherent in the act of "paying the debt."

Dopamine is released at times of sexual arousal before the act itself. It intensifies the libido, facilitates orgasm, and has a central role in activating the pleasure areas in the brain. Medications that increase dopamine release or imitate its action by connecting to its receptors might increase sexual desire.

There are additional examples of the importance of pleasurable situations (reflected physiologically in "spurts of dopamine") in terms of gene survival. In the society we live in, a large financial profit involves improved survival chances for the single organism, for obvious reasons, and a higher life expectancy, which increases the chance of a higher incidence of sexual intercourse. Laughter and music improve socialization and the chances of forming relationships.

The lives of many of us are managed as a search for the golden fleece of the master key to the doors of happiness, which is, in fact, a renewable chase after the next dopamine spurt.

In the course of a natural, normal life that does not involve use of psychoactive substances, we aim at this much-longed-for moment of dopamine spurt and enslave numerous moments of our life in order to reach it. It often involves a loss of precious time. It seems that this was what Freud meant when he claimed the libido, which directs our behavior toward reaching the next dopamine spurt involved in sexual orgasm, has an intense driving force in our lives.

A major part of our conduct can be seen as a desperate chase after dopamine spurts and the blessed serenity of endorphins, similar to a drug addict who craves his next trip. We should be aware of the fact that a major part of our behavior derives from our desire to "break the dopesick for dopamine." In this gloomy sense, we are nothing but biochemical marionettes dancing to the sounds of the hidden chords of the dopamine harp.

Being aware of this fact might loosen the unseen grasp dopamine has on our brain a little bit. We sense a feeling of inner tension when the biochemical-electrical pendulum of central neural pathways (such as the neural mesocortical pathway, with its three-dimensional networks between the limbic system and the frontal lobe) deviates from the point of balance. Such a deviation, above a certain threshold, creates an inner motive (urge) for action. This action is mediated by dopamine, which brings the pendulum back to the point of balance.

The above mechanism shares similarities with Freud's pleasure principle in terms of release of tension and aspiring for homeostasis.

The point of balance is not meant in the sense of ataraxia (stoic tranquility), since, under the influence of dopamine concentrations, we are often directed at innovation-oriented behavior, excitements, and even dangerous challenges.

The craving for the new is mainly dopamine-driven. An innovation that makes our heart beat is at the basis of our attraction to fairytales, new books and films, and the surprise component similar to a punch line of a good joke. The dopamine, as aforesaid, is at the basis of our yearning for innovations. Studies point out that a medium level of novelty is more attractive than absolute novelty, which is perceived as threatening due to its lack of familiarity, and it is also more attractive than absolute familiarity, which is perceived as boring and lacking excitement.

The initial attraction between spouses is created thanks to the shower of dopamine at the reward system in the brain. These showers are the start switches of the courtship-matching-parenthood software wired in our brains.

Fred and Wilma from *The Flintstones* (a very nice animation series about a human family living in the Stone Age) acted under the influence of the same potion flooding our brain.

Subjects of an experiment were introduced to scenarios that presented various versions of "the prisoner's dilemma." During the experiment, their brain was scanned by means of an fMRI. It was found that among the subjects who tended to cooperate more with others, the dopaminergic neurons pathways that run the reward system in the brain were highly active.

These areas demonstrate intense activity when people look at attractive partners or experience contact with them, when their tongue meets their favorite dessert, and when they are under the influence of cocaine. Dopamine activity is a central component in these reactions, and it is also a prime cause for the fact that love is addictive, trust is pleasurable, and cooperation is relaxing.

Sounds as Agents of Pleasure

Music can shape an emotional state, which increases various aspects of thought flexibility. The mysterious path of music in terms of forming emotions is based on the link between audio-processing centers and the limbic and reward systems. This connection is at the basis of the ability of the fingers of sounds to play the keys of emotions in the brain.

Music causes changes in the neural pathways of the reward system with the help of dopamine and opioids, similar to the changes that are caused by familiar pleasure generators such as laughter, play, sex, and drugs.

Sounds are active in our brain as effect agents. For example, in a wine store, it was found that while French music is played in the background more French wines are sold, and when Italian music is played more bottles of Italian wine are sold.

Dopamine Coordinators—Brain Performance

A higher amount of dopamine at the prefrontal lobes is found to be in correlation with better performance of a great number of cognitive tasks. In the brains of young people, the average content of dopamine at the prefrontal lobes is higher compared to that of aging brains.

High levels of dopamine in the brain lead to a good mood and improve cognitive processing. Learning as a result of positive or negative reinforcement depends on the level of dopamine.

The dopamine is a central service tool in the mechanism of the "crystal ball" in our brain. Its dosage at the designated neural networks enables us to predict rewards. The correlation between the actual reward and the predicted one affects its level. The changes in the level of dopamine determine our level of satisfaction from the result. A wide gap between expectation and realization, however, when the actual reward greatly exceeds our expectation, bounces the level of dopamine. An expectation of a significant reward increases the level of dopamine, and it is also increased when the actual reward is more satisfying than the predicted one.

Neurons whose communication signals are transferred by means of dopamine mediation constantly create prediction pattern such as "if-than." Correlations that are validated by reality manifestations reinforce the neural pattern at their basis, which means the network of neurons that constitutes the structural infrastructure of the predicting pattern that was realized. A fulfilled prediction rewards the neuron network that predicted it with a spurt of dopamine, which creates a sense of pleasure and reinforces its structural infrastructure. An unfulfilled prediction causes a sharp decrease in dopamine discharge, which is the biochemical parallel of the sense of disappointment, and the structural infrastructure at the basis of the unfulfilled prediction becomes loose. A reward that exceeds the prediction causes the greatest spurt of dopamine—three to four times higher than the basic level of dopamine—and, accordingly, it leads to a maximal sense of pleasure.

The dosage of dopamine and the level of arousal it induces are correlated with the gap between our expectations and the actual events. The wider the gap between the expected result and the actual positive result that exceeds expectations, the higher the amount of dopamine secretion. In this sense, dopamine is the trademark in the profile of "a good surprise."

Prediction is not only related to the quality of the expected result, but also to its timing within a certain window of time. When the reward's quality and timing match the expectation, the level of dopamine is retained and even increases ("the pleasure of being just"). On the other hand, a reward that is lower than expected significantly reduces the level of dopamine.

The brain area called the anterior cingulated gyrus is in charge of weighing the production of emotional brain areas and rationalistic brain areas and their mergence. Such an activity enables the "lesson" our brain is supposed to learn, in light of the unfulfilled prediction, so we will not make the same mistake over and over again. An accurate intuition results from laborious trial-and-error practices whose lessons are written with the dopamine pen.

The level of dopamine in certain brain areas is translated into our tendency to initiate certain acts. If we know the level of dopamine at certain brain areas, we will have a tool with which to assess to what extent a certain person is likely to initiate daring acts and take risks. In other words, there is a correlation between the level of dopamine and the level of daring in various circumstances, such as investing in the stock exchange or even initiating romantic relationships.

The world is represented in our brain as a scope of reward areas of various degrees (positive, neutral, and negative). Our brain navigates through this gradient of rewards in an attempt to reach an area with a high concentration of rewards and draw away from areas with low or negative concentration of rewards. The main compass used by our brain in its maneuvers is the level of dopamine.

Chapter 6
Psychonautica

One can find evidence for initiated use of cognition-modifying substances in almost every human culture. It turns out that the smoke of the frankincense resin, which curled upward during various rituals in the Middle East in ancient times, has an antidepressant psychoactive effect.

The Aztecs and Mayans used to use enemas that contained extracts of deliriant seeds during various rituals. This pathway of achieving euphoria is quicker than better-known patterns pathways, such as the pathway of the esophagus (drinking) or the pathway of the trachea (smoking), and its side effects are fewer.

According to mythology, the gods talked from the throat of the Oracle of Delphi, who apparently owed her richly descriptive prophecies to earthy intoxicating substances rather than to divine inspiration.

African tribes' members used to use the hallucinatory drug iboga.

Opium dens were common in the past in the Far East.

It was 1897 when researchers probably started researching deliriant substances in an organized, scientific manner. This was the year in which the chemist Arthur Heffter isolated mescaline—the psychoactive substance found in the peyote cactus—and, by doing so, took off to the sky with diamonds in a scientific rocket of psychoactive substances.

Out of This World—Heaven's Breath and Angels' Dust as Cognition Expanders

When laughing gas (nitrogen oxide) was discovered, and that upon meeting our brain it improves our mood, it was termed "breath of heaven." Cognition-modifying substances were termed breath of heaven and angels' dust, since their fans thought they were capable of making our cognition soar from the ground level of earthy cognition to divine heights of improved cognition.

Freud described his real-time thoughts, sailing his stream of consciousness while experiencing the use of cocaine, to his fiancée, Martha, in a letter he wrote to her on February 2, 1886. The rapidly changing mix of emotions under the influence of the drug is reflected in his words. He also sent her a small portion of the drug in order to "strengthen her and redden her cheeks a little." At the time, the use of cocaine was not prohibited. Later, he studied the medical usages of the drug and wrote about his experience using it in an essay called "*Uber Coca*" ("About Coca").

Aldous Huxley experienced the use of the mescaline drug and documented his experiences in his paper "*The Doors of Perception.*"

The chemist Alexander Shulgin[23] might be considered the Mendeleev of psychedelic drugs - the founder of the periodic table of psychedelic drugs. Shulgin wrote some very thick books about psychedelic substances. He arranged encounters between his brain and some of these substances and described the implications meticulously and scientifically.

Psychedelic Hallucinatory Drugs

Psychonautica is a like a hitchhiker's guide to the universe of cognition. Some people consider psychoactive substances a hidden path to the "forbidden city" of cognition, a type of heavenly Tree of Knowledge apple that is available to residents of Earth to bite and enables them to "exit inside."

Hallucinatory drugs are described as spaceships traveling in expeditions across the brain galaxy as the psycho-nautic universe opens up for them without need of a compass or road map. Some give excuses and describe their use of hallucinatory drugs as an attempt to open the windows of cognition and let the "light of pure reality" in. They describe their experience with these substances as a journey to places in which the limits of time and space dissolve; places where the wind has color, and a field of flowers of every scent, shape and color, never encountered before by our senses, stretches out to the horizon.

Psychoactive substances like amphetamines and cocaine increase the amount of dopamine at the synapses (communication junctions between one neuron to another) in various neural pathways. This mechanism is central in relation to the effect of many hallucinatory drugs. An additional important action mechanism in the effect of psychedelic substances is activation of A2 receptors of serotonin. These receptors greatly affect the processing of the senses' input and perception processes.

It was also claimed that musical sounds at unique frequencies and rhythms, downloaded from various Internet sites, might affect the level of consciousness beyond the known emotional effect of music. In other words, the fingertips of frequencies, which can be seen as the morphological aspect of sounds typing on the keyboard of consciousness, and the aspect of overall general acoustic harmony, which can be seen as the aspect that is more related to content, seem marginal in this context. This type of effect on brain activity has also been defined as a "digital drug."

Some people claim that they managed to go deeper into the depths of consciousness when they were "exhausted after spending a night of too much of everything." As "rain pilots" who perform cloud seeding using silver iodine at stormy nights and "milk" from the clouds more showers than they would voluntarily give, it seems that certain psychoactive substances "milk" our brain for more than it is willing to give us under normal circumstances.

Some consider the effect of psychoactive substances as the incantation that opens the Ali Baba cave of cognition, which is normally hidden from consciousness. Sometimes, they are used as "consolation pills" or biochemical escapism. In his dystopian novel *Brave New World*, writer Aldous Huxley describes a "soma pill" taken by people to allay their mental suffering.[24] Like Huxley, researcher Susan Blackmore believes that certain psychoactive substances are "consciousness expanders and tuners."

Some people zigzag between exiting sanity and going back to it (brief, controlled exits) by means of psychoactive substances, as a repetitive pattern, in the name of "cognition expansion."

There is no dispute over the fact that multiple risks are involved in psycho-nautic journeys in the universe of consciousness: psychoactive substances are justifiably sensitive legality-wise and health-wise.

Brain ambrosia at every street corner involves the risk of inflicting our brain with madness, which would be reflected in temporary or permanent functional disruption.

The Inner Skull Drug Cartel

There are highly complex relationships between the various ingredients that compose the biochemical cocktail that floods our brain.

The brain is full of streams, falls, and pools of compounds that mediate its activity. There is constant dynamic balance among the concentrations of various psychoaffective compounds, which compose the soul-potion cocktail, that greatly affect our emotional and mental state at every given moment.

The brain is a large laboratory in which self-produced drugs are constantly made for its own usage (see dopamine, testosterone, serotonin, acetylcholine, adrenalin, noradrenalin, etc.). We might claim that all of us are naturally drugged by substances mixed by our brain. The periodic table of human brain drugs has not been completed yet; further, strenuous research is needed in order to improve our understanding regarding the order and regularity of the mix of potions that flood our brain.

The number of potions made at the lab located inside our head and the distribution patterns of their receptors determine, to a great extent, the nature of the "cognitive creatures" we will become, in the sense of our typical thinking style, the emotional soundtrack that is usually played within our skull, the preferred processing pattern, our tendency to prefer a certain type of information to another, etc.

Some might claim that, as time does not exist in a mechanical watch but can be inferred from the pattern of the hands' configuration, "high emotions," such as affection, do not "exist" in the brain but are, rather, formed in a pattern of "inner interpretation" of a configuration of typical activity of neurons' networks (such a view involves the existence of the "experiencing self," and it resides in the twilight zone of body–soul relationship). In this spirit, we can compare the activity configuration of the neurons and the hands of a clock, and it derives, inter alia, from the mix of neurotransmitters, which mediate the activity. In addition, the neurotransmitters also have a role in interpreting the content granted by the "experiencing self" to the morphological aspect of the activity pattern of neural networks. According to a rival approach, the activity configuration of neurons' networks directly creates the modes of consciousness, which means that the "experiencing self," which supposedly resides at a higher layer of cognition, is not really required.

Tailor-Made Emotions—Knowing Your Momentary Trip

The crucial effect of the "inner drugs" (the neurotransmitters), the array of receptors, and the pathways of their transmission are hidden from the eye of consciousness; we usually sense only the final result of their activity.

Is it possible to get to know the ingredients of the potion that bewitches our brain at any given moment?

Looking through a narrow peephole, and from a simplistic point of view, we might be able to claim that the central mediators of signals that compose the Morse code of the emotions in our brain are serotonin, which induces relaxation and peace of mind; dopamine, which initiates expeditions in search of novelty; cortisol, which induces stress and distress; opioid, which induces tranquility; oxytocin, which derives a sense of emotional satisfaction from contact with others; adrenalin and noradrenalin, which add energy, focus, and vitality; and the acetylcholine, which serves as lubrication oil in the wheels of the loom that weaves the threads of our memories.

And the endorphins, natural painkillers, are similar in their structure and effect to opioids, inducing relaxed tranquility and good mood.

Each emotion is accompanied by a typical mix of neurotransmitters; for instance, the biochemical fingerprint of emotional distress is a descent in the levels of serotonin, dopamine, and oxytocin and an increase in the level of the stress hormone cortisol.

The brains of men and women may be different with respect to the mix of neurotransmitters, even in a similar emotional climate. A central component in the "pink cocktail"—estrogen—thus causes an increase in the levels of dopamine and oxytocin.

The oxytocin inducts a tendency of relating to others and developing intimacy and is considered to be the "drug of closeness," in women's brain in particular.

On the other hand, vasopressin is the main hormone in charge of turning on the lamps lights in the lane to the fellow man in men's brains and inducing affectionate emotions toward others.

An increase in the levels of oxytocin and vasopressin takes place in the brain after taking the drug "ecstasy," which encourages tendency toward "artificial love." In this sense, it might be seen as the "love drug," which provides contemporary support to the old insight according to which "love is addictive."

DHEA is a spring from which both testosterone and estrogen flow and is considered the Zeus and Hera of hormones.

Androgens are hormones related to aggressiveness and intense interest in sexual intercourse, and, on the other hand, to reduced empathy toward others among both men and women. The main ones are the testosterones, DHEA, and androstenedione. The level of these hormones is maximal at the age of nineteen among women, and twenty-one among men. In the male body, testosterone is produced at a rate ten times higher than in the female body; thus, its influence is more prominent in average male behavior. The level of androsterone increases during the second and third week of the female period. Women who take birth control pills, which depress the activity of the ovaries and make them discharge fewer androgens, are less susceptible to the effect of androgens on their behavior.

One of the most familiar effects of the hormonal potion flooding our brain at a given moment is related to attraction between men and women. For example, a study pointed out that, on most days of the month, women prefer a male face characterized by symmetric, round features; during ovulation, they prefer more prominent, manly features that emphasize gender differences, such as a "square" face and prominent jaw.

The Sunset of Sunrise—The Janus Faces of Drugs

The issue of using psychoactive drugs is a charged one and requires cautious walking on a rocky lane. (A comic illustration of the complexity is Homer Simpson saying, "I'm too drunk to drive—wait a minute! I mustn't listen to myself when I'm drunk!")

When the ship of our thoughts is carried on the waves of hallucinatory drugs, it is prone to cross the equator of rationality and reach illusionary destinations that do not appear on the maps of logic.

Should we be afraid of using psychoactive substances to the point that we avoid using them altogether? The answer to that is simple and unequivocal: yes and no. The difficulty is in determining the borderline for the admissibility of psychoactive substances from an external source (admissibility in the sense of nondestructive impact on brain function).

Our brain knows how to deal with self-produced drugs, but some of the drugs that come from outside of our body might disrupt our brain's function irreversibly. On the other hand, among these drugs there are some that might better brain performance and improve certain aspects of its function. When we consider using them, we must adapt a selective, realistic approach.

Throughout history, man has been trying to search the kingdom of psychoactive substances for a panacea—a universal magic medicine, a sort of magic bullet that will hit the core of sadness and desperation and has the power of curing all existing diseases of the soul. Unfortunately, it has not been found yet.

Psychoactive hallucinogenic substances that are considered psychedelic, such as LSD and magic mushrooms, might be used as therapeutic substances. There have been attempts to combine them as part of therapy in cases where patients do not respond to conventional medical treatment in cases of chronic depression, post-traumatic stress syndrome, and drug and alcohol dependency.

There are "legal drugs" that are socially accepted nowadays and might have a good effect on brain function. Drinking coffee, for example, was found to reduce the risk of dementia, and it seems that nicotine itself, isolated from other chemical substances that are usually attached to it in tobacco and make the smoking more damaging than beneficial, might have a good effect on thinking functions. In this context, a significantly lower frequency of Parkinson's disease was found among smokers. The reduced risk is usually ascribed to the energizing impact of nicotine on dopamine-producing brain cells. In light of the above, researchers are considering the option of prescribing isolated nicotine, not attainable by smoking.

A weighty argument of those who object to non-therapeutic usage of psychoactive substances is often presented as a question of whether we were meant to turn into "zombies" and sacrifice our precious, ephemeral sobriety for the sake of momentary delight.

The functions of our soul are vulnerable to the impact of external drugs: our conscience might melt in alcohol; our short-term memory might be damaged as a result of lasting use of marijuana; LSD might resurrect hallucinations performed on the stage of our consciousness, accompanied by dazzling pyrotechnics, even years after taking it.

The peak of the mountain of "artificial happiness" is white, and its slopes are dangerous: the brain profile of a person who uses cocaine is similar to the brain profile of a person who experiences divine happiness, but the re-encounter with reality, after the dream of happiness evaporates, might be cruel and lethal. Among other things, cocaine increases the risk of stroke.

Hallucinogenic substances might produce, in the halls of our brain, an illusion show, which is actually a synthetic scenario that does not match reality. In this spirit, people say jokingly that one of the ways of knowing that you are drunk is when you lose an argument with standstill objects. A person who takes a psychoactive drug is present in his body, but his spirit sails on the sea of imagination. Under certain circumstances, the lack of compatibility might cause a person to make decisions that are valid in his fictional world but can risk his life in reality. There are psychoactive substances that initialize apoptosis (from the Greek word for exfoliation) among brain cells—a process of death that sends cells to the gallows.

There is always the risk of getting caught in the spider's web of an addictive substance. Addiction causes structural changes in the brain, some of which remain for life. The main features of addiction are obsessive craving for the addictive source; developing tolerance, which requires higher levels of stimulation in order to experience a sense of satisfaction; and difficulties in making do without the addictive substance.

Dopamine is sprayed from the ends of dopaminergic nerves in response to an achievement that involves effort. Addictive substances provide the "reward" by infusing dopamine that we achieve in an effortless manner.

The wave of dopamine that stimulates excitement shapes and reinforces the nerve network that encodes the activity that caused its creation. A feedback mechanism is thus created, which reinforces the link between dopamine and the addictive behavior. Sometimes, in order to retain the fragile balance between the concentrations of compounds of neurotransmitters, which constantly change, our brain performs "initiated adaptive changes" with respect to the concentrations of these substances. For instance, when a substance like cocaine meets our brain and increases the concentrations of dopamine at once, especially at the structures of "the cycle of pleasure and reward," our brain, as a means of creating balance, reduces self-production of dopamine. Brain imaging through PET scans shows that, in the brain of a person who uses cocaine frequently, there is a decrease in the basic amount of dopamine when the person is not under the influence of the drug. Such a decrease, as in a vicious circle, increases the craving for cocaine.

Chapter 7
Memory Functions

Memory Classification in the Mirror of Time

A conventional classification of memory is done according to information preservation on the time axis.

At first, reality manifestations meet our sensory organs, and the changes in these organs resulting from this encounter are called "sensory memory" or "very-short-term memory." These are the perception impressions, as they are perceived by the sensory organs and the primal input areas in the cortex, at their raw, initial state, prior to brain processing.

These changes are like the pattern of a circle perimeter that expands when a drop of water hits the surface of the puddle.

The initial visual input is called "iconic memory." Its "life span" is long enough so that a modern film, which is projected at a speed of more than twenty-five pictures per second, will be seen as a continuum. The previous picture still "remains in the eye" when the next one is projected. Old movies from the "silent era," however, which are projected at their original speed of fifteen pictures per second, seem segmented, since the icon of each of the pictures starts to evaporate before the next picture is shown. The retention of primary auditory input is called "echoic memory." The retention of the liveliness of the sound beyond its actual occurrence enables one syllable to echo in our mind until we hear the next one, and we recognize the meaning of the auditory input by means of the context.

The same is true for music. "Echoic memory" enables the retention of one sound until the next one appears, and the interface between sounds that touch each other's heel makes music come into being.

"Present-Momentary Memory" or "Working Memory"

The second meeting point between the sensory impressions and the brain along the time axis generates the short-term memory. There is controversy about the borders of this type of memory in the kingdom of time, but most experts claim that its range is between several seconds to thirty seconds.

Those who believe in the motto "business before pleasure" might prefer the term "working memory" as the term describing short-term memory, since it enables us to make lightning-fast decisions regarding pieces of information that are piled up on the desk of consciousness, subject to time and capacity restrictions—which will be further processed and doomed to oblivion.

Working Memory Characteristics

Our working memory preserves perception impressions on the screen of consciousness once they are perceived and for a short time afterward and, by doing so, enables us to process information consciously. This type of memory is formed and exists primarily at the frontal part of the frontal lobes (the prefrontal area). A secondary contribution to its formation derives from activity at the parietal lobes and the brain area called anterior cingulate gyrus.

The prefrontal areas serve as a sort of working desk on which we pile the perception impressions so that we can manipulate them mentally and process them. In this sense, these areas serve both as a storage place and a processing workshop.

The short-term memory has a passive aspect—the perceptional recording—and an active aspect that includes thinking about the input in a conscious manner or memorizing the input in order to retain it in our consciousness. The active aspect represents voluntary involvement "from within" regarding contents that mostly derive from the outside.

The Outlines of Real-Time Memory Capacity

Studies have shown that the capacity of working memory (the channel of existence of momentary consciousness) is seven items of general verbal information on average. The average capacity of spatial-visual information is about four items of information. In other words, we retain in our real-time memory seven "general information" items, and when it comes to spatial information (the location of objects in space), the average capacity is about four pieces of information. The volume of our real-time consciousness changes according to the type of information retained in it.

Two central aspects of the working memory are the spatial-visual memory and the verbal-tonal memory, which is also called the "phonological loop."

In our brain we retain names of figures in a resonant phonological loop (the "phonological loop") with a two-second life expectancy. The longer the name of the figure is, in the language our brain thinks in, the shorter the sequence of numbers recalled by our brain will be. A long name of a figure extends the encoding time necessary for its preservation and leaves us less time to encode the next numbers in the two-second window of time before the expiration of the phonological loop. A study showed that Chinese speakers of the Cantonese dialect are able to recall a longer sequence of numbers compared to speakers of other languages, since they blessed with names of figures in their language that are especially short and catchy.

This is another example of the fact that the language in which our brain "speaks" casts its shadow over "pure" thinking functions as well.

In the spirit of time, some may compare the transition of information from the working memory to the long-term memory to the transition of information from the computer screen to the hard disc.

The vitality of our working memory in terms of verbal information is extremely frail. In most cases, it exists in the frontal lobes for several seconds. The preservation of its vitality is enabled due to the bioelectrical resonance induced among the neurons. When the resonance is over, the information fades away at once. In love terms, some compare working memory to a one-night-stand—when the future of the relationship is vague and uncertain. On the contrary, our long-term memory bases its status on structural changes that create consistent footprints and may be compared in the same spirit to a life-long marriage ("Catholic memory").

Working Memory and the Screen of the Shadow Memory

People who are well familiar with the secrets of fairytales know that one can identify a witch or a demon on a sunny day, since they do not cast a shadow. On the other hand, in museums of science, there is popular exhibition called "shadow memory." The exhibit is formed due to a flash of light projected on the visitors, and their shadow, which is projected on the screen, remains on it even after they draw away from it. The shadow lives on its own and gradually fades out. The magic derives from the fact that the screen is made of phosphorescent substance, which preserves the "shadow memory" and creates the delay in its fading out. Similarly, our working memory, as with the phosphorescent screen, preserves the footprints of sensory input for a short period of time after its actual disappearance.

According to a common classification, the long-term memory (LTM) includes recent and remote memory.

The LTM is a "Methuselah memory"—in the sense that it has long life expectancy.

The biological substrate of the LTM—its material footprints—are the structural changes at the junctions of interface between the neurons (the synapses) involved in memory preservation.

LTM is essential to "deep-perspective" thoughts, since the narrow window of time during which working memory exists does not provide enough time for building a high tower of thoughts.

For instance, a tomato is perceived in our sensory memory as a round, red entity with a smooth texture. In order to conceptualize it as a tomato, it is essential that this information meets our LTM, which will contribute the necessary information and complete the task of identification.

The volume of the LTM, in the sense of the volume of pieces of information stored in the brain of an average human being, was estimated by researchers as five hundred times bigger than the information included in all volumes of the *Britannica Encyclopedia.*

The Weight of Memory

A short story demonstrates the heavy weight of the LTM. Two monks were about to cross a stormy river, and, along the bank, they met a beautiful girl who begged them to help her cross the river. One of the monks took it upon himself. He carried the girl on his back and crossed the stormy river with her and with his friend, the monk. On the other side of the river, he put down the girl, and she thanked him and went away. The monks continued the journey to the monastery quietly. Twenty years later, the monk said to his friend, "I will never forget that twenty years ago, on our way to the monastery, you broke the monk's vow and touched a woman's body while carrying her on your back to the other side of the river." The other monk answered, "I carried the girl on my back for a few minutes and forgot all about her afterward. You have been carrying her in your memory for the last twenty years and never forgot her. The weight of your memory is much heavier than her physical weight."

Memory Classification According to Their Interface with Consciousness

Many brain researchers support the assumptions that most memory processes are formed and preserved out of consciousness's jurisdiction. These memories are described as implicit or concealed.

In many cases the impressions of our experiences affect our behavior and the principles of our faith without being accessible to consciousness. The lack of accessibility to conscious recollection does not mean a person does not know. In this case, his knowledge is implicit or concealed. Moreover, it does not mean his behavior will not be affected by that knowledge that is owned by his brain. Such "dumb knowledge," which is present but "does not speak itself" in the bustle of the stream of consciousness, is still heard. Implicit memory impressions are involved in information processing in the brain routinely and affect our behavior at any given moment, even if they do not "float" on the surface of consciousness while they exist in the layers of unconsciousness or preconsciousness. The penetration of memory into the domains of conscious experience is not essential to its effect on our behavior.

Aspects of the Structural and Operational Infrastructure of Memory

Our memory is like a loom that weaves threads connecting the representations of perception impressions of the various senses. The connecting networks are scattered, and their winding paths create integration of inputs derived from single senses, such as the visual and audio input.

Memory can be defined as a mental state anchored in a biochemical bedding.

The patterns of neural activity are at the basis of mental conceptualization.

Each material or abstract concept we think about is represented in a unique neural activity pattern, which virtually resurrects it.

Our short-term memory is created by means of repetitive cycles—groups of neurons that are mutually linked and operated together in a circular continuum of bioelectrical signals; one signal activates the other, and vice versa, in a loop-like pattern.

The preservation of the mutual activity pattern forms the active recollection—retaining information in momentary consciousness.

The intensity of the repetitive resonance decreases as time goes by unless the pattern of resonance is preserved and reinforced.

Once a certain loop-like cycle, which is a three-dimensional networking pattern between neurons, has been created, its destiny, and, as a result, the destiny of the information it contains, is determined according to the intensity and duration of its activity. If it crosses the crucial threshold, its chances of becoming a LTM will greatly increase. In such a scenario, the pattern of loop-like activity intensifies itself from time to time by increasing the linkage between the neurons at its basis. The increase of linkage is formed according to the rule that is known as the Hebb Rule, named after physiologist Donald Hebb, who explained the intensification of the connection between neurons through repetitive resonance. As early as 1888, Freud came up with the supposition of reinforcement of the connection between two neurons when the neural potential (the bioelectrical signal) is projected at the same time. He defined the phenomenon as the "rule of connection that derives from simultaneous timing." Later on, Hebb developed and elaborated the supposition, which was later named after him, and characterized it as the profound pattern of neural learning.

The repetitive resonance—the pattern of repetitive sequences of bioelectrical currents that are at the basis of initial memory (working memory)—is not a passive code that is available for the conscious mind from time to time but, rather, an active code that constantly changes and is involved in constant dialogue.

Our working memory is also a "resonant memory," since it is based on preserving information by means of resonance. In fact, the resonance loop is the basic mechanism of our working memory.

The destiny of most perception impressions that exist as resonant loops at the representation stage can be compared to a candle in the wind. Most of them are extinguished and fade away before they have a chance to become permanent lights.

The physiological infrastructure at the basis of the short-term memory is not completely clear, but it is known that its nature is different from the nature of the infrastructure at the basis of the LTM.

"Anagram" is the term describing the change in the structure of junctions between neurons, which makes the links between them—the structural-biological layer of our memory—durable. This change grants our memory the ability to be preserved for the long term. The anagram can also be defined as inscribing information in neurons, the pattern of material footprints of memory.

Before our memory codes are encoded in the solid slate of the anagram, they exist in a highly fragile state, in which they "echo." The echoing is done through a repetitive, lasting activation of the same neural paths that preserve them in a repetitive activation pattern of a "loop." Most "resonant loops" tend to fade away before they manage to cross the river of memory and reach the bank on which the anagram is created. This stage is susceptible to fading unless it is preserved above a certain critical threshold or the information preserved is presented repetitively on the screens of "present consciousness" of our working memory. The moment of formation of a structural memory footprint, which is the moment of anagram creation, constitutes the point at which our echoing memories have reached the safe shore of LTM.

When our perception impressions are young and only echo, the weight of the structural changes is light. The moment of encoding involves changes in the structure of connecting junctions between neurons, and the latter become more sensitive and accessible to the transition of bioelectrical action potentials. This is the moment when the status of our memory changes, and it goes up the ladder toward the layer of LTM.

The result of the structural changes that intensify the connection between the neurons is the transformation of a short-term memory into LTM. It seems that this process is composed of two stages; at the first stage, the changes are only related to function. Such changes involve, inter alia, changing the tendency of certain openings on the surface of the neuron's membrane. These openings, which are called "ions channels," enable the entrance of molecules with electrical charge.

This change is followed by the second process, a more sustainable one, which is the change at the structural level—change in the physical connection patterns, which actually means changes in the wiring between neurons.

The lasting mutual activity of neurons that are interconnected leads to changes in wiring, which means changes in the structure of the connection pattern between neurons by means of creating new neural communication junctions (synapses). These junctions are added to more senior junctions that inducted their creation. The LTM is created by means of activation of genes in the nucleus of the neurons that are encoded into "connection strengthening" proteins. These proteins improve the transmission of signals in the existing synapses and, in addition, induct the creation of new synapses (a process called "synaptogenesis"). These proteins lead to prosperity of the neurons that are part of the information encoding network and upgrade the wiring between them, which results in expanding the "bandwidth" of the communication pathway between them and improving its durability.

At certain nerve junctions, genetic mechanisms are activated within the cells and encourage the growth of additional junctions between the involved neurons. Now the cells connect at more durable nerve junctions, which are more suitable for a lasting signals' transmission with fewer disruptions.

Eric Kandel won the Noble Prize in Medicine in 2000 for this discovery when he exposed the fact that repetitive cycles and resonant loops between neurons that constantly echo have a trophic and permanent induction on the cells that compose them.[25] This effect leads to the creation of a thicker cerebral tissue, and it depends on activity and lasts throughout life.

Memory registration on the whole is reflected at the material level, as a bulb of ultra-thin threads (axons) that are interrelated and create a sort of "three-dimensional cloud" composed by neurons whose connection pattern is the infrastructure of the structure of specific memory.

In order to transform short-term memory into LTM, sleeping genes must be awakened.

Deepening the footprints of memory is accomplished by diving from the surface of echoing waves into the depths of the cell—into the DNA. The DNA encodes reinforcement proteins to the nerve junction, such as the protein called CREB.

CREB is the name of a protein that has a central role in reinforcing the inter-synaptic connections between neurons and thus contributes to the transformation of a temporary connection (short-term memory) into permanent connection (long-term memory). CREB is like a wedding ring that changes the status of a relationship from temporary to permanent.

Flies that were inducted, by means of genetic engineering, with intensive production of CREB became "superflies" that demonstrated superiority with regard to performing learned tasks in comparison with nonengineered flies.

The process of learning new information is manifested in the design of specific communication pathways in the neural network in the brain. Certain communication pathways enjoy reinforcement, and electrical signals pass through them more easily compared to communication pathways that were not prioritized by means of reinforcing the connection between their components. The reinforced pathway constitutes a structural representation that is in correlation with the information it represents. Studies show that not only the intensity of the electrical message is reinforced in pathways that became prioritized through a learning process, but more complex processes that create a correlation between the frequency of the transferred stimulus and the intensity of the reaction also take place. This process shares similarities with the mode of fine-tuning of a radio station to a unique frequency of electromagnetic waves. This mechanism allows for "selective reception": the addressee neuron receives only the messages it is tuned to and deletes the "noise" (the components of information it is not tuned to absorb) from its worldview.

The positive effect of constant learning and memorizing lasts throughout all seasons of our life.

The secret of lasting memorizing lies in the fact that it lengthens the candle thread and improves the chances of the flame of information to become a permanent part of LTM.

The action of conversing information with short life expectancy (the working memory that exists at the prefrontal cortex) into information encoded for the long- term, mediated by the hippocampus, is the core of the learning process.

The hippocampus, which means seahorse in Greek, because it resembles the shape of a seahorse, is a magnificent conversion station that constitutes the gate to the kingdom of LTM.

It seems that the hippocampus is a type of dynamic supermap that contains the locations of brain sites in which different aspects of experience impressions are kept and, during recollection, upon recollection, create the memory of the entire experience.

Legendary Memory

Information that captivates the heart of our attention beam will be carried in the carriage of the senses to the brain palace; will enjoy a short, jiggly dance with the short-term memory; and, if the short-term memory approves, will win the right to proceed toward the LTM boulevard. There, its recording will be imprinted in the boulevard of anagrams in the brain, as the palms of Hollywood superstars are imprinted at the Walk of Fame.

Four-Dimensional Preservation

It seems that our memory is also characterized by a unique shooting pattern (a bioelectrical signal that has typical shape and duration) that is formed at the three-dimensional networking that encodes them. In this sense, one may claim that memories are encoded in four dimensions: in the three spatial planes and in the plane of time (in this sense, the term "chrono-architecture" fits in).

Memory is, to some extent, a web of addresses that helps retrieve the various memory components spread throughout various graded scales (strips of continuity). For instance, a memory of a trip to the South Pole might include the following data: the measure of fatigue on the first day was seven, the level of internal arousal was eleven, satisfaction from food was five, and so on. The main emotional imprint of the trip will be a changing mix of the various measures. Different components of the experience are mixed and merged into a composition of the overall experience and the emotional imprint it leaves behind. Thus, the unique impression of the experience is preserved.

Strips of continuity that are hidden in the links between cells in a three-dimensional pattern constitute the structural infrastructure that encodes the recording of experience, and the links encode "a continuum of intensities" for the different aspects of the experience.

Whenever we invoke a certain memory and the recollection process takes place, we reunite the components of the experience that were scattered to various brain areas.

Recollection is formed from a continuum of links between the various measures of an experience. For example, the concept "cold" in the archive of a person who is fond of journeys might invoke this continuum of impressions in his brain: the most intense cold—while climbing the Everest, medium cold—during the trip to South America, heat wave—on the trip to the Sahara, etc.

Studies have shown that the process of preserving an LTM is an ongoing process that requires constant maintenance; it's not a one-time process. It seems that there is a sort of molecular engine that preserves the change in the neural activation pattern caused by learning at the nerve tracks. The disruption in the operation of that molecular engine, copies of which exist in all of the neurons, disrupts the preservation of memory.

The Lego Blocks of the Memory Palaces

Prevalent memory components become modular; they are constructed with a type of universal building blocks, that can be used for creating memories from new experiences.

This economical pattern omits the need for encoding a huge amount of perception impressions that share similar characteristics individually.

The stages of memory consolidation, preservation, and maintenance are characterized by unique and typical electric activity at the junctions of the neurons that share the "secret" of the specific memory.

At the stage of memory retrieval/remembering, structures that are mostly located at the thalamus and the ventral frontal lobe put the coded perception impressions into patterns so that they become more coherent and make more sense. This is how a memory that matches the picture of perceived reality and fits the directionality requirements of the arrow of time (from past to present and from present to future) is reborn.

While we are sleeping at night, the memory impressions that appear in our dreams do not pass the stage of being put into logical patterns that is typical of wakefulness, and it seems that this can explain the surreal nature of dreams.

Brain Areas—The Land of Memory

It seems that archives of mental maps that enable us to recruit different brain areas to the task of assembling memory (the process of recollecting) are located mainly at the cerebellum, at structures of the central core of the brain called the basal ganglia and at the hippocampus. These areas contain supermaps that contain, in a nesting manner, the maps of memory-details preservation sites.

The hippocampus has a central role in turning experience recording into personal, long-term memories.

The "hippocampus spider" weaves a three-dimensional spider's web between concrete entities representations, and it seems that it has the role of the "great integrator."

A major supermap that is located at the hippocampus enables it to compose the holy grail of memory out of the various pieces of information.

The hippocampus interweaves the pieces of information, composing memory as a whole as beads that make an "anagram necklace." The hippocampus is in charge of coordinating the time and place between the various brain areas that store memory components. The manner in which such a magnificent coordination is performed is still not sufficiently clear to brain researchers.

The creation of personal memory of experience, which stores the details of our personal autobiography (also referred to as "episodic memory"), mostly relies on the hippocampus and the structures of the temporal lobes. The creation of procedural memory (memory of "how to perform"—for example, how to brush one's teeth) mostly relies on the basal ganglia at the central core of the brain and the cerebellum. In various types of brain injury, the procedural memory shows better survivability compared to the episodic memory.

The meta-memory is the information we have regarding our overall memory capacity—the knowledge about the knowledge. People who have suffered an injury at the frontal lobes have difficulties in evaluating which knowledge is stored in their memory and which is absent. It is thus likely that the capabilities of the meta-memory derive from the activity of the frontal lobes.

The Address of Our Memories

"The future memory" is a term that refers to our ability to see future scenarios in our mind's eye, and it mainly relies on extrapolation of existing information that is stored in our memory.

This ability is formed through the activity of the temporal lobes.

The amygdala is in charge of the emotional aspect of memories. The memory of beliefs and habits is centered at the cerebrum cortex and the basal ganglia. The nondeclarative, procedural memory of the sequence of actions required for performing a task is probably located mainly at the cerebellum and the basal ganglia.

The G-Spot of Memory

Each memory has its own G-spot—a point in time in which it is "at its peak."

After a certain amount of time from the time of encoding, the memory fades away as a result of aging and lack of reviving. There is a time peak on the timeline in which the memory is at its best. Afterward, it might gradually fade away when not used.

Studies have shown that there are glucocorticoids (hormones in charge of the metabolism of carbohydrates, proteins and fats, such as cortisol, which takes part in regulating the level of glucose) receptors at the hippocampus. These receptors are activated whenever there is a high concentration of stress hormones, like cortisol, at times of distress. This finding might provide an additional supporting explanation for the claim that mild arousal improves memory function, while intense stress disrupts such functions. This pattern of effect can be graphically represented as an inverted U-curve (this principal is called the Yerkes-Dodson law after researchers Robert M. Yerkes and John D. Dodson, who presented it in 1908). It is assumed that, at times of distress, the higher levels of cortisol decrease the efficiency of the hippocampus' activity and, at the same time, increases the amygdala's activity.

The Yerkes-Dodson Law:
The Connection Between the Level of Arousal and the Level of Performance

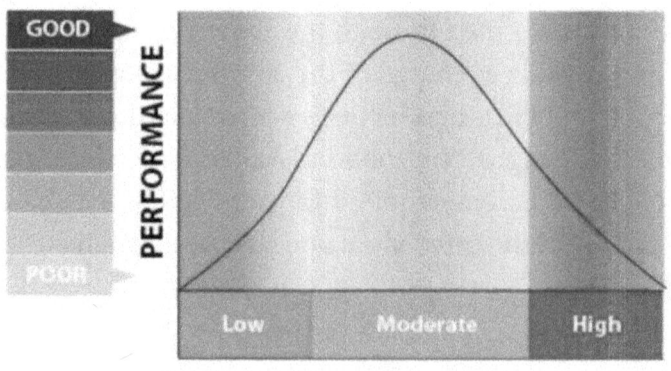

Forgetfulness at the Service of the Genes

The blunting of memory takes place naturally with regard to labor pains. It seems that the oxytocin that is discharged during labor and induces contraction of the uterus muscle blunts memory with regard to the aspect of physical agony in labor. Some may say that by doing so, it lessens the worry about additional labors.

Desirable Blunting of Memory

The painfully glaring lighting of memories that derive from traumatic experiences may be dimmed through the use of drugs that belong to the pharmacological family of the adrenergic beta-blockers. Such drugs turn down the sounds of the emotional soundtrack attached to the experience and soften the scraping of the traumatic experience. Such an effect is possible because the activity of adrenaline and noradrenaline, which are discharged in stressful situations and encourage emotional arousal that deepens the traces of memory, are blocked.

On the other hand, it is possible to intensify memory or to deepen its traces in the substrate of brain cells by giving adrenaline or other substances that imitate its activity.

There is the "blessing of forgetting"—a "forgetting pill" that is given to a person who has experienced a traumatic event and is intended to erase painful memories from consciousness can be compared to a drop taken from the legendary River Lethe that can erase past memories.

The opportunity to get rid of "painful" memories we are all doomed to have was tragically seized upon by the unfortunate lovers in the movie *Eternal Sunshine of the Spotless Mind*.

Medications that will enable us to broaden the limits of our memory raise the concern that we might become trapped under an avalanche of details that will roll all over us at any given moment. Will we become prisoners of overly detailed memory recordings that will compromise our ability to process information and plan our future moves? Will all veins of every autumn's leaf become engraved in our memory, as in the story "Funes the Memorious" by Jorge Luis Borges? Will the effect of the sometimes-blessed oblivion be doomed to disappear from our life, although it has a survival role in the adventurous book of the evolution? The frail interrelation between memory and oblivion might be interrupted and make the brain deal with a challenge it has not encountered before.

Universal Pattern of Memory Organization

Four main activities are related to memory processing in our brain: encoding, storage, consolidation, and retrieval.

Encoding is the acquisition of new information. Storage is the preserving of information. Consolidation is the middle stage, which combines encoding and storage features. Retrieval is the stage of recollection. Each activity is organized in our brain in a different pattern. It seems that there are similarities between the patterns of memory organization in the brain of human beings and other animals.

Procedural memories (which encode performance skills of sequences of actions) are partly engraved in the DNA; thus, for instance, horse embryos perform galloping movements in their mothers' uterus.

Memory researcher Eric Kandel, who was greatly involved in studying the biological substrate of memory, chose an Aplysia sea slug, which has about twenty thousand neurons, many of which are visible. Kandel assumed that the nervous systems share similarities across all hierarchies of living creatures' kingdom. He conditioned the sea slug so that its body being touched led to wincing of its gills and moving away from the area of stimulation. Afterward, he followed the recordings of the neurons' activity. It was found that the intensity of the signals that pass between them doubled once the behavior was internalized.

Kandel's insights, based on his studies, aim to prove that there are many similarities in the pattern of information processing and encoding among animals along the height of the pyramid of evolution.

Memory as Woven in a Nesting, Modular Pattern

The neurophysiologist Charles Sherrington once called the brain a "magic loom" due to its ability to weave together multiple pieces of information as necessary. The footsteps of our memory are scattered—they are not located at a specific location in the brain but gathered and woven together into a coherent memory from different brain areas.

Long-term memories have considerable preservation ability, inter alia, since their components are scattered and derive from different brain areas.

Recollection usually has a nesting, modular nature. The nesting pattern of recollection tends to flow from the general to the specific (e.g., from a "trip" to a specific trip to the summer camp in the woods).

The modular pattern of recollection relies on the fact that recollection is an anabolic process (a process during which a complex structure is formed from smaller, simpler units) that summons shreds of experience from various brain areas to the memory-composing workshop. These shreds of experience, once merged, become an overall memory experience.

Remembering How to Forget

Time weeds shades and sounds from the garden of our memories.

The changing streams in the river of our life enforce a constant drift of oblivion, which shapes the route of our memory channels every day anew.

A breakthrough in scientific research about the pattern of memory fading occurred when the researcher Hermann Ebbinghaus published his finding regarding the process of recollection. The "lab mouse" in Ebbinghaus's study was Ebbinghaus himself. With Sisyphean persistence, he repeatedly memorized multiple lines of meaningless syllables, to the beat of a metronome, and followed the pace of their fading from memory.

His findings showed that the curve of natural fading of memory is not linear.

He found that about 40 percent of the syllables were forgotten about twenty minutes after memorizing. One day after memorizing, about 60 percent of the syllables were forgotten, and later on the pace became slower. A month later, he could still remember about a quarter of the syllables he memorized. These findings led him to conclude that the lion's share of forgetfulness takes place soon after learning and that the pace later becomes slower. This was an empirical finding that served as evidential infrastructure for the memory-fading curve he later developed.[26]

A reservation regarding Ebbinghaus's fascinating findings is that by choosing to focus on meaningless syllables as the object of memory, he actually neglected important essential qualities of memory, such as emotional charging (which is absent in the case of emotionally neutral syllables) and personal meaning. These are two fundamental measures that greatly deepen the traces of memory in our brain.

When we study "transparent" information that is not colored in the colors of emotions or personal meaning, the result is likely to be similar to the one reported by Ebbinghaus.

Ebbinghaus published his findings in 1885 in a 123-page-long book called *Memory: A Contribution to Experimental Psychology*, which became a constitutive classic in memory research. Following the discovery of his findings, he coined the term "the forgetting curve" to represent the finding, according to which the main part of forgetting takes place in proximity to learning and its pace becomes slower afterward. The validity of his conclusion was also confirmed in temporary studies.

It seems that, although forgetting among all of us follows the general rules discovered by Ebbinghaus, the fading pace of specific memories depends on individual factors, the most important of which are the memory's emotional charging, its relevance to contemporary circumstances, the level of personal involvement in the experience, and so on.

Some of the methods used to slow down the pace of fading are constant memorizing of the information, or coloring the information with the paintbrush of emotions in order to make it more "rememberable."

Another great contribution from Ebbinghaus's studies is the identification of the "spacing effect"—using specific time intervals when memorizing results in deeper traces in memory.

Ebbinghaus's studies show that there is a window of time in which repetitive memorizing of information yields "maximal memory dividend." The type of information we wish to remember is an essential factor in drawing the borders of this window of time.

There is computer software that operates according to the "spacing effect" and calculates for us the best time intervals to memorize information. One might claim that letting a machine supervise this component of our spiritual life should not be perceived as lifting the burden of thinking off the shoulders of our brain but, rather, an intelligent use of the silicon brain for the sake of improving the performance of the protoplasm brain.

Despite the impressive empirical proofs of Ebbinghaus and his followers, it seems that the "spacing effect" has yet to reach its appropriate status as a central insight related to the human learning process.

It seems that we still have not used the full potential of these insights as a tool to accelerate the learning curve and redesign learning methods.

Gone with the Wind—How Do Memories Fade Away?

The locks of memory become rusty as time goes by and are more difficult to open.

Or, alternatively, the candle wick of memory usually becomes shorter as time goes by.

A well-known joke describes two very old men who sit on a bench in the park. An attractive woman passes by. One of the old men asks his friend, "Where are the days we used to chase those? Remember?"—"Yes, I remember we chased them," says the friend, "but I don't remember why."

Many of us believe that a future of forgetting awaits us all and that, ultimately, the lights of memory will turn off and we will go back to darkness. Whether this is our destiny or not, there is no doubt that the sharp teeth of time blunt our memory.

The Czech writer Milan Kundera said that there is death within forgetting. In his opinion, the scariest aspect of death is not the loss of the future but, rather, the loss of the past.

Memory loss of life experiences normally does not occur at once; it is usually formed in a pattern of "graded diminution." Details become less and less clear and gradually fade away. The levels of resolution decrease, and, as memory is fading away, the picture of experience becomes more granular and less accurate.

As for episodic memories, which contain our personal life story, when the emotional soundtrack that accompanied the experience becomes blurred, the shades of the memory of the overall experience tend to fade away (as "out of the heart, out of the memory"). From a distance of time, the eyes of memory have difficulty noticing the exact details of the experience. When the emotional melody is played in a minor scale, the voice of personal memory sounds weaker.

Memory Consolidation

The consolidation mechanism shows lasting selectivity and continues the sieving process at the time when memory impressions have already started leaving their signs on the slate of permanent memory (long-term memory). The dripping of perception impressions, classified as "small fish," from the sensorial memory (very-short-term memory), the working memory and the short-term memory into the depths of the ocean of oblivion is possible due to the consolidation mechanism, which continues the sieving process within the memory storage systems. The consolidation mechanism has a "Janus face"—one aspect allows for assimilation of information that survived the phases in which short-term memory becomes long-term memory, and the other aspect channels information that is classified as inappropriate for becoming long-term memory out of the borders of the memory kingdom.

There is an important distinction between passive forgetting—disappearance of impressions—and active forgetting—repressing perception impressions.

Selectivity is thus manifested not only in the stage of input and initial encoding, but also in the consolidation stage. The process of forgetting at the consolidation stage takes place through passive forgetting (disappearance of impressions) and active forgetting (repressing perception impressions). The consolidation process is like a refinery that refines memories and preserves their essential core.

Abstract and general (generic) representations are more resistant to the harms of memory fading compared to concrete memories, which lack personal meaning. In other words, the more the memory represents a more unique piece of information that lacks a personal emotional soundtrack, the more vulnerable and prone to fading it is.

The frequency of using generic memories, which are naturally connected to mental representations of multiple entities, is higher than the frequency of using concrete memories, which constitute mental representations of single entities.

The level of resistance of a generic memory to the drifting forces of the process of dementia contradicts the great vulnerability of a memory of a single event that lacks personal shade. Generic memory is the mental representation that shares common patterns between different memories, and these patterns create an interface that serves as an anchor that prevents their drifting toward the waterfalls of oblivion.

Resistance of Memory Types

Memory disruptions that occur with different brain injuries constitute a wide scope for inferences regarding processes of memory processing in the brain.

In the 1880s a French psychologist, Théodule Ribot, discovered an important connection pattern in the complex relations between brain injury and memory. He noticed that when brain injury occurs, the "younger" memories are more likely to be lost and the more senior memories are more likely to be preserved.

The memories that are more prone to fading away after brain injury are those that were created in proximity of the time of injury and encoded hours, days, weeks or months prior to the injury. The memories that were created a long time prior to the injury are more resistant. This phenomenon takes place as a result of the consolidation process, which implants the memories as time goes by. The more lasting the process is, the deeper the traces of memories are in the material storing bed (anagram). This consolidation process might be seen as an extension of the encoding process.

In most cases of brain injury, short-term memory is damaged more than is long-term memory. Remembering the name of our third-grade teacher does not guarantee the overall effectiveness of memory.

After a sharp brain injury (such as head trauma), memories from distant events are more resistant than memories that were formed more recently. It seems that the roots of old memories are more deeply implanted in the soil of our brain, thus explaining their high resistance.

Invisible Survivor

In cases of brain function injuries, the unconscious memory shows higher resistance, and its survival chances are higher than those of the conscious memory.

The procedural memory (which encodes how to perform motor tasks) has better chances of survival even among those who suffer from severe memory disorder reflected in other aspects of memory.

In this sense, procedural memory is like the memory of the muscles, and factual knowledge is like the memory of the mind. For example, among Alzheimer patients, many times the "memories of the muscles" are preserved as the "memories of the mind" that fade out.

Following electric shock treatments, which are mainly used for relieving depression symptoms in cases where medications do not help, the "forgetting of the previous" is common. This is a type of lasting forgetting and, at times, permanent forgetting of the previous chain of events that took place prior to treatment.

A person who suffered from lasting memory loss after electrical shock treatments described meeting himself, as he was, in various places only through external documentation (such as photos, and stories told by acquaintances) and not through guidance of his internal memory.

The Memory and Me

Memories are the bricks used for building of the self. The answer to the question "Who am I?" lies, to a great extent, among the volumes in the library of our private memories.

Memory is the bedding from which our personality sprouts. It provides narrative consistency in time and space, which is necessary for the existence of self-identity.

In the absence of personal memories, "our story" evaporates. We disconnect from the chains of narrative gravity that link us to our past and become "nameless" in the land of memories and in our being, one moment is merged into the next in seamless, nondescript unity.

The answer of philosopher John Locke to the question "What is the self?" was that the self is a solid succession of memories. My uniqueness as myself derives from the solid succession of memories in my brain.

As a rearview mirror that allows us to look backward, our memory is like a historian of our journey across planet Earth.

Personal memories carry latent information in them—information known only to the person who was present at a certain event (or to someone who examines the various aspects of that event in retrospect). For instance, a criminal is familiar with pieces of information from the crime scene, which are known only to him and to those who thoroughly investigate the event. Metaphorically speaking, all our personal memories are latent details known only to our brain or to the brain of people who are close to us or shared the same experiences with us.

The recollection process has a personal shade.

The manner of processing personal information means, inter alia, our "personal fingerprint" in the recollection process. For example, different people grant a different level of importance to the various types of input coming from the senses.

We add part of our worldview and part of ourselves to each of our memories, even if it is done unconsciously. Each memory is, to some extent, in our image.

The hippocampus, which is essential for the creation of long-term memories, is a private mint that produces coins with our personal portrait on them.

"The Rashomon phenomenon" refers to the facts of the same event as they are reflected in the memory of different people who witnessed it. Often the facts are reflected in their memory in different, even contradictory, versions.

In this sense, reality is "in the eyes of the beholder" and does not have "universal validity." This phenomenon raises questions regarding the ability to perceive reality as it is (the event itself) and reinforces the feeling that each memory is, to some extent, in the image of the person who remembers it.

Memory and Emotions

As time goes by and we draw farther away from a certain event, the sounds of the accompanying emotional soundtrack become weaker and vaguer. What is left is a type of aphorismic extraction of the experience.

The concealed emotional mark of the components of the experience that do not reach consciousness pours its impressions into the cocktail whose averaging constitutes the emotional mark of the overall experience. The averaging weights the contribution of the components of the experience from the conscious and unconscious layers.

Invoking a past experience and raising it to the conscious layer sometimes triggers experience-compatible emotions, and, in fact, we retreat to our "emotional past" while living the experience in real time. Restoration of the emotional climate through recollection induces a similar emotional climate in the present, thus the "lapse in time" that allows the recollection.

The Hippocampus–Amygdala Duet

As alcohol content in alcoholic beverages varies, our memories carry various levels of emotional content with them. The amygdala is the main bartender who pours emotions into memories. Memories that contain emotional components rely on a functional duet that includes the brain areas of the hippocampus and the amygdala, combined as central components.

Our personal memory is characterized by the resurrection of the original soundtrack that accompanied the experience when it was etched in our brain. The amygdala allows for the resurrection of the activity pattern formed in it when the brain "met" the original perception impressions of the remembered event.

The Dial of Emotions

We experience our life as a narrative continuum embedded with "high amplitude events" in terms of emotional lurch. These events leave more prominent records compared to routine events, which are mostly "flat-amplitude."

Numerous studies have shown that we remember information (such as from photos) that triggers strong emotions better than information that leaves us emotionally indifferent. Moreover, memories that trigger negative emotions are encoded at a more detailed level compared to memories that trigger positive emotions; they serve as warning signposts.

Thus, for example, behavioral skill acquired by rats in a stressful condition proved to be extremely resistant to eradication and became almost a "non-erasable imprinting."

Some claim that exceptional events, in terms of the emotional lurch they induct, are recorded in the "body memory" as well as in our brain—this is reflected in muscle pains, abnormal heartbeats and a tsunami of adrenaline—as a mute engraving etched in body tissues, which is claimed to be retrievable and verbally conceptualized by specific methods of therapy. On the other hand, some argue that this claim is just romantic, wishful thinking that lacks factual support.

The Emotional Prism of Memories

Recollection is the invoking of past impression. It has the power of reconstructing serenity in moments of turmoil, and, on the other hand, it allows us to dig in the exhausted pile of an emotional memory whose flames once touched the clouds.

Sadness is a mood in which we are likely to "visit the land of memories" and leaf through emotion-triggering memories recordings within the book of the heart.

The emotional value of memories is reflected facing the flames. It is a known fact that many people whose house went up in fire chose to save their photo albums and other items containing memories first.

Aspects of Memory Failures

The old insight that the strength of a chain is measured according to its weakest link is also true with respect to mental functioning: some people are prone to failures in memory assimilation, some to failures related to maintenance of memory, and some to failures related to retrieval.

An overall deletion of memory contents almost never occurs when the person is conscious.

In certain extreme situations, events that cause strong emotional lurch, such as traumatic events, are not encoded in memory, or are encoded in a thin pattern that requires great effort in order to resurrect the memory; thus, the product might be only partially reliable.

Our memory contains past effect impressions that continue to exist in present. Each recollection leads afterward to re-encoding and, thus, the danger of "blurring the original."

"Changing of the judgment backward" is like diverting the river of memories to a certain direction, which is different from the real course, due to new information that dims the reality-compatible information the memory contained in the first place.

For example, the way a person perceives himself in the present greatly affects the way he remembers his old self.

The level of reassurance regarding the reliability of our memories and, alternatively, the level of reliability we ascribe to information that conflicts with our memory, determines to what extent we are prone to "changing of the judgment backward."

One of the reasons people tend to underestimate it is related to slips of memory in old age. As we grow older, the load of tasks and information on the shoulders of our memory systems becomes heavier. The load of insights accumulated by our brain becomes heavier, and the act of navigating among them, as in a thick forest, becomes more and more difficult.

What's Your Name?

The most embarrassing memory failure, in many cases, is failing to remember people's names. A plausible explanation for that is that such a failure shames the person who forgot in the presence of others. The aspect of publicity intensifies the distress. A failure that is not exposed to the public, such as a grocery item we wanted to buy but forgot all about, lacks the embarrassing public aspect.

H.M.—Prisoner of the Eternal Present

There is a macabre cliché according to which the science of medicine develops from one funeral to the next. Similarly, brain science was also developed as a result of studying tragic cases.

The case of the person who has been known in scientific literature for many years only by his initials—H. M.—has become a milestone in the search of the golden fleece of understanding the formation of memory. Henry Molaison was nine years old when he fell off his bike. He was badly injured and started to suffer epileptic seizures that involved frequent loss of consciousness, which greatly disrupted his quality of life. At the time, medical treatment was insufficient and did not help him much. In 1953, at the age of twenty-seven, he underwent an operation in an attempt to better his condition and remove the rebellious brain tissue that caused the epileptic seizures. During the operation he underwent bilateral amputation of the hippocampi and adjacent structures at the middle layer of the temporal lobes on both sides of the brain. In many senses, H. M.'s soul was imprisoned that afternoon of September 1, 1953, while he was undergoing surgery through which both his hippocampi were removed. Once they were removed, the gate of his memory kingdom was locked to any new information.

Following the operation—once his mind was distracted, and the immediate perception impressions were washed away by the waves of a new experience—the recordings of the previous experience were deleted for good.[27]

The worldview, as it was reflected from his brain, was an island of the consciousness of the present, which was eradicated and recreated from one moment to the next and was surrounded by a sea of eternal oblivion. Following each blink of the eye, he perceived a whole new world. H. M. lost big parts of his past (and his future) but involuntarily received a place of honor in the pantheon of brain research. His tragic case clarified the critical importance of the hippocampus in terms of assimilation of new information and its encoding as a long-term memory.

Pseudo Memories

"Merging of false memories" is a phenomenon in which false memories are created based on structuring the memory backward, which means gathering pieces of information about a certain event after it occurred and merging it into personal biography in retrospect. Often there is a real, factual basis for the memory, but, since it was constructed as a patch quilt, certain patches are "false patches." Sometimes they are connected to the "factual patches" in crude stitches that are easy to identify, but sometimes they are sewn as if by an artist, mostly inadvertently, and thus more difficult to distinguish. It might occur while creating a recording of an experience out of real, direct experience recordings and perception impressions mediated from other sources of information, which were not directly experienced by the self, such as photographs and family stories. Thus, when sewing the quilt of memories, we must be very careful and try to distinguish between real patches and false ones.

A reliable memory is supposed to match the memories that interface with it in terms of time and place. In case the details of the memories are incompatible, "memory dissonance" is created and a review process is formed, which "straightens" the memories' impressions according to a chosen outline.

A false memory is sometimes as accurate an imitator as a glittering zircon.

"I am sure it was like that, but reality thinks otherwise." The level of subjective sureness, with regard to the truthfulness of memory details, is not a reliable measure of its truthfulness.

Incorrect memories share brain areas with real memories; this is the reason subjects feel that these incorrect memories truly reflect past reality. It is similar to imaginary "thinking images" that share the same brain areas that are activated when we observe real visual input.

A memory that includes many pieces of information that are based on sensory input is likely to reflect a real experience. There are exceptions to this heuristic rule, however.

Induction of False Memory

It is possible to create a false memory so that the person who remembers it is entirely certain of its truthfulness. An example is in the case of a list of words that are related to the familiar context of the "target word," which is not included in the list.

Words such as "tire," "steering wheel," and "gear stick" were read to subjects who were later asked whether the word "car" was included in the original list. Most subjects said it was, indeed, included in the list. The words related to cars made the subjects prone to making the mistake. They assumed that the word that is related to the context of the words they heard was also read to them. The memory drained into the drainage basin of the word "car," although this word was never mentioned explicitly.

Sometimes our memory tends to distort under the bending arms of imagination, like a spoon in the hands of a magician.

It is known that a person who is familiar with the secrets of the soul is capable of inflicting on another person, mostly inadvertently, persuasion and even a false sense of truthfulness with respect to false "past experiences" that did not actually occur. It can be done by means of suggestion methods such as hypnosis, age regression, trance induction, and guided imagery. According to memory researcher Elizabeth Loftus, even the most subtle suggestion affects memory.[28]

Testimony

Memory and truth sometimes meet and sometimes walk in parallel paths. In other words, memory sometimes aspires to reach truth as in centripetal attraction, but at other times it moves around it, not touching it, as in a centrifugal course. In between, there are numerous situations across the spectrum, on the ends of which total truth and total falsehood reside.

The Rashomon phenomenon, which was mentioned earlier, is named after the famous 1950 film by the Japanese director Akira Kurosawa. The film describes the different versions of a description of the same event as they are reported by the different people who witnessed the event. Rashomon is an example of competitive narratives, and it is a daily phenomenon in our lives. The importance of this phenomenon is intensified in court rulings, which are mainly based on testimonies of witnesses with regard to specific events.

The probative weight of testimonies of different people should reflect the limits of human memory in general and the specific limitations of the witnesses.

In many cases, the witness expresses his beliefs, in the sense of ideas and insights that have already taken root in his brain, and those are merged in the goblet of facts as he experienced and witnessed them in a type of alchemy.

Thus, the great difference between versions is not due to the mere facts that are reflected from the witness's consciousness but, rather, his beliefs and perceptions that are added into the goblet of his memory.

We tend to preserve the generic perceptual impression of an experience, which is formed from the collection of details and the relations between them. It is a type of holistic pattern that we remember better than each of the details separately. When composing a facial composite, however, the witness is required to dive into the depths of his memory into the "abyss of small details."

The number of facial components in the reservoir used for composing suspects' facial composites is about 560, from which sixty-three billion different facial composites can be composed.

Witnesses might fall into the trap of relative judgment. A general impression of "familiarity" might lead them to identify a person who shares the most similar characteristics with the figure planted in their memory as the figure itself.

Wide gaps between the acquisition environment and the retrieval environment will make the process of recalling and following the trail of elusive memory through channeling clues more difficult.

Exceptional Memory Skills

Under the Wings of Hypermnesia (Enhanced Recall)

The paths of human culture are embedded with wondrous memory performances. Folk singers from northern and southern Europe sang and recited their songs by heart. Homer caused wonders by remembering the lines of *The Iliad*.

In Islamic tradition, a person who memorizes all 114 chapters of the Koran is called a Hafiz, which means "guard." An ancient Indian tantric drill trains the trainees to give two written speeches at once, orally, one sentence from each of the speeches at a time. As practice advances, the number of topics the trainee is required to talk about simultaneously rises.

Solomon Shereshevskii, who had extraordinary memory and who was discussed in a book by the Russian neurologist Alexander Luria, *The Mind of a Mnemonist*,[29] demonstrated incomprehensible memory capacities—capacities that can be attributed, at least partially, to the synesthesia (merging of the senses) he suffered from (or was blessed with?).

After he had heard a list of words, Shereshevskii recalled more than seventy words accurately (an average person recalls five to nine words in similar conditions). He said that, at the phase of assimilating the information, his thoughts wander the familiar streets of Moscow and he "scatters" the words among the familiar sites. At the stage of recollection, he goes back in his mind to those streets and "picks up" the pieces of information from the sites he anchored them to.

Shereshevskii's memories were characterized by exceptional accuracy and vividness, but it seems that, at a certain point, his brain collapsed under the weight of his exceptionally detailed memories. Toward the end of his life, he wrote down many of his memories and then set them on fire with a metaphoric hope that they would evaporate from his brain as well.

Among children who suffer from autism, there were a few who were found to be able to say the exact weekday on which each one of the days in the recent century took place.

Genius is reflected as an oasis in the midst of a desert of limitations in the case of the British painter Stephen Wiltshire, who was diagnosed as an autist. In his case, the hypermnesia is reflected as an amazingly accurate photographic memory. Stephen puts on paper detailed urban sceneries accurately after having observed them only once. He has a particularly high level of iconic memory skills.

Brad Williams, a radio announcer from Wisconsin, sails in his memory to the seashores of his childhood, and the sailing log in his brain is amazingly accurate. His memory of past events was compared to archival documentation and was found about 90 percent accurate.

Jill Price from the United States also demonstrated unbelievable recall performance. Her performance was limited to defined fields of interest, however. One of those fields is her personal life. In her fifties she played on the strings of her memory the unique past melody of each day of her life from the age of fourteen. Another field in which she excels is the history of television shows and aerial accidents. Her mastery of the details is wondrous and draws the outline of human memory: accurate and detailed memory abilities are not comprehensive but focus on fields of interests, to which numerous "brain hours" were dedicated.

Magnus Carlsen from Norway, who is called "the Mozart of Chess" and who won the title World Chess Champion before the age of twenty-three, demonstrated superior memory performance as a child. It was said, for example, that at the age of two he remembered all the car brands he had seen. At the age of four he remembered the names of almost all 430 municipalities in Norway, including their symbols and number of citizens. During his legendary career, he played against ten competitors simultaneously with his back to the chess board, not seeing the location of the 320 pieces on the chess board (in each game of chess, each player has sixteen pieces; thirty-two pieces on the whole) and relying solely on his memory for virtual demonstration of the location of the pieces on the board and for giving instructions as to how to move them accordingly until he won the games.

The Sweatshops of the Brain

Maintenance of memory skills is worthy of becoming a routine procedure, exactly as many of us engage in physical exercise as a means of maintaining and improving physical health. Often, various types of physical activity are also beneficial in terms of maintaining and enhancing thinking skills.

British rowing champion James Cracknell, who crossed the Atlantic Ocean in a rowing boat (with a partner), once said, "If I had eight hours to chop down a tree, I'd spend six sharpening the axe." The tool we use for chopping down the trees of reality is our brain, and one cannot stress enough the importance of constantly sharpening our brain.

The masters of sharpening the brain axe and maintaining memory skills gather once a year at the brain sweatshop of the World Memory Championship.

At the air-conditioned hall, in which the competition has taken place annually from 1991, brain muscles are stretched silently, and virtual puddles of brain sweat cover the floor.

Ben Pridmore, a British accountant, was the 2009 World Memory Champion.

In a journalistic interview focusing on his wondrous memory performance, Pridmore complained that his memory always betrays him with regard to remembering names and faces.

His performance was extremely impressive; he remembered the order of cards in a standard pack of fifty-two cards after looking at it for 24.97 seconds.

The task in a competition called "binary 30 minutes" was to remember a sequence of the figures 0 and 1 that appear in a random pattern. Maximum memorization time was thirty minutes and maximum recollection time was one hour. Pridmore achieved the world record for 2009 when he recalled a sequence of 4,140 binary figures.

In the "short distance" competition, the task is called "binary five minutes"—memorization lasts for up to five minutes. Pridmore holds the record for recalling a sequence of 930 figures.

At the brain athletic stadium, other competitions take place simultaneously, such as "cards in an hour"—the order of cards in the maximum number of packs the competitors remember after an hour of memorizing.

Pridmore's world record in this competition is twenty-seven packs of cards. In other words, he remembered the sequence of cards in twenty-seven standard packs, in each of which there was a sequence of fifty-two cards, which adds up to a sequence of 1,404 cards!

When Pridmore was asked about the secret of his success, he said that, unlike many of his colleagues, who convert each card to a single mental symbol, he converts in his brain any possible sequence of two cards into a single mental image. For example, a combination of the two of clubs followed by the ace of hearts is translated in his brain into a fishing net. Pridmore weaves a story in his brain according to the sequence of mental images, where the chronological arrow of the plot represents the order of cards.

According to Pridmore and his senior German competitor, Dr. Gunther Karsten, their memory performances do not depend on innate supertalent but, rather, on an acquired skill. During a journalistic interview, they said that the way to enhance memory skills depends on converting sequences of cards into a suitable mental image and memorization. The performance curve improves in accordance with the duration of memorization.

This is a known approach that is based on converting one mental symbol (a sequence of figures or cards) into another mental symbol that is more accessible to recollection (visual mental images). The mental symbol is easier to remember as a sequence by means of weaving a plot based on a time arrow. Pridmore developed a "private language" whose grammar rules are clear only to his brain, and it enables simultaneous translation of the figure language and sequences of cards into the language of mental images from which the plot is weaved.

Memory Characteristics

The term "memory" refers to various, different soul functions.

Some define it as the lasting footprints of an experience, past effect impressions that continue to exist in the present, or a bridge over the river of time that connects the past bank to the present bank. The fingertips of memory allow the present to "touch" the past and sometimes to try to touch the future.

According to an anonymous saying, "beautiful memories are the paradise from which no one is able to expel us" (at least not until the dreadful scenario of overall memory deletion, as described in several science fiction movies, takes place).

A memory whose impressions live within us, even if it is of an event that involves people who are no longer with us, such as a memory of a talk with a person who has passed away, is like observing starlight coming from a star that has already vanished by the time its light reaches our eyes and our consciousness.

Plato talked about the Greek word for truth, "althea," in its etymological sense, which is "without forgetting." He thought that memory was a solid documentation of past events. Freud, on the other hand, claimed that the memory was a cocktail of facts and heart wishes.

Although we generally aspire to create memories that are as reliable and reality- compatible as possible, it is more accurate to perceive memory as a biographer of the brain than to perceive it as a biographer of a reality experience as it is. Memories are incomplete reconstructions of our experiences. Our memory lives in the meeting point between complete accuracy and fiction.

Memory as Commemoration of the Essence of Experience

We attempt to grant the profile of our experiences an eternal face in our memory. Memories tend to minimize into a refined extract. Their refined core is preserved, while "peripheral features," which are not perceived as the essence of memory, tend to fade away and vanish.

Unlike copying information from a computer file, which results in a new file that is totally identical to the original, our brain does not remember the experiences with total accuracy. The brain outlines an encoding pattern that preserves the "essential core" of the experience and the essential relationships among its various components. "The spirit of things" is prioritized over perfect matching.

The various memory manifestations—encoding of experience impressions, consolidation of memory and rebuilding it at the time of recollection—are all formed based on preserving the essential core of memory ("essential" as perceived by the brain).

Preserving the essential core of an entity takes place in our brain with respect to all types of sensory inputs.

Learning and memory are like inseparable Siamese twins.

Memory is the bedding on which new learning is based. Thus, our memory is not only a historian; it also sends its arrow toward the future in the sense that our conceptual infrastructure and many of our capabilities are based on pieces of information stored at the cellars of our memory.

A desire to cover the scope of solutions, which is an important layer with respect to solutions offered by digital computers, turns out to be a poor strategy with respect to many tasks in our life. Adapting a "computer approach" in dealing with problems in our daily lives might severely decrease our life expectancy. As illustrated by the example that was previously mentioned in this book, the attempt to identify the type of a snake accurately and wonder about the level of its toxicity when coming across it accidentally might turn out to be the last taxonomy investigation made by the "computer approach" strategist.

Mythological Memory

According to Greek mythology, the love between Gaia, the goddess of Earth, and Uranus, the god of the sky, created Mnemosyne, the goddess of memory. Indeed, memory, like its patron, is both divine and earthy.

According to Greek mythology, at the moment of death we cross the Lethe River in abysm, and the ones who drink from its waters forget all their memories at once and wander to the lands of not-knowing in the absence of memories. Thus, memory loss is one of the manifestations of death. The blessed ones get to taste the water from the lake of Mnemosyne, the goddess of memory. According to this myth, memory preservation is the earnings of the "illuminated."

According to Jewish tradition, "Seeking to forget makes exile all the longer; the secret of redemption lies in remembrance" (Baal Shem Tov).

And from the dark depths of tragedy, a request rises as a testament: "And remember me" (*Hamlet*, Shakespeare).

Constant Change of Memory

Our memories undergo change and reshaping, weaving and ripping as a routine procedure.

The surface of the various aspects of the sea of memory—encoding, preserving and retrieval—changes as the surface of every sea. It depends, inter alia, on winds—an external factor to the sea—and on depth currents—an internal factor to the sea. Metaphorically speaking, it depends on external phenomena and on phenomena related to our "internal weather"—within our body. Thus, we have only partial control over the depth currents of the sea of memory.

The image of our memory changes with time and depends on multiple factors, some of which are subjective (such as our mood, our age, and our health) and they might subtract from the "objectivity" of memory. For example, recollection at a time in which we are in a gloomy mood is likely to darken the shade of memories. The weathercock of memory casts its shadow on the image of memory.

We have a gallery in our head—pictures of our life experiences hang in the corridors of our brain. Some of the figures in the pictures become bigger and their colors become more intense as time goes by. Some of them become smaller, and their colors fade away, until it is hard to notice their outline. Metaphorically speaking, people who were once central in our life might lose their importance in our life, and vice versa—they might become even more important, as would, accordingly, our memory of them in our brain.

In the philosophical sense, each change in our brain—and such changes occur all the time—leads to changes in the entire capacity of information in our brain. Thus, a plausible inference is that the shades of our memories constantly change (the memories' chameleon), even when we do not process them consciously. Memory, in this sense, is like a view from a kaleidoscope that constantly changes its shape. For instance, information with which our brain is confronted after the experience takes place constantly affects the components of the memory of the experience as a lasting dialogue.

Evolution and Memory

Memory, as a reliable documentation system, is essential in terms of survival. As George Santayana once said, "Those who cannot remember the past are condemned to repeat it." Making the same mistakes as a repetitive pattern, especially those that involve life risks, is a suicidal behavior pattern. The struggle for survival will not be sympathetic toward a creature whose reality documentation is not "sufficiently" loyal to the facts.

Mental density is the manner of preserving information from past generations and ancient areas of evolution ("evolution memory"), which preserves patterns and insights of knowledge spread across the ages.

This density is reflected at the structural level by dense neural structures that are linked together in a compact pattern and that hold the wisdom of all the generations that preceded us for millions of years. According to a common estimate, in the approximately 250 thousand years since our species, Homo sapiens, appeared on the face of Earth, about ninety billion men and women of this species were born and died. The products of their brains, and the brains of those who preceded them in the evolutionary chain, are etched in the evolutionary memory, which is encoded in us in a genetic pattern.

The amygdala (an almond-like structure at the core of the temporal lobes, close to the hippocampus) is a dense concentration of representations of these ancient patterns, and is like a refined capsule of evolution memory.

Each and every one of us stores within us an evolutionary bottle that is capable of bringing out ancient demons directed at various behavioral tendencies. These demons reflect survival-related lessons that were bought in blood, sweat and tears by our primeval ancestors in various ages of evolution.

The Archetypical Memory—The Womb in Which Myths Are Formed

Conditioned reactions to unique world stimuli, which improve survivability, are, in fact, memory etched in our DNA. This is a specific repertoire of behaviors that exists due to an automatic protocol that is genetically dictated and is essential to survival. The final design of such a memory pattern requires appropriate environmental stimulation, which completes the fixation of the pattern.

Possible examples of the memory that is etched in our brain from birth: the tendency to be deterred by fire, fear of snakes, fear on the brink of an abyss, rejoicing in view of sunshine, a sense of tranquility triggered by the sound and view of running water.

The British ethologist Desmond Morris claims that our tendency to be covered derives from being "naked monkeys" who remember their hairy evolutionary past. The primeval yearning for body fur is, claims Morris, at the core of the urge to cover ourselves during night sleep.

Carl Gustav Jung, Freud's student and one of the minds behind the collective unconscious concept, described how the concept regarding the collective unconscious popped up in his brain following a dream. In his dream he found himself touring his house, and, when he descended the stairs to lower floors, the furniture seemed more antique and typical of earlier periods. When he reached the basement, he found a cave that contained pottery and human skeletons, which symbolized the age of prehistoric man. It seems that the consciousness of prehistoric man puts on its play on the fringe theater stage at the basement of our consciousness as human beings, and more than once the "fight-or-flight" cry in the caveman version derives from the basements of our brain.

Like our childhood sceneries as individuals penetrate our being in a pattern of our personal memory treasure, our "childhood sceneries" as a race also penetrate our being. These sceneries form the basis of collective memory, assimilated in us from the moment we are created. Along childhood sceneries in our individual mother's lap, our brain stores the childhood sceneries as a race in the African savanna.

According to this view, man is the collection of his experiences as well as a collection of the experiences of past generations encoded in him. The memory of our species' history (phylogenetic memory) is etched within us and contains representations of our ancient forefathers and foremothers, even before we became human—in other words, before our species, Homo sapiens, showed up on the stage of life. In this spirit, said Ralph Waldo Emerson: "Every man is a quotation from all his ancestors."

"*Anima mundi*" ('the world soul' in Latin) can be used to refer to the collective memory of our entire evolutionary journey etched in our DNA strands.

The archetype memory as genetic memory saves experiences that are beyond the personal. This is a collective memory, hidden as an innate gift among the folds of our brain, and, like the Internet, constitutes a reservoir of memory and knowledge available to every human being at any given moment.

Jung focused on the concept of the collective unconscious and characterized it as containing universal human contents, which are reflected as primeval archetypes and myths. As such, they share numerous similarities and are weaved into the different human cultures.

A hint of the previous memories that are part of us from the moment of birth can be found in the Aggadahs (fables) of the Sages of Blessed Memory (Chazal), which tell of a baby boy in his mother's womb. The embryo in its mother's womb is like a folded notebook, a glowing candle on its head; knows the entire Torah (Pentateuch); and observes the world from beginning to end. When the time comes for it to be born, an angel comes, slaps it on the mouth, and makes it forget all chronicles. Thus, according to the story, we are born equipped with previous knowledge that we were made to forget. Our role on the face of the earth is to remember it all over again. According to this story, life is a process of remembering.

Historical Memory

For ages, prior to the invention of the alphabet, human memory used to be the slate on which the chronicles of history were written.

Ideas of primeval myths were passed from father to son as oral traditions and marched from one generation to the next as an army of determined ants in a pattern of ideological heritage: the Deluge, the Epic of Gilgamesh, etc. Each generation poured into the plot the medicament of "spirit of time" from its era.

Prior to the invention of the alphabet, whatever was not told orally disappeared from the collective consciousness and sank into the depths of oblivion.

The Iliad epos, which included sixteen thousand verses, was told orally by Homer. This myth challenged even the memory of the residents of Olympus, and even more so the memory of plain people. Mnemosyne, a Greek goddess by profession, and the patron of memory, by her specialization spread her wings over Homer, who had a reputation as a storyteller who consistently, wondrously, and heroically (following the behavior pattern of some of his heroes) remembered the plots of all his stories.

Herodotus is considered "the father of history." He was a chronicler and a wandering storyteller. He used to tell stories about the chronicles of the capitals of the realms of ancient Greece until he decided to document the information he stored in his brain in writing and to contemplate the meaning of the described events in writing. By doing so, he established the documentation of history.

The Griot mnemonist who preserved lineages among tribes in Africa carried in his brain the core memes (ideas skipping from brain to brain) at the basis of tribal culture and orally passed them from one generation to the next.

The myths that were common in Middle Age Europe and were characterized by spectacular deviation from the familiar, such as dragons, giants, and dwarfs, reflect the clear connection between the grotesque and exaggerated to the memorable.

Memory and Technology

Externalization of a high volume of information to memory-auxiliary devices, such as palm-top computers, cell phones, and the like, changes long-lasting memory strategies of humankind. In certain cases, students feel there is no need to memorize by heart or learn the multiplication table, or how to read the analog clock, since readily available computers and digital clocks provide the answers quickly and reliably. The question we have to ask ourselves is, can these memory devices be considered our forgetting devices?

We rely more and more on external memory devices and might lose significant skills of recollection, knowledge and operation. For instance, people who remember telephone numbers by heart mostly belong to the older generation. Nowadays they are referred to as archaic, since the address book in our mobile phones frees us from the burden of remembering telephone numbers. A more complex aspect is reflected in children's perceptions regarding the need to learn by heart the multiplication table, in light of the user-friendly calculator. We should not consider the aforementioned a Luddite saying (the Luddites were members of a social movement that opposed the industrial revolution in England of the early nineteenth century). Information technology improved human welfare, but at the educational level we should consider the issue of cognitive core skills that should be taught to the next generation, in light of the technological devices that are available to us and become more advanced from day to day.

Various memory skills are being neglected, since technological devices preserve the information in a friendlier, more reliable manner. We become dependent on these devices, and, in this sense, our memory devices can also be considered our forgetting devices.

The temporary self is partially a "digital self" in the sense that our computers store an increasingly large part of our private and collective memory. They constitute a type of extension of our brain, or an auxiliary brain for our body. We digitally invoke (on the computer screen, as an example) pieces of information and perception impressions that used to be mentally invoked by our brain.

Recollection as Rebuilding of Memory

Recollection is summoning perception impressions documenting a past experience to our consciousness.

Memory is, in fact, created only at the moment of recollection. Until then, it is just a "potential entity" whose forming requires the art of coordinated and timely interweaving of multiple pieces of information that, together, form the designated memory.

Memory impressions just lie there, aimlessly, and wait for an opportune moment, which might never come, until they are urgently called, at the time of recollection, to take part in the act of assembling a conscious and unconscious memory experience.

Memory impressions are scattered in our brain in a cloud that sometimes seems chaotic. The GPS that locates the locations of memory impressions stores the map that enables memory building. Memory impressions spend most of their life in the darkness of unconsciousness. Nevertheless, their impact on our behavior is firm and abiding, even when they are at the unconscious layer.

Each memory is composed of multiple details that contribute to the recollection of the overall experience. The details' impressions are scattered across different brain areas, and in each recollection we create new links between them. Sometime not all of them are linked, so the memory is partial and does not reflect everything our brain remembers about the experience.

Retrieval from memory is a process of restoration and reconstruction.

Remembering means reconnecting the components of a certain experience.

The word "remember" contains the secret of memory consolidation, since it is composed of "re" (again) and "member" (component)—in other words, reassembling of the components of an experience and making them available to consciousness.

Memory is more like restoration than exact reproduction.

In the process of recollection, we do not throw our fishing rod into the sea of oblivion in order to catch the fish of memory but, rather, create it anew by connecting the fishbone to the tail and the tail to the fin. It seems that there are essential core components that constitute the backbone of memory, and the memory cannot be built without them. When they are found, however, it is easy to reconstruct the entire memory.

The modularity of memory and its reassembly are essential aspects of the process of recollection.

Thus, in the midst of recollection, in order to retain the experience true to the original, we should act as if we are archeologists reassembling ancient pottery. We must remember the balance of powers in the memory mix so that, at the time of re-encoding, following recollection, the original balance of powers of the components of the experience is retained. We should be accurate in terms of the dosage of the accompanying emotion and its type, as well as in terms of the sensory inputs that led to the perception of the experience, etc.

When we bring up memories of our grandmother, we do not activate a single neuron toward which all related memory impressions are channeled—a sort of "grandmother cell," as some brain researchers believed in the past. In the process of recollection, we trigger an array of neurons netted across different brain areas while each area adds its contribution, such as our grandmother's voice, the flavor of her dishes, the look of her profile, and the smell of her cookies. It is all done for the purpose of reconstructing our memory of her.

The retrieval process is an active building process and not a passive gathering from the memories' archive.

The retrieval process assembles the perception impressions encoded in our brain into a cohesive memory and pours them into a baking pan that matches common sense according to homogeneous rules such as time and space compatibility and matching other memories encoded in our brain.

Memory Production Line

Recollection is mostly managed in a narrative pattern.

At times of resurrecting a personal memory, a "voluntary activation" of the amygdala (the central generator of emotions in our brain) takes place in order to reproduce the nuances of emotions. Before the frontal lobes inscribe a reality-matching pattern that is loyal to the direction of the time arrow on the perception impressions, while we retrieve them from memory, the perception impressions act like anarchists who do not obey reality testing.

Only at the retrieval stage are compatibility to reality, rationalism, and correct chronological order inscribed as an added value at the production line of memories.

"The logic machine" in our brain, whose central part resides at areas of the ventral frontal lobes, matches memories with chronological logic and reality. When its function is disrupted, raw materials, which are illegitimate as memory components in normal times, suddenly become legitimate.

Acquisition Environment and Recollection Environment

The importance of reconstructing the acquisition environment, in which the original experience was acquired, as a trigger for recollection, is demonstrated in the story of writer Jean Cocteau in his book *Diary of an Anonymous*. In this book, he describes his return to the street in which he spent his childhood. He regains his childhood habit and drags his hand along stone fences, gates and bushes that encircled the line of front yards of the houses in the street of his childhood. The sequence of textures that "flowed" through the palm of his hand does not trigger anything, however. Then he realizes that, as a child, his height was shorter. He bends to a child's height and drags his hand again along the various objects that encircled the front yards, and all of a sudden the memories come back magically. Just as the needle moves across the bumps of old vinyl records, the melody of his childhood came back to life through the palm of his hand, and he rediscovered the texture of his schoolbag, the names of his friends and teachers, his grandfather's voice, the smell of his grandfather's clothes, etc.

Reconstruction of the environment in which the experience was acquired is a very powerful retrieval cue for resurrecting its memory.

There are crooked paths, like goat paths in the mountains, and, alternately, fast highways between the stimulus and the memory it evokes. Dragging one's hand on the wall at a "childish height" constituted the fast lane to memory—a lane compatible to the acquisition environment in the childhood. A journey to memory in which one walks in different lanes that are not compatible to the acquisition environment is often winding, and slower, and sometimes involves getting lost and not reaching the desired memory destination.

Matching the recollection environment to the acquisition environment is also the mechanism that stands behind the attempt to reconstruct the stages of a crime by the suspects at the crime scene.

The emotional environment also affects recollection; events that a person experiences at a certain emotional state will be better remembered at a similar emotional state.

"Instrumental memory" considers an inanimate object as containing a living memory. In a poetic spirit, some might claim that Cocteau's memory was kept inside the stone fences, the bushes, and the gates, like the musician who asked his friend if he wanted him to play a famous melody since it was waiting at that moment inside his guitar. Or, as an optional exercise for guitar players: let your guitar remember your music for you; imagine that your instrument of expression absorbs your thoughts.

The Resolution of Recollection

We are capable of performing a memory bounce, in the sense of skipping between various levels of resolutions of the experience's registration: recalling the overall experience or focusing on the small details that compose the entire memory. Recollection of a loved one may include, for instance, the memory of his voice, his profile, typical phrases he used, and the typical emotional imprint he used to leave on us—in other words, perception representations that are mapped in mental maps at different brain areas.

Each time we recall a personal memory, our consciousness focuses on a single detail out of the entire recalled experience and it "fills the screen momentarily." The overall detailed experience is played in the background, however, and other details show up and fill the screen of consciousness in matching timing according to the order in our memory narrative.

A memory of an experience is often the averaging, or a merger, of similar experiences.

During each recollection the experience impressions are re-encoded and redesigned. Recollections are sometimes "hybrid products." Most of the time, we will remember "averaging" of experiences of similar nature—not a single chess game with our grandfather, but several games played with him whose shared impression is merged in our memory, unless a certain game involved unique characteristics with a personal touch that made it particularly rememberable.

We spent about nine months in the eighth grade. If we challenge our brain to come up with memories from this period, we will sketch the "outline" of the experiences remembered from this period, recall extreme experiences (extraordinary events from this period), and, on the whole, we will work out an average of the experience and the average emotional soundtrack that accompanied the experience of being an eighth grader.

The relationship between time and memory are complicated. As time goes by, the image of memory fades away. Similar to a figure that stands close to us in space and we are able to notice its minute details, so it is in the case of a memory that is close in time. And, similar to a figure that stands far away from us, and its details are blurred, so is the case of a past memory whose impressions have faded out. We tend to remember the recent past and imagine the near future in much more concrete terms compared to remembering the remote past and imagining the distant future. A memory of the remote past is sometimes like an apple that shows the teeth marks of time.

Encoding and Formation of Memories

The Actuarial Science of Memories

The frequency of using memory is the main actuarial indicator of its survival. Its emotional weight is another actuarial indicator, which depends mostly on the level of the arousal of the amygdala linked to the perception impressions experience that becomes a memory. A stimulation of the amygdala is like a stimulating drug in the marathon race of our memory toward the finishing line of the anagram.

Thus, we can claim that the intensity of our memories is mostly determined according to two components: the frequency of use and the emotional soundtrack that accompanies them—in other words, the level of vibration of the amygdala's strings.

Emotional intensity is a significant prognostic component in terms of the life expectancy of memories and of the level of compatibility to the reality in which it is encoded. During an event that inducts an extraordinary emotional reaction, we mostly encode the various aspects of the experience with accuracy that exceeds standards events. But it seems that it is an "on-the-verge situation," since, beyond a certain verge of emotional charge, the emotional turmoil might compromise the quality of the encoding.

In "healthy" dosages, stress might blow wind into the sails of recollection.

It is a challenging arousal, since it is a sort of "healthy stress." Adrenaline in a certain "dosage window" might assist in focusing attention. On the other hand, excess dosages of stress (and adrenaline) will make it difficult for us to encode the impressions of the experience and to retrieve them at the recollection stage.

The "exposure shutter" of our memory camera determines the quality of the experience photos. Prolonged exposure usually results in deeper traces of memory.

The environmental conditions at the time the information is acquired are environmental noises, lighting, temperature on site—all of these factors affect the quality of the input from the senses.

A failure in the function of sensory organs, such as dysosmia (disruption of smell) or dysgeusia (disruption of taste), dims and distorts the face of experience. A disrupted input might lead to a disrupted, partial memory like a patchwork quilt of memories sewn in crude stitches.

The basic worldview of the observer also affects directly and indirectly his perception of reality.

The relevance coefficient for us is another important actuarial indicator. The more the information is relevant to our current situation, the more priority it gets in terms of processing and input by means of focusing our attention on it and shifting our attention from phenomena that are not as relevant.

Present tasks that share similarities with past tasks will be encoded more easily. Once a complex mental skill is acquired, similar skills will be acquired more easily ("more of the same"). The brain will use the neural network of coordinates that is already in place in a modular status in order to encode the new task.

Our brain is better at remembering a sequence of events that are linked in a causal manner than a random sequence of events.

The directionality of the arrow of causality signals the direction of recollection, draws its outline, and facilitates its formation.

We are all anxious about the fading away of memories of precious moments and try to commemorate them by external documentation means, hoping to retain the intensity of emotions we experienced at a certain moment, which naturally decreases as years go by. It can explain our need to document precious moments by taking pictures of them. In this sense, we are all similar to Leonardo, a character from the film *Memento*, who tattooed information on his skin for fear it might fade away.

Aspects of Memory Functions Enhancement Methods

Sometimes the ripples of memory softly touch the shore of consciousness, and the encounter creates the recollection. At other times, it seems that breakwaters as high as dams stand between the sea of memories and the shoreline of consciousness and prevent any encounter between them.

In a situation in which certain information that we know exists in our brain refuses to float to our consciousness when we need it, we might choose to act according to the approach that says "let the brain be a brain" and not expect that all phases (and not even most of them) of problem solution will take place at the conscious layer. Apparently solutions are cooked in various layers of our consciousness. When the memory of a certain detail does not float to consciousness when we want it to, as in a case of a word we cannot utter but feel at the tip of our tongue, a common approach is not to insist on bringing it to the conscious layer here and now. Such insistence often does not yield the desirable results. Instead, according to the "let the brain be a brain" approach, we should stop searching and allow "thinking time" for the unconscious layers of thinking. This insight is reflected in one of José Saramago's sayings,

"Answers do not always come when we need them, and many times the need to simply wait for them is the only possible answer."

This approach sometimes leads to the desirable result—the exposure of the rebellious piece of information that is finally found in the dark cellars of memory. As with the Phoenix—the legendary bird that goes up in flames after five hundred years and is reborn out of the ashes—so do memories that seem to be forgotten often come up from the ashes of oblivion.

Retrieval Clues

Retrieval clues help in resurrecting ancient memories. Sometimes they serve as a hidden door that leads to the cellar of forgotten memories and improve our ability to skillfully leaf through the pages of our memoir.

"All roads lead to…" It is possible to reach a memory from different paths: its emotional charge (i.e., the emotions that were involved in it); its configurational features; sensory aspects that were involved in it; and, sometimes, external memory assistants such as old pictures.

"The hinting intensity" of the retrieval clues depends on their nature.

Retrieval clues are more efficient in memory retrieval when they are linked to content. The configurational features of memory usually provide retrieval clues such as color and the like, whose hinting intensity is lower.

Enhanced associative networking is a familiar retrieval assistant. A multitude of associative links increases the availability of experience impressions to our consciousness at the desirable time.

Nicknames also serve as a familiar assistant to recollection, and it seems that this is the reason children tend to invent nicknames for each other. It seems that it is related to the emotional charge linked to nicknames.

In Homer's epos, which were memorized by heart, all the heroes' names contain adjectives, which make them easier to remember.

Sometimes, linking a body movement to an informative item helps in remembering it and producing an appropriate behavior pattern.

It is illustrated in the following case, which I happened to witness personally. A woman who had a malignant brain tumor was operated on and underwent radiation and chemotherapy treatments. She became totally mute, but when the cellular phone rang and was given to her, she took it in her hand, put it to her ear and said, "Hello, this is…" and nothing else. This repeated itself numerous times; beyond that, all attempts to encourage her failed.

The Art of Retrieval

The recollection test—i.e., the availability of information to consciousness at a desirable moment—is the important test for our memory.

Retrieving memory recordings often seems like the weakest link in the chain that is being tested, routinely, in everyday life.

The stopwatch of recollection changes according to the retrieved information and the identity of the person retrieving it. Thus, for example, ten seconds are what we need on average in order to retrieve an autobiographic memory.

Young subjects "pull out" words much faster than older subjects. It was reported that old subjects are slower in retrieving a memory following a word that describes emotions (eleven seconds on average) compared to retrieving a memory following a word that describes an object (seven seconds on average).

As in the Wild West, it is the speed of retrieving information from memory (and the precondition is the existence of retrieval) that sometimes makes all the difference. "The fastest gunner in the West" is sometimes the winner who takes it all.

Memories that are concealed from the eye of consciousness at a given moment make us fear that they might fade away, but, in fact, they might be kept in a layer that is concealed from the ordinary retrieval mechanism and pop up in front of the eye of consciousness at another time. The delay in the availability of information, however, might lead to a condition termed "staircase wit." It refers to delayed recollection of information that, at the time we needed it, was not available to our consciousness but pops up later, when it is no longer needed. Retrieval clues might assist in preceding the thought of "too late."

Means of Improving Recollection of Information

Mountains and valleys, as in a topographic map (and, metaphorically speaking, weak and strong points), exist in the scenery of our personal memory; some of us are good at remembering names but do not remember faces well, and vice versa. Some remember jokes but have difficulties with spatial memory, etc. Each person has weaknesses and strengths in terms of his memory abilities related to different types of information. Self-mapping, or mapping assisted by external sources of the profile map of personal memory abilities, will expand the boundaries of meta-memory—the ability to navigate between various memory sites across the vast memory map. Thus, we might be able to come up with an improvement plan for dealing with the valleys we encounter in the map of our memory.

promotions that make the information more catchy, such as editing information in an aphorismic pattern, triggers emotions, amusing, clear, and organized—lengthen the shadow of memory.

Using the maximal number of senses for the sake of recalling an experience enables us to encode it in an auto associative manner, which is easier to remember.

The film *The Jazz Singer*, which was premiered in 1927, contained, for the first time, a soundtrack of speech and singing and heralded the transition from the era of silent films to speaking films and, thus, contributed to an enhanced experience. Similarly, encoding of an experience while registering its impression on the various senses deepens the traces of memory.

It seems that the synthesis gives an advantage to certain recollection abilities due to the multiple-sensory conceptualization inherent in them. One can practice synthesis conceptualization through conscious encoding of the projections of an experience on various senses, and not only on those senses on which the projections of the experience are projected in an intuitive pattern.

Paying attention and granting meaning to information makes it rememberable.

Focusing on the unique components of the experience improves its recollection.

When we remember an experience as unique, we mark it. In other words, we encode its unique characteristics on the background of ordinary routine experiences. Memories that lack uniqueness tend to be forgotten.

Improvement of old techniques might enhance memory performance—using, for example, the "Roman Room" method, in which pieces of information we wish to remember are mentally linked to items that are stored in our memory and are highly available to it, such as pieces of furniture in our living room. (Shereshevskii, the magnificent mnemonist discussed earlier, used a similar method). An imaginary tour in a familiar room according to a certain sequence helps resurrect matching pieces of information, which are linked to the objects in the room. While using the principle of mental linkage, one can convert the imaginary tour to a tour through the words of a familiar song. The words of the song will serve as the anchor, or the hook on which the new pieces of information we wish to remember will be hanged.

Our brain tends to arrange short audio input items in a pattern called "phonological loop." This pattern preserves the representation of the sounds of information in our working memory.

In preliterate societies (in which reading and writing were not yet used), oral texts accompanied by music were used—a preferred method to bequeath memes (ideas that pass from one brain to another) to the young brains.

A familiar song can serve as a "memory trick"—in different languages there are children songs that present the sequence of the alphabet in a memorable fashion, accompanied by catchy melody, which facilitates assimilation of the alphabet in children's memory when they learn to read. Music also has a role in terms of assisting recollection among adults. Homer's *Iliad* and *Odyssey*, for instance, had rhyming and rhythm, which made them singable and, thus, facilitated memorization.

Chapter 8
Thoughts About Thoughts: Climbing The Mountain of Thoughts

Thoughts and Thermodynamics

Some think that our brain is a thermodynamic system (in the sense of a system that contains energy and undergoes transformation from a certain energy manifestation to another) that aspires to the balance point, which is manifested in rather low entropy (disorder, in energy terms).

As a clarifying analogy, which is far from being perfect, one may say that our brain is an open thermodynamic system, which means that energy and substances from the environment enter it and leave it. Our brain operates out of a (permanent?) gap in the thermodynamic balance that exists in its environment. This is the environment with which it has constant exchange relations in terms of substance and energy. The brain's operation is characterized by relative stability due to the "low entropy condition," which it preserves through energy exchange with its environment.

In an attempt to preserve the low entropy condition, the brain's systems have constant energetic dialogue with their environment. Our brain has exchange relations of import/export of energy with its environment, as is dictated by the circumstances.

The metaphor of the thermodynamic balance point of our brain usually refers to a condition in which the brain is in a state of an activity profile whose final purpose is inducing satisfaction and relaxation on its owner. It is still so even when the means to achieve this goal go through adventures and novelties.

Chemical medications and physical means of treatment, such as electroconvulsive therapy, magnetic stimulation, light stimulation, barometric medicine, and the hyperbaric chamber, are intended to divert the balance pendulum toward a more balanced point.

The condition of the brain pendulum may be of increased entropy or decreased entropy. In both conditions, the aspiration is to divert it toward a more balanced point—entropy that is on the spectrum of desirable values in terms of the function of the system—as it was "programmed" to be.

The dopamine oscillator is a central entropy generator in our brain, and it is probably a central regulator of the pendulum movement toward the balance point and from it.

The daily dose of escapism is almost a necessity, since rest is needed after the (tumultuous) tango most of us dance with reality. In this spirit, one may say that the secret of watching an action film is the escapism experience it provides. At this time, our brain is not required to make immediate decisions, and the amygdala practices emotional accompaniment of information that does not have a direct or immediate effect on us. While we are under the shelter of an escapist experience, our thoughts are less reality diverted.

Maximal assimilation in the present, detached from past scars and concerns about the future, is an effective escapism approach. Children have a natural ability to experience the present with their entire being—when they are immersed in play, for example. They are thus sometimes referred to as the "masters of the moment." The devotion of the being to this moment, focusing on the here and now, total dissolution in the river of the present that is flowing right now, reduces concerns and temporarily eases the burden of reality we carry on our shoulders.

The Theory of Relativity at the Sunflower Field

The philosopher Arthur Koestle[30] raised the question, Is Einstein's universe closer to reality than van Gogh's sky? Is the value on which the dial of reality-compatibility points at a higher place on the scale with reference to Einstein's surprising universe or van Gogh's starry sky?

In the first surreal manifesto, which was written in 1924 by Andre Breton, surrealism was defined as follows: "Psychic automatism in its pure state, by which one proposes to express…the actual functioning of thought…in the absence of any control exercised by reason, exempt from any aesthetic or moral concern."

Some might define such a way of thinking as thinking at the intermediate level, between the upper course and the lower course. Others might see it as fetal, raw thinking, mostly amygdala driven, and which mostly derives from a subconscious source.

Among patients who suffer psychotic episodes, it seems that their thoughts do not obey the gravity of reality but, rather, float in the weightless universe of their hallucinations.

The attempt to conceptualize "humanity consciousness," in the sense of consciousness of the entire humanity, is a theoretical thinking exercise that has numerous philosophical aspects. If we could merge the consciousness of living humans on Earth into a single, inclusive consciousness, what would it look like? Would citizens of poor states pour a more pessimistic worldview into the collective consciousness due to their dire straits?

And if we merge the consciousness of all living creatures (humans and animals) into a collective consciousness, what would that consciousness look like ("the consciousness of biosphere")?

Features of a Thought

Basic Thinking Characteristics—Managing Abstract Symbols and Conceptualizing Them as Mental Symbols.

Some refer to thinking as a symbolic activity whose aim is personal structuring of information. In other words, thinking is organizing and creating meaning and, in a broader sense, can be seen as an activity that is directed in the opposite direction of the thermodynamic fall of increasing entropy, since it involves ordering and creation of meaning in raw pieces of information in which order is not built in.

Vectorial thinking—i.e., thought directionality—is important if we aspire to process focused information, since in the absence of guidance, our thoughts tend to spread out and mingle one with the other and, as a result, to lose their uniqueness. Often an energetic deprivation (the "material fatigue" in the brain version) is responsible for distracting the directionality of thought and for spreading out its beam.

The big questions in life, such as the essence of relationships with people, the various aspects of love, various aspects of human existence, the end of our universe, and the like—topics from which telenovelas are made—are regular visitors in the lounge of our thoughts while the birds of thinking freely fly within our skull.

The Thinker and Rodin—as a Single Entity

We sculpt and are being sculpted at the same time. We create thought, and it creates us. A new insight born in our brain mutually changes (in a pattern of mutual influence) its wiring due to the new contexts it weaves.

Thought in Do and Re—A Hidden Magic that Turns Substance into Spirit

Among those who support the view that the entity called thought will forever avoid the grip of actual-facts science, there are some that compare thought to a musical piece. We can decode and name the notes, we can view each sound as a musical entity (such as the typical air vibration frequency created by the pitch), but the musical content of the piece that appears in our consciousness cannot be dismantled, because, as other consciousness manifestations, it cannot be externally, directly observed. It is perceived in our consciousness as a subjective experience rather than an entity that can be measured objectively.

Sometimes a high ladder is required in order to go down the depths of thought or, alternately, climb to the top floors of the thinking tower.

The fist of our perception finds it hard to grip the evasive entity of thought and what seems to be its Janus faces—its material appearance, on the one hand, and spiritual appearance on the other hand. Its material appearance, which is formed in the neuron-wiring pattern and the bioelectric signals patterns that flicker among them, follows the rules of physics, but is the spiritual aspect of thought an obedient citizen in the familiar land of physics? Let us consider the possibility of superposition: the existence of two bodies at the same spot in space is not possible with regard to substance. But can thoughts that are the daughters of the spirit employ the same spot in the space of our brain?

Ceaseless Thoughts

Galileo Galilei dared, at a young age, to challenge a scientific axiom that had been dominant from the days of Aristotle, according to which the speed of a falling object is directly related to its weight. In a series of experiments, courtesy of the Pisa Tower and gravity, he threw a light weight and a heavy one and found out that, contrary to common belief, both reach the ground at the same time. Thus, in his twilight days, following numerous hardships and in the shadow of a great loss he experienced—the death of his beloved daughter, Maria Celeste—he was quoted as saying, "My troubled brain will never cease grinding." In the spirit of his words, he never ceased to explore and search for new knowledge until his last day.

There is a famous old saying, according to which man needs to know a lot in order to realize he knows so little. Indeed, although we are required to look as deep as possible in search for the secrets of our world, the knowledge that the horizon and the depth seen by our thoughts are limited and partial should be built in into our consciousness, including our thoughts about our thoughts.

The Complicated Creation of Thought

We tend to see various mental functions as a single conceptual category when they are, in fact, a varied group of skills. This partially derives from language limitations, which weaves the threads of information into a uniform conceptual layout. Thus, for example, sometimes subconscious insights and those interwoven with conscious insights are included in the same conceptual category.

Many researchers of thinking processes have adapted a graded model of thinking processes, on the bottom of which there is a low-graded thinking skill that is based on raw information as is. An example of this type of information is the retrieval of a raw piece of information from our memory, or memorization of information in a simplistic manner. On the other end of the spectrum of thinking complexity, there is high-graded thinking, in which complex processing of information takes place, which brings new information that is added to the existing database.

Conscious and unconscious layers are woven into the complicated creation of thinking.

Numerous cognitive activities are based on a mix of unconscious and conscious thinking. Thinking processes at the unconscious layer, which usually consume a smaller amount of mental energy, allow for simultaneous channeling of brain energy resources to processes of conscious thinking.

The Geology of Thinking

Some compare the earth's crust, which is exposed to sunlight, to conscious thinking processes and the inner part of Earth, which sizzles and storms below the surface, to unconscious thinking processes. The crust is as thick as our consciousness, and it is thinner than the depth layers that represent the depths of unconscious. The lava flows, which sometime split the thin crust, might be suitable to represent "behavioral slips"—situations in which we act automatically, in a manner that is uncontrolled by conscious filtering mechanisms.

Acknowledging the importance of unconscious thinking processes, acquiring communication skills between conscious and unconscious thinking layers, and creating a constant interface between them might reinforce our cognitive ability.

Freud is the known cartographer of the unconscious world. He compared the different layers of consciousness to the surface of Earth and specified deep valleys, high mountaintops and in-between plains, hills and craters. The mountains, which are illuminated by the sun, represent the contents we are aware of. The valleys represent the darkness of the unconscious. In between the mountains and the valleys, there are plains, which represent the subconscious. The subconscious contains contents that are on the edge of consciousness but have yet to cross it (preconsciousness). Sometimes they "climb" up and are exposed to the glowing sun of consciousness, and sometimes they remain in the shadows, untouched by the rays of the sun of consciousness. Some relate the effect of contents from the plains area to the psychological sense of knowing that the information exists in our brain but is not available to us at this specific moment, as in the "tip-of-the-tongue" phenomenon.

Experimental psychology is like an inverted periscope that enables us to glimpse, under the water surface of consciousness, into the unconscious depths of our soul.

Hunger Burns a Hole in the Pocket—an Example of the Subconscious in Action

One should not go hungry to the supermarket. This street wisdom derives from the fact that when we are hungry, we tend to estimate our needs inaccurately and purchase many more groceries than we really need. Our conscious decisions are affected by our physiological state, which triggers unconscious thinking, which trickles into our conscious thinking.

The Invisible Hat

Edward De Bono, who wrote with considerable talent about thinking processes, compared six different types of thinking to hats of various colors that we are wearing while we are thinking. The white hat represents fact-based thinking; the red hat represents intuition-based, emotional thinking; the black hat represents critical thinking that emphasizes shadowed differences; the yellow hat represents positive thinking that emphasizes illuminated aspects; the green hat represents creative thinking, which "shatters conventions"; and the blue hat enforces the rules on the various thinking modes. A seventh hat may be added: the colorless hat as a representative of unconscious thinking. As this type of thinking, it is invisible, thus colorless, but it most certainly exists.

In order for the seventh hat to truly represent the wide scope of this type of thinking, the hat should be wide brimmed, like a sombrero.

The Bricks Composing the Tower of Thinking

The inner layout, which enables us to think, is mostly invisible to the eye of consciousness. Usually, we are aware of the external products of our thinking, such as ideas that are uttered out loud, written messages, actions (think on the end before you begin), and problem solving. The bricks from which the tower of thinking is built are made of different raw materials. Some of them are made of words, some are made of images, and some are made of senses that are hidden from the eye of consciousness.

The allegedly nonmaterial thoughts leave a distinct material signature behind them.

The composition of the coalition of the brain parliament, which formulates the thoughts, constantly changes (even more frequently than the Italian government).

The thinking bricks are piled in a modular pattern—the fruits of contribution of different brain areas.

The first experiences build mental scaffoldings for the conceptual worldview. The emotional experiences we experienced at the beginning of our life leave their imprint, which is learned intuitively, such as the verbal and nonverbal dialogue between mother and baby.

"A rational city of thought" is built in accordance with some rules. One must follow the building rules so that the towers of thought will not collapse under the heavy burden of reality. Freethinking allows for more liberal rules, so that even "castles in the air" are sometimes built.

"No, no, you are not thinking, you are just being logical," said Niels Bohr to Albert Einstein.

Although it is clear that Einstein's brain was a master of thinking, and not only of logic, this saying reflects the insight that thinking is more than using one's logic. An additional, central component of its formation is emotion, which is mixed with the products of logic and is weaved through them, usually inseparably.

We tend to "speak thoughts"—to conceptualize verbally. Words are the central bricks of the towers of conscious thinking, but they are not the only bricks.

The weight of non-verbal thinking in the cognitive sphere is controversial, though no one denies its existence.

Sometimes, the products of thinking are improved as a result of using both conceptualization of words and conceptualization of visual images and images perceived by other senses such as the auditory sense, the sense of smell, and the sense of touch.

Verbal conceptualization is suitable for representing certain conceptual worlds. Nonverbal conceptualizations are more suitable for other conceptual worlds.

The Stork's Nest of Thoughts, or How Thoughts are Born

Thought is created in the virtual gap between the neurons—the thinking cells.

The manifestation of thinking products is a behavioral output (externalized product), a pattern of silent thought (internalized product) or various combinations of the two.

In a common scenario of multilayered thinking, the initial conceptual skeleton often represents the valued core—for example, altruism-oriented thinking. On top of this layer, an oppositional layer, which considers cost-benefit analysis, is laid. The behavior is the result of a clash—the vector that wins the argument in the brain's parallelogram of forces.

While we manage to preserve conceptual vectoriality, in the sense of directionality and focus, the threads of thinking are woven in the loom of our brain to create a thinking cocoon that represents a regulated, thorough, and reasoned process.

Genealogy (generations' lineage) of thought traces the thoughts from which it was born, thoughts that are its relatives, and its offspring thoughts.

Usually it is easier to become a collector of thoughts than a producer of thoughts. All of us represent a mix of the two accumulating-knowledge approaches. Thoughts collection is an approach based on thinking creatures that were born in another brain's conceptual womb and which are collected in the archive of our insights. Self-production of thoughts is a more demanding, difficult process, and our brain tends to minimize it.

Practical Wisdom

The practical aspects of wisdom (guiding knowledge) are based on knowledge of the world (descriptive wisdom). Wisdom provides us with devices that we use to solve problems, in the sense of mental tools to identify patterns. On the basis of methods of coping that proved to be successful and, alternately, in light of the lessons that were learned from methods that proved to be unsuccessful, we shape our thoughts and responses to given stimuli.

We Think as We Act

Organization of thinking processes is similar to organization of movement. The brain, as a frugal artist, relies on structural and operational infrastructure, which is similar to thought in the spirit, and to the act as a material manifestation.

In Parkinson's disease, there is failure in the output of the basal ganglia, which are structures of clusters of neurons in the brain core. The disrupted output reaches the motor cortex, in which plans of performing movement are formed, and leads to prominent slowing down of movement. Thinking processes that are slower than the norms (bradyphrenia) might also be part of Parkinson's, in addition to slow movements (bradykinesia), which characterize the disease. This phenomenon possibly supports the assumption that thought, as movement, also contributed by the motor cortex. The flow chart of performing a motor task probably also serves the formation of thought as a conceptual sequence, as a motor task is formed from a sequence of separate movements.

The Brain's Chimera

We are hybrids: a little reptile, a little mammal, a little ape, and a little human, in the ideal sense. The relative proportions of the mix ingredients change from one person to the next, or, as the joke goes, the ape is the ancestor of mankind, but some people demonstrate it more than others.

Our thoughts are also heterogenic and derive from the various ingredients in various mixes.

A clash between conflicting thinking patterns is like a clash between two tectonic plates. It creates an earthquake at the halls of thinking, a situation that is known to us as "cognitive dissonance." Other people, like us, are not made of one piece, and they are also trapped in a jungle of conflicting desires that routinely fight one another in the arena of their brain, as they do in the arena of our brain.

The Eternal Flame of Thoughts

The fountain of thoughts never ceases to flow, like the stories of Scheherazade; they never end, since her life depends on them.

A thought follows its former thought in an associative course, and a conceptual slalom is created in the tracks of the feedback loops of the neural network. One association resurrects the other in a repetitive pattern.

Our brain moves across the continuum between associative laxity and strict adherence to reality (over concretization).

The mention of the smallest piece of information might activate the galloping train of thoughts, whose railways are created in real time while it gallops. Most of them cross the territories of the unconscious, and a few of them cross the territories of active attention. This process was well described by writer José Saramago: "one muse leads to another, and many a time we reach the next muse without paying attention to the path that connected them. Like walking from one riverbank to the next on a covered bridge. We walk without seeing around us—crossing a river without knowing it exists."

Sometimes the maps of our thoughts interface, edge to edge, and enable us to navigate. At other times, our thoughts are formed as a recursive reflection of one another when a basic thought is contained within a more complex, detailed thought.

The Time Arrow of Thought

The choreography of the thoughts' dance takes place in the dimensions of time and space (the involved brain areas and the timing of their operation).

The birth of thinking products does not take place simultaneously with the stimulus that allows their creation. Thought is formed at the interval between stimulus and response: between the water flowing from Archimedes' bathtub and the crown, which is not made of pure gold—between the test question and its answer.

The chronoarchitecture of thought (the pattern of its development along the time axis) is unique to each of our thoughts. Constant practicing of information processing of a certain type, however, will allow us to learn lessons from the past and shorten the processing processes when a similar problem is presented again.

Through thinking, we try to grant order and meaning to the reality representations in our head. Sometimes our brain does that by enforcing patterns, such as single directionality of time and cause-and-affect relations on the world of phenomena.

The consistency of the flow of thoughts is like legato playing, in which the toe of a certain sound touches the heel of another. Sometimes it seems that our thoughts follow one another, as in a ceaseless continuum.

In the spirit of hegemony among the various clocks' hands of our body, there is a method of relaxation that aims to synchronize the rhythm of thinking (the part which can be consciously guided) with other selected body rhythms, such as the pace of breathing. An example is a thought that is designed to last three cycles of inhaling–exhaling and the like.

When the stork of thoughts does not reach its destination, the thought it carries will not be born.

In most cases, we cope with thought in its completed form, but our worldview does not include those cases in which the thought is not visited by the stork, which is supposed to carry it to the world of completed thoughts. This type of thought remains unripe, and sometimes it is dropped from the womb of thinking as a stillborn, and we are not even aware of the attempt to form it.

The art of writing involves the art of erasing, and, as a metaphor, contemplation of a deep thought involves desertion of numerous unripe thoughts that did not turn into a deep thought.

A Thought as Promoting Itself

A neural pathway (neural network) becomes stronger each time a thought passes in its route and each time the behavioral skill to which it is encoded is performed in practice. These are as metaphorical feet that tread its course. The more frequent the marching is, the wider it becomes, and it might turn from a narrow goat trail into a broad, paved highway.

These findings might bear moral complexity. The thought police might claim that thinking about a felony is the first step of performing it in practice, since it relies on the same neural infrastructure, and that it might be required to monitor our thoughts in a preventative manner. Such a disturbing argument might be raised in the future, when there will be machines capable of decoding our thoughts.

Attempts to ask people to consciously block thoughts about a certain subject often lead to, in a paradoxical manner, an amplification of the flow of thoughts about this subject. In a study in which a group of people was asked not to think about white bears, and to describe, out loud, the thoughts that appear on the screen of their consciousness in real time, five or six white bears walked peacefully along the subjects' field of thoughts. Thus, for instance, the neuroses entrench themselves in the trenches of our brain as a neural map that deepens its roots by force of habit.

Thoughts and actions in practice intensify the tendency to think them or repeat them in the future. This fact constitutes a neurologic basis for the saying "Examine your thoughts—they become words. Examine your words—they become actions. Examine your actions—they become habits. Examine your habits—they become your personality. Examine your personality—it becomes your destiny" (paraphrased from the *Upanishads*—the Hindi scriptures).

Thinking Etched in DNA

Thoughts shape genes and genes' products (proteins), and vice versa.

The pattern of activation and inactivation of genes, which exist in the nucleus of the neuron, and as a result of the genes' products, the proteins, is affected by our thoughts and our actions and by our various experiences in the world. The thought and the accompanying emotional stimulation are capable of using their fingers to press the activation and shutdown switches of different genes and, by doing so, to change our brain's anatomy from the micro to the macro level. Constant dialogue takes place between "soft" thinking and the strict genetic information etched in the DNA. As a result, thinking as an action is capable of stirring up dormant genes and, alternately, to bring about unemployment of active genes (thus the epigenetic impact on genetic agents).

Thoughts have biochemical and anatomic manifestations. In this sense, thoughts that supposedly lack materiality have a clear material manifestation. Such an understanding leads psychology to graze in the pasture fields of biology.

In the animal kingdom, there are many examples for behavioral conditionings etched in the DNA. For instance, the cuckoo nestling that has just hatched immediately pushes away the other eggs from the nest that is forced to host it. In the twilight days of its life, the salmon fish, impressively, navigates itself to the exact same stream in which it was born, traveling thousands of kilometers on its way.

Studies that focus on human behaviors show that a considerable part of our behavior derives from genetic encoding at the DNA level. In fact, we are, in a sense, evolution robots encased in a biological straitjacket—thus the importance of emphasizing the non-programmed (or, at least, less-programmed) functions and maximal usage of our unique human capabilities pertaining to our intelligence.

The Wisdom of the Swarm—as a Whole Bigger than the Sum of its Parts

In the case of life forms that live in colonies, such as ants, the wisdom of the whole (in this case, the entire swarm) is bigger than the sum of its parts (the wisdom of individual ants).

The collective wisdom minimizes the levels of freedom of individual thinking.

Preferring innate rigidity over acquired flexibility characterizes the behavioral patterns of colony insects like ants and bees. The role of each individual at the colony is dictated from birth. The rigid programming, etched in the DNA, does not allow any freedom of independent behavior that deviates from the genes' dictation.

For example, the cooperative lifestyle in burrows of ants' nests and beehives' cells is a result of accumulation of behaviors etched in the DNA of these creatures, which structures their brain according to this unified path. The gathering ant is not equipped with reproductive organs, and it is destined to act as patroller, gatherer of food, warrior, or nanny of the queen ant's offspring, having no offspring of its own.

"The ant's handcuffs" partially handcuff our brain, as well.

The wisdom of the collective serves, in a way, as mental handcuffs that reduce the levels of freedom and decrease behavioral options. The ants and the bees, in many senses, are DNA programmed. Their wisdom is mainly systemic wisdom.

The genetic programming is, in this sense, a very tight biological straitjacket, and, at least to some extent, we are also born wearing it.

Reincarnation and genetics: Various folk traditions around the globe refer to reincarnation, such as the reincarnation of the grandmother in her granddaughter. Some will say that, in western societies, it is referred to as genetic heredity.

With respect to pairs of identical twins, the level of identicalness partially depends on the stage at which the common fertilized egg was split. The question is, to what extent does a single soul look at us through two separate sets of eyes? To what extent do they share brotherhood of thoughts?

Adaptation that was preserved due to neural Darwinism, such as preserved representations, is also reflected in action modes related to the function of memory and thoughts.

As for the preeminence of man, some claim that the uniqueness of mankind is the meta-knowledge: the knowledge about knowledge, about the partiality of the knowledge, and even its lack.

The apples of the Tree of Knowledge, which we picked while climbing the evolution tree—insights that were etched in our brain while we climbed to the treetop of intelligent creatures on planet Earth—are like the collective memory of the entire evolutionary journey etched in our DNA strands.

Some may disagree with the phrase "climbing the evolution tree," which may ascribe teleological reference to evolution, and say that the process of evolution is better described as a random wandering through the paths of life. In other words, they might say that a more suitable graphic description of evolution is a tree with several branches; our species is at the end of one of these branches, rather than on top of a ladder. However our evolutionary journey is described, its lessons are etched in our brain and have a great impact on our behavior.

Language as a Central Conceptual Tool

Language is a manner of defining world patterns. It is a tool for classification and diagnosis—a taxonomy that categorizes the multiple impressions that flood our perception. The world of phenomena constantly throws at our consciousness perception impressions, which change as in a psychedelic kaleidoscope.

Language equips our brain with an aphorismic tool, which enables us to summarize the overall impression of an experience without having to retain all of its shades in our memory. This is the source of the spell it casts on our brain, which is in charge of aphorism as a central tool of memory and thinking.

In language, there are no "built-in truth filters," and yet it is a means of patterning our world, because the rules of language match the ecological niche and the accumulated life experience of humankind. The language fences and isolates sayings that contradict human logic, such as the saying "I tripped and fell up."

Language rules create pattern and order in the world of phenomena around us. In this sense, language assists in patterning our world and also serves as "cognition designer." Language allows conceptual flexibility, which does not "break the rules."

Language is not only the spokesperson of logic. Sometimes it is the spokesperson of emotions and often presents various mixes of the two. Language contains fingers that skillfully press the buttons of emotions in our brain.

Describing a subjective experience and matching it to the linguistic categories might be a type of *Taming of the Shrew*.

At a young age, thought is formed alongside the momentarily sensory input, and it is extremely tangible and "earthy." When language skills are acquired, the ability to conceptualize abstract concepts or fictional occurrences that are not related to the material, momentary experience is acquired, too. Language gives us wings, and yet some see it as restricting the ceiling of thoughts. In this context, philosopher Ludwig Wittgenstein once said, "The limits of my language mean the limits of my world."

Our verbal competence might be affected by hormonal changes in our body. For example, a study showed that women are better at performing verbal tasks on the second week of their menstrual cycle. It might be ascribed to the effect of estrogen, which induces the creation of synapses in the hippocampus and is at its peak then. (Thus, women are recommended to conduct an unavoidable argument with their spouses especially on the twelfth day of the menstrual cycle, on which their verbal competence is at its peak.)

And, again, we will repeat as Cato the Elder: words and thoughts have a chemical manifestation in our brain. In other words, words and thoughts are reflected at the level of material molecules, and they magically cause changes in the mixes of neurotransmitters and the patterns of bioelectrical potentials among brain cells, which are translated into emotional and conceptual motivation.

Although thoughts are spiritual creations, they cause material changes. It is also true for the words of our language—in many senses, words equal actions.

Monological Dialogue—The Inner Speech

Some see thinking as a type of "inner discourse." Language and thinking are strongly connected. Some see thinking as the source from which the spring of words flows, or a cloud from which a shower of words derives and which is formed in a pattern of "inner speech."

Very often the language skills grade the capacities of this inner discourse. On the other hand, there are thinking layers that are not verbal; thus, some might claim that the saying "The limits of my language mean the limits of my world" is too comprehensive.

Fuzzy Logic—The "Roughly" Thinking

Fuzzy logic is an approach that ascribes to truth a relative value that ranges from zero (total lie) to one (total truth). Most cases are in between zero to one and can be defined as "partial truth" (or, alternatively, "partial lie") in accordance with the level of proximity to one of the ends of the continuum between one and zero.

The fuzzy logic approach is different from the Boolean logic approach, which is based on a dichotomous-binary approach and ascribes absolute values to truth and lie values of one—lie—and one—truth—without any intermediate values in between them. This is, in fact, the ordinary theory of logic in which a claim can only be true or false. In fuzzy logic, graded, qualitative rulings are used, such as "more," "less," "large," "small," and "good" instead of absolute verdicts, such as "lie" and "truth."

Some may define fuzzy logic as an accurate method of thinking in terms that are not accurate.

Fuzzy logic enables us to deal with the difficulty of quantifying qualitative data whose values are relative rather than absolute. The importance of "roughly" is that using terms such as "in most cases" and "usually" enables us to grant a relative value to truth. Although the conclusions of fuzzy logic are not as absolute as those of accurate logic, they often provide a satisfactory solution while covering a wider range of situations.

The famous saying in the world of medicine, "In medicine, as in love, there is no always and no never," is based on the medium between the two absolute ends.

In an attempt to quantify a qualitative measure, I may privately think that the beauty of the neighbor who lives at the end of the street is 0.84 (a thought I will not share with my wife, of course…).

Language is a classification tool in the service of soft, fuzzy logic. It seems that the need for the scale of relativity of the fuzzy logic was an important trigger for the development of language, which, in turn, broadened the scope of distinctions of the soft, fuzzy logic.

Soft, Fuzzy Logic—The Brain's Favorite Logic Language

Our brain zigzags between using soft, fuzzy logic and using Boolean logic in accordance with the circumstances (the soft, fuzzy logic relies on the definitions that do not convert the qualitative characteristics into a numeral value, as in standard fuzzy logic, but, rather, grant a relative value to them, thus the term soft, fuzzy logic), but it seems that it uses soft, fuzzy logic much more. Thus, for example, the rules of thumb (heuristics) that guide numerous aspects of our behavior derive from using soft, fuzzy logic. Our brain uses soft, fuzzy logic as a result of constraints, such as lack of information or lack of time, which exist in many situations in life.

Common sense, which serves us in our daily life, in which numerous situations are ambiguous, is primarily based on soft, fuzzy logic.

The component of relativity is inherent in fuzzy logic, and it serves us well, since most of the situations in our life are painted in shades of gray rather than the black and white of totality. The relative terms allow the differentiation that characterizes the resolution of our daily life (thus, for example, our neighbor is tall compared to most people we meet on the street, but not compared to the Maasai people, who wander in Eastern Africa and whom most of us do not meet in our routine life).

Fuzzy Logic in the Service of Artificial Intelligence

An ordinary computer program that tries to reach a decision based on certain source data and ordinary logic rules is capable only of absolute rulings and cannot cope with "almost" or "roughly" situations.

In order to spill "common sense" into the "silicon intelligence," computerized expert systems, in which human knowledge is assimilated, need to use fuzzy logic in order to cope with reality manifestations that are not perceived as unequivocal.

Using fuzzy logic enables the expert systems to use prediction tools that are true in most situations (heuristics). This provides an answer to a variety of situations in which information is not complete and absolute.

Fuzzy logic enables computerized systems to reach conclusions in uncertain conditions through an inference engine, which relies on values that are relative to truth.

Intuition

Some believe that our "gut feelings" merge, in various degrees, voices from the emotional brain with the voices of subconscious insights. We tend to accept or reject their recommendations according to the circumstances and to our basic approach toward "nonrationalistic" arguments.

Our gut feelings, which are not based on evidence, at least not in a conscious manner, are partially channeled toward the brain areas called the insula and the anterior cingulate gyrus. It seems that the insula is a brain area that has a central role in the processing of such feelings. The cingulate gyrus, which, on average, seems bigger and more active in the female brain, tends to add to our mental climate a cloudlet of skepticism and might cause us to perceive reality through a lens in a dark, impartial pair of figurative glasses.

Logical inference and intuition are two main thinking approaches in our brains.

The systematic inference, which is based on rules of logic, is formed in the cortex. Intuitive identification greatly involves the amygdala.

Often, gut feelings are not some abstract emotional conditions that echo within us but, rather, intensive feelings that direct our behavior. Goose bumps usually serve as a behavioral compass. Some might claim that whenever we let the instant judgment of the intuition have a great impact on our decision making, we live according to the manifestations of our goose bumps.

Our natural tendency to find order and common sense around us, and to identify causal relations between events, sometimes leads us to the wrong conclusions. A common trap is the perception of noncausal correlation as a causal one, or an erroneous link between events that are not related in a causal manner (such as the link between a rain dance and the amount of rainfall).

Children tend to perceive noncausal correlations between world manifestations as causal due to the lack of sufficient database and immature brain structures in charge of forcing the causal, logical connection (primarily, prefrontal lobes). This is how charming "childhood insights" appear, such as the request of my young son not to take off his glasses while he is asleep so that he can watch his dreams clearly. There is also the example of the boy who left his parents' house at a late hour of the night and started coughing. When he was asked how he felt, he said, "Everything is fine; I just swallowed some darkness." There is also the girl whose aunts used to pinch her cheek while calling her "Honey." When she asked her mother to kiss her, she told her to kiss her on the "honey," which meant her cheek.

Intuition is a survival tool that allows rapid interpretation, which bypasses rationalistic, tiresome analysis. Sometimes the price we are required to pay for these shortcuts is erroneous conclusions.

Yet, sometimes it is logical to be "illogical." When we are about to make a decision based on very partial, or doubtful, knowledge, not infrequently it would be wise to make a decision based on our gut feelings, which supposedly lack the logic component.

A common perception of intuition is as a "wondrous" thinking leap, which sometimes seems puzzling, does not plod along the tiring path of information processing in a logical, systematic manner, and does not require thorough processing and analysis. Some see it as a process that completely lacks conscious analysis, or as one that takes place at the preanalysis stage. Some brain researchers, however, believe that intuition is, in fact, a concentrate of complex analysis processes that are compressed into a sort of refined capsule of analysis—a product that may be defined as an "extract of patterns." According to this view, intuition is, in fact, a stage beyond the ordinary information analysis (a post-analysis stage).

In this context, Einstein had a surprising insight: "There is no logical way to the discovery of these elemental laws. There is only the way of intuition, which is helped by a feeling for the order lying behind the appearance."

Aspects of Thinking Failures

The Danger of Permanent Residency in "La La Land"

The sweeping tendency to interpret reality appearances through "rose-colored glasses" is like sprinkling powdered sugar on our memories. This tendency is also called the "Pollyanna syndrome," after the literary character who never notices a dark cloud in the sky of her life. Such a tendency might carry calamity seeds of self-deceit. Gambling agents are grateful for this tendency every day.

In a light spirit of wishful thinking that tends toward too-optimistic interpretation of reality, a common, humoristic saying in bachelorhood tradition is "A woman who walks with a dog is lonely, while a woman who walks without a dog simply forgot it at home"—therefore, the duty of every decent bachelor is to try to ease the widespread loneliness.

In previous times in history, life presented to our species a much narrower scope of options. The amount of information we needed in order to solve our day-to-day problems was smaller than today. The complexity of required processing was also usually lower compared to what is required from the brain of an average twenty-first century Homo sapiens.

Complex probabilities were not usually required in previous times in history; thus, our brain is not equipped with a set of mental tools designed for complex statistic processing.

Often, probabilistic discretion does not come naturally to our brain, and so the task of statistic inference is sometimes difficult for it to perform.

We have difficulties with rational probabilistic thinking in uncertain conditions. When we consider our alternatives, we tend to make decisions according to heuristics.

An anecdote that can illustrate the obstacles of predicting the future and the capricious nature of destiny is told about a doctor who shared with his colleague his conflict related to terminating a woman's pregnancy. It happened at the end of the eighteenth century. The woman's husband had syphilis, and she had tuberculosis; their first child was born blind, the second died at a young age, the third one was born deaf, and the fourth had tuberculosis as well. The reaction of the colleague was to say, "I would terminate the pregnancy." The doctor then said, "If I had, I would have sentenced Beethoven to death."

Our brain is directed at the sensational and dedicates to such events attention and processing resources that are very much exaggerated according to statistic logic and risk management philosophy, which is based on common sense. All types of journalism thrive as a result of this tendency of our brain.

Often we are caught in the trap of illusionary correlations, and then noncausal correlations are perceived as causal ones.

"In-depth interpretation" sometimes adds an artificial dimension that does not exist in its absence. Many of our deeds, especially habitual reactions, lack the symbolic depth that is sometimes linked to them. Often our actions are the result of a momentary condition of the brain's parallelogram of forces, when the action patterns of neurons that were resurrected at a certain moment lack the depth dimension that is sometimes added to them in retrospect as an interpretive act.

We must not confuse between what is eloquent and pleasant to our ears and the factual truth.

Subliminal priming (information that affects us before we make a decision, which we are not aware of since it does not reach our consciousness) is one of the factors that are hidden from the eye of consciousness and might cause the perception of noncausal correlation as a causal one.

Central components in our behavior derive from subconscious processes that evaded consciousness. Often, such a behavior is justified teleologically in retrospect. We do not understand many of our responses, especially those that are colored in intense shades of emotions, and sometimes we are required to explain ourselves to ourselves. Sometimes the retrospective explanation is erroneous, since it creates an erroneous cognitive-emotional correlation.

At the time of occurrence, events develop from beginning to end, while, when we consider them in historical view, we observe them from end to beginning, in inverted direction. Thus, it seems that our bias of retrospective wisdom is built in, which colors the events in a more deterministic shade and darkens the shades of randomality.

Sometimes our brain inverts the direction of the arrow of time, and we err by inverting cause-and-effect relations.

Entire life philosophies are based on so-called identification of "order patterns" within the white noise of randomality. These philosophies build, through selective gathering of convenient findings, an evidence-based infrastructure that supports their worldview. Whole lives move along paths of "absolute truths," which are nothing but white noise. Or, alternately, whole lives are based on using claims deriving from an axiomatic doctrine that precedes the facts and is not affected by them. This is conceptual paleontology—fossilization of thought, in the sense of adhering to a conceptual dogma in an uncompromising, inflexible manner that lasts even in the absence of factual support, or even when the facts contradict it.

Our basic tendency is to trust. This is the default of our perception. It is easier for us to trust than to doubt. A claim we hear, and do not have any previous knowledge about, will be considered true until proven otherwise.

It is also true for a claim that is not an axiom; if our brain meets it repetitively, it has a very good chance of being considered an axiom, since it became familiar knowledge. This tendency of our brain is called "the illusion of truth." It is broadly used, cynically, for propaganda purposes.

Reduction, when it involves overflattening of the depth and multidimensionality of phenomena, might lead to oversimplifying, which substantially distorts the nature of the phenomena.

Each logical inference has a "window of cognitive scope" that matches it. There are very limited windows, and there are windows that reach the horizon of cognitive view, and their validity seems universal.

Deviation from the valid borders of a logical inference "beyond the verge" might lead us to absurdity.

Often our brain prefers to focus on the short or immediate term (the present) and, by doing so, to mortgage future goals. We often prefer to feel good here and now rather than considering the possibility of improving our condition in the future—thus the difficulty related to postponement of gratification. We have a built-in apprehension about mortgaging the "safe" present on behalf of the unknown future.

The comforting embrace of routine is sometimes a bear hug—it prevents us from looking around in a process of self-interpretation or, in other words, to think reflectively. Such thinking relies on staying at the point where our consciousness replaces the exclamation mark with a question mark. When we hurry through the routine race of life, our brain is prone to thinking fast in response to the immediate challenges of reality. Often, the process of reflective thinking, in which we "look at ourselves from the outside" and perform "second-level thinking processes," is missing. Intelligent thinking derives from an appropriate mix of reflective thoughts and reactive thoughts.

"The Status Quo Bias"

The status quo bias refers to our tendency to identify the existing with the desirable. The old insights nesting in our brain are probably at the basis of the status quo bias. The traditional supposition our brain tends toward, according to which "the existing is the better," relies on the old memes.

Thus, for example, the German philosopher-mathematician Gottfried Wilhelm Leibniz wrote in 1710 that, despite the disadvantages of our planet, "We live in the best of all possible worlds." (Time will tell if he was right.)

The existing has the highest priority only because "it is there." In a lighter tone of loose association, the mountaineer George Mallory answered "Because it's there" when he was asked why he wanted to climb Mount Everest.

The Sin of "Performance Bias"

If, on certain circumstances, it is customary to "take action," then not doing anything is perceived as a greater failure than the failure that is caused by taking an action, even if that action does not go well. In other circumstances, it is the other way around—taking an action might cause greater regret if it fails. This behavioral bias is applicable to numerous domains of our life. For example, capital-market experts claim that performance bias, which leads to frantic overactivity in the stock market, is usually prone to cause losses to the ones who adapt it.

Social Suggestion Subjugating Inner Truth Perception

One of the most fascinating studies about social influence was "Asch's conformity experiment," which was carried out by researcher Solomon Asch in 1951. The study showed that, in the absolute majority of cases, subjects tended to change their opinion and match it to the "consensus," even though the data they were shown presented a totally different picture.[31] The wisdom of the crowd often wins over our better judgment and we surrender to it, even when our consciousness interprets the situation entirely differently.

The World as My Image—the Projections of the Ego

We tend to use excuses that please our ego. It is difficult for a person to take out the "self" from himself.

Thoughts, cognitive tasks, and memories create ego projections. We tend to pad the ego and protect it by using shock absorbers, which are built in within us, in an attempt to facilitate its drive along the bumpy paths of reality. Thus, when an operational task is ego-friendly it is called "ego-syntonic," and when it leads to a sense of discomfort it is called "ego-dystonic." In fact, each piece of conscious information casts a shadow of "ego projections," and the magnitude of the projection (the shadow) changes in accordance with their relevancy to the ego.

Numerous pieces of so-called "objective advice" we offer are related to consideration of ego and profitability (how does it affect us?), even at the unconscious level. An attempt to pinpoint such effects enables us to delete them in order to offer the fairest advice. In order to be able to cope with ego projections, we must first be aware of them. Our ego casts a long shadow over each of our thoughts—and being aware of it enables us to weight its proportion in the end product of our thought, or, alternately, to let it be or try to minimize it.

The Good Deeds Quota

A common behavioral tendency is reaching a "quota of good deeds" (we have had this tendency from the days we recited the "Scout oath"). The level of the tendency is determined by each individual in accordance to his beliefs and values. Each person has a personal "reference point," and he dynamically calibrates his behavior according to it. When the personal line is crossed, as we perceive it, the behavioral bias orders us to find the average for our actions by exhibiting less moral behavior, and vice versa: when we feel we have exaggerated with unfair conduct toward others, the behavioral bias guides us to act with more decency.

This is a common behavioral tendency, although partially undesirable. Obviously, it would be better to do the good constantly and consistently.

It seems that our tendency toward decent behavior matches "the momentary level of decency" in which we position ourselves. It depends on our tendency to position ourselves at a certain point, which is often the point of balance between doing good as a value and the energetic price we have to pay in order to reach a certain threshold and retain our position above it.

Clarifications About Art

Some see art as a connection with dimensions of experience that are beyond the personal—as broadening our consciousness from the personal experimenting dimension to the collective experimenting dimension.

Anima mundi (the soul of the world) refers to an experimenting layer that we all breathe—i.e., all of us experience it because all of us live in this world. The artist mines the impressions out of this common medium, and his brain processes them in a unique way that is both subjective and objective.

Part of the aesthetic experience is the sense of excitement that is formed when we identify a familiar pattern. Whenever we meet a new creation, we are in need of the familiar in order to be able to enjoy the unfamiliar.

The fingerprint of the artist's brain is imprinted on the reality manifestation of the work of art.

Thus, a painter from the Impressionist movement dismantles the vision of his eyes into separate specks of color on the canvas—as specks of light on the retina as they appear in the sensory, iconic "vision memory"—in an attempt to transfer the raw input of the vision experience. The "inner logic" of the creative brain, however, always merges with virginal view of the "reality as is." We will never be able to find two identical paintings of the same "reality manifestation" painted by two separate Impressionist painters. The personal fingerprint of the brain is an inherent part of the supposedly raw input.

And, as art teachers like to stress, positioning a standstill object in front of art students often turns their work into self-portraits of each and every one of them.

In a sense, composers also reflect the inner rhythms of their brain in their music. Thus, some claim that they could hear the approaching paces of insanity that overtook Schumann in his late compositions.

A great artist as a great scientist—both are sometimes perceived as a Promethean figure who exposes "ordinary" human beings to the secrets of the high spheres.

The Impressionist painters, therefore, were seen as the artists who documented the invisible energy (van Gogh's sunflowers scream loudly,) as magicians who deal with "bewitching physics laws" through art.

In art there is no such thing as "no such thing" (or, as the country singer Paul Brant puts it, "Don't tell me the sky is the limit when there are human footprints on the moon"). It expands the limits of human perception, and it breaks limitations and barriers.

Aspects of Thoughts and Emotions

Emotional ethos or conceptual patterns that are formed frequently are manifested in material changes in the brain tissue and, sometimes, reflected as prominent changes in brain structure. For example, a study shows that the hippocampus in the brains of people who suffer from long-term depression is about 15 percent smaller compared to its size in the brains of those who do not suffer from it.

Behavior researchers who belong to the behaviorism school of thought raised the banner of observational science—i.e., science that focuses only on what can be seen, quantified, and measured. They avoided the subjective aspects of exploring emotions and desires, but it seems that the gap that this approach left in the puzzle of understanding human beings is too wide.

Compatibility between physical senses and thoughts and emotions is required for the formation of a mental scenario. The thinking processes in our brain are related to the overall condition of our body. It is difficult to think about compassion when we are hitting a punching bag. It is difficult to imagine a sleepy afternoon on the shore of a tropical island when our fists are closed tightly and our jaws are fastened. On the other hand, imagine yourself pushing your car, which is stuck in the middle of the street, when your body is in a status of conscious relaxation and your muscles are loose. Compatibility between brain and body is an essential condition for the ability to create a reliable emotional-mental image.

A fascinating experiment showed that when our elbow is bent, as happens when we bring an object closer to us, and we are asked to imagine a list of objects that is read to us, the accompanying feeling is more positive compared to a situation in which the same list is read to us when our arm is extended, as happens when we make a rejection movement, when we push a certain object further away from us. It seems that the position of our body creates a certain tendency to a typical emotional mix.

The values of the scale of "intensity of mental representation" are compatible to the dimensions of the anagram (neural networking) that encodes it. Various insights in our brain have different levels of representation. When we imagine some of them, it creates extremely loud seismic noise and a tsunami in the sea of our brain. In their Richter scale of mental representation, a higher value will therefore be registered compared to representations that make weaker waves in the sea of our brain. The intensity of the emotional charge of the experience greatly affects the intensity of its mental representation.

High Spirits

The Pace of Thinking is also the Pacemaker of our Emotions

Rapid thinking, which is sometimes defined as a state in which the pace of producing thoughts is faster than the ability to express them in words, sometimes improves the inner emotional climate. Studies found that rapid thinking triggers high spirits and usually induces a sense of mental vitality. This finding might be related to the "run of thoughts" and high spirits related to a manic condition. On the other hand, "slow thinking" is compatible to a gloomy mood. These findings suggest that inducing a faster pace of thinking might be a therapeutic tool to improve people's mood.

A common assumption among brain researchers is that our brain is wired, on average, to a positive bias of reality—i.e. to perceiving our world in rosy hues—without disregarding the fact that among people who suffer from depression, the pendulum of emotions turns toward negative bias with respect to reality. As aforementioned, however, researchers generally tend to think that the common human tendency with respect to predicting the future is optimistic, with all the reservations that exist with respect to this comprehensive declaration.

High spirit is an emotional trampoline that swings us upward, softens our fall, and often induces a conceptual ethos that strikes like the soles of a flamenco dancer in the midst of a passionate dance.

Just as the Australian plant banksia needs a fire in order to spread its seeds, the flames of enthusiasm might inflame the wick of thought related to a great idea.

It is claimed that Handel wrote the "Messiah" within three weeks only, when he was in high spirits, or even at a manic state. During a manic state, ideas might flow so fast until they "bump into" each other.

High spirits increase the speed of thought and are in a linear-like correlation with the output of the products of thinking. This renaissance of imagination takes place up to a certain point—the point of mania, from which the correlation is distorted and even turns over—then the products of thinking are found trampling along the quicksand of chaos.

At a manic state, the thoughts buzz like bees in the busy center of the beehive. Thinking strings of various lengths are formed—some reach the end of the nose, and some reach the horizon. All great life enigmas stand in line and demand full and immediate answers.

Regret

The director Woody Allen once said that the only thing he regrets in his life is that he was not born someone else.

Most people, however, refer to regret in the sense of actions they could choose whether or not to perform. In this context, there is a common distinction between two types of regret: the regret of doing, which derives from an action that was performed, and the regret of not doing, which derives from regretting not doing something that could have been done. Studies found that, in the short term, we are more concerned about the undesirable outcomes of actions we performed, while, as time goes by, in the long term, we become more regretful concerning actions that we could have taken and didn't.

This tendency was not found as gender-dependent, but it was found that the regret of not doing increases as we become older.

Bonnie Ware, who worked as a nurse in a hospice for terminal patients, described in her book that one of the most common regrets among dying individuals, men and women alike, was not expressing their feelings toward others more frequently. Thus, a person who adopts the approach of Mr. Spock from Vulcan and seldom expresses positive feelings toward others might regret it toward the end of his life.

Elegiac Thoughts

"The inevitable misery of every-day life," as Freud mentioned, often triggers sorrowful thoughts about the journey of our life.

Sometimes the strident accords of the melody of our mood are formed into a nihilistic tune that sounds like a whistle of a train. This is a march that is heard by those who believe that we spend our days on a train that is headed nowhere.

Studies have found that the predictions of those who tend to provide more strict interpretation are more accurate. The term "depressive realism" connects the rigorous emotional ethos with more accurate and reality-compatible interpretation of reality appearances as they are. We often suffer "emotional phantom pains," however. This is the anxiety we feel due to reality scenarios that never take place but, nevertheless, cause us emotional distress, because we think about them and are preoccupied by them.

Sometimes knowing causes suffering. A refreshing example of this painful subject is the example of Jerry, the cat, from the cartoon *Tom and Jerry*, who continues to run even when there is no ground under its feet. But after a short time "air running," once it finally realizes that it's walking in the air, it drops down at once like Newton's apple. The "knowing" summons gravity to perform its familiar magic. Indeed, knowing can bring about suffering.

The Abyss Within

The "Lucifer Effect" refers to a situation that makes "normative" people act wickedly.

The French philosopher Jean-Jacques Rousseau is the prominent representative of the view that says that human nature is basically good (the imagination of man's heart is good from his youth).[32] On the other hand, there is the view of the English philosopher Thomas Hobbes, who claimed that human nature is basically bad (the imagination of man's heart is evil from his youth).[33] It seems that each of the two worldviews is partially true.

The Russian writer Aleksandr Solzhenitsyn, who was well familiar with human cruelty, wrote, "The line dividing good and evil cuts through the heart of every human being."

Pandora's boxes also exist outside the Greek mythology, and one of these boxes is hidden within each and every one of us. The dark side can materialize in each and every one of us. In order to prevent the seed of evil from developing within us, we must constantly struggle not to let this aspect of our being control our behavior throughout every moment of the day, throughout all our life. We must prepare ourselves for this constant struggle according to the time perspective of "The Hundred Years' War." We should aspire to select the compassionate action pattern from the repertoire of behaviors unfolding in front of us, as our favorite performance option at any given moment, knowing that, unfortunately, our human essence will probably not always allow this optimal scenario, but, nevertheless, we should always aspire to it.

Several parallel channels of performance options that derive from the momentary brain climate run around in our brain at every given moment. Some of the options are unfair, or even cruel, with regard to others, and we are required to stand guard in order to minimize their expression and even to eliminate them as much as possible.

"Your Wish is My Command"—the Genie Aspect of our Personality

In the spirit of the submissive obedience of the genie from the stories of Aladdin, the obedience-to-authority tests that were carried out by psychologist Stanley Milgram in 1961 showed that two-thirds of the people who participated in the study obeyed the instructions of an authoritative figure, to the point of causing severe damage to another person, even after distress signals appeared and the suffering of the other person was evident.[34] The fear of authority, which seems to derive from emotional motivation whose source is primarily the amygdala, blurs the image of the other in the mirror cells of our brain (which are in charge of reflecting the emotions of the other) and covers up the voice of our conscience from the prefrontal lobes.

A possible claim deriving from Milgram's experiment might be that the circumstantial context, the place, time, and social norms are very powerful factors that greatly affect human behavior.

Opsimism

Some claim, somewhat cynically, that sometimes the difference between an optimist and a pessimist is that the pessimist knows more.

We are all optimist to some extent, with different mixes of optimism and pessimism ("opsimism") that depend on our natural tendency and the different circumstances. The echoes in the cave of our thoughts move from bright to dark interpretations, and vice versa.

A circular bias sometimes characterizes the pattern of the opsimism mix.

For instance, from personal experience I know that during night sleep, dark sandbars sometimes sprout above the water during the low tide between three and four o'clock in the morning and fill up the cave of thoughts (skull cavity) with gloomy echoes. Afterward, by the time we wake up, a comforting wave of high tide appears, refines the contents, and reinforces our self-confidence as if preparing us for another day of coping with reality. In my view, this pattern has a personal component but, possibly, also a universal component.

Thoughts About the Journey of our Life

Our thoughts are the soundtrack that constantly accompanies the "movie of our life." Some of them are claims regarding the movie itself (critique) within the movie. The music of life is mostly played at very low volume, but sometimes the volume suddenly soars upward and shakes the ears of our being.

At any given moment of our existence, we send out the ship of our life toward the capricious waves of circumstances whose direction and height we are only partially able to know.

This pattern of existence summons the painful understanding that the world is run according to its internal algorithms of probability and cause and effect, and human hope is not mentioned in its rule book.

From time to time, life bestows upon us cruel skits; they offer us possibilities that enchant our heart but, at the same time, prevent us from reaching them—like meeting the person who seems the most appropriate to be "the love of our life" the day before we die.

Friedrich Nietzsche once said that the night is deeper than the day could possibly imagine. this idea shares some similarities with Albert Einstein's words. He said that the world is more bizarre than we can possibly imagine.

We have to understand and come to terms with the fact that our ability to comprehend many phenomena in our world is rather limited. The complexity of many of the phenomena in our world is beyond man's range of understanding. The insights his brain gathers throughout his lifetime do not allow a deep understanding of many of the phenomena in our life, but they provide a certain level of understanding of some of them. Accepting the concept that the level of complexity of numerous phenomena in our world, and the interaction between their various components, is beyond the scope of our understanding, even if we persist constantly in discovery's journey during all our life, is a painful, yet sobering, insight. The scope of our life is limited, and the horizon of our understanding is limited, and we must match our expectations regarding our understanding of the world to this insight. All of these bring about acceptance regarding the built-in imperfection of life.

Experiences in our life are sometimes similar to a Rorschach test, and they are a sequence of occurrences that are always suboptimal as far as man is concerned. We aspire to reach the kingdom of perfection, and sometimes we come very close to it, but it seems that we will never be able to actually enter its gates. Our brain is the means that might bring us closer to the essence of this aspiration, but we cannot achieve it entirely. And in a spirit of (too) gloomy and severe interpretation, some might claim that our life is a "dystopia that aspires to utopia."

The size of the straitjacket of the reality of our life derives from the fact that our time on this planet is too short, too much confusing information surrounds us, and the sources of our brain, as wondrous as it is, are, in this context, too limited. We must run our life within the limits of this straitjacket of the reality that surrounding us.

Views in Life

Life is a continuum of various dosages of suboptimal situations. Imperfection is built into them. Various situations in life are at different distances from the optimum point, and in extremely lucky circumstances they come closer to it in an asymptotic proximity.

There is a Buddhist saying that when we are born, we receive a key that can open the gates of both heaven and hell. Some interpret this saying as emphasizing the importance of the hand that holds the key that determines which gate will be opened. Others see it as a metaphor for our world, which is both magnificent and horrible. It seems that most of the time we are somewhere in the middle—in a hellish heaven that contains a changing mix of the two components in accordance with the circumstances.

Life as a parable: Marcus Aurelius, the Roman emperor who fulfilled Plato's vision of a philosopher king, said that life can be compared to a war journey in a foreign land. Using the same tone, we might even say that, sometimes, life seems like a survival journey across a hostile planet.

Our mental life sometimes seems like an interplay of turning a blind eye between naivety and critical observation.

The circumstances of our life require coping patterns that range between an ethos of macho audacity, on the one hand, to caution that might be perceived as hesitancy, on the other hand. Sometimes it seems that the circumstances of life require what is non-euphemistically called "King Kong balls." In other circumstances, however, it seems that the best coping requires "ping-pong balls," in the spirit of the hedgehog version of the *Kama Sutra*: make love "slowly and cautiously."

Sometimes we are both the prisoner and the jailer, with respect to the choices we make or choose not to make in our life.

From the moment we are born we are doomed to freedom, as philosopher Jean-Paul Sartre once claimed. Hence the constant necessity to have doubts and make decisions, which grips our brain in virtual pincers and never lets go as long as we are alive. Some believe that the loosening of the grip of this tiring and sometimes-paralyzing necessity is the secret magic of the worldview according to which metaphysical entities beyond ourselves share the burden of responsibility for our destiny with us.

All of us sometimes long to be "freed from the freedom of choice"—in the sense of minimizing the options, wishing to rely on consistent and arranged hooks of reality in the midst of the chaotic carousel on which life forces us to spin. An arranged set of life rules has worked magic on the human soul from time immemorial, and the psychological serenity that is involved in it is the reward for adhering to it, as among religious people.

The processing processes in the brain, which involve the constant need to have doubts and make decisions with respect to daily matters (reflection of the freedom that was forced upon us), are probably the hungriest consumers of energy resources in the brain (each thought has an energy cost). Our brain, as a tough energy banker, aspires to minimize the thinning of its resources as much as possible; it prefers to rely on structured behavior protocols that ease the burden of contemplation and making decisions and that require lighter brain processing and, as a result, reduced energy consumption. It might explain the source of attraction to customs that are common in various countries and outline arranged patterns of behavior that are more economical in terms of brain energy. The freedom that is forced upon us, according to Sartre, and the accompanying freedom of choice have a high energy price. Perhaps the wish to reduce brain energy consumption is at the basis of our desire to be "freed from freedom."

Thoughts in the Service of Survival

From an evolutionary point of view, it seems that our brain is designed to improve our ability to survive and multiply and is not intended to serve as truth detector regarding our world. While pursuing the discovery of truths in the world of phenomena around us, we channel some of our capabilities' resources toward the aspiration to discover the truth.

On the dark side of the thoughts' moon, there are thoughts that were intended to improve the survival chances of the individual at the expense of the survival chances of others.

Reality is sometimes a "rain of falling knives." Turning a blind eye to the sights of reality's battlefields is like teasing the goddess Fortuna or taking a nap on a railway.

It seems that our brain tends to make friends with the tangible and to avoid the abstract.

Similarly, reality manifestations with a high relevancy coefficient are prioritized during processing: our brain prioritizes images that are perceived as more relevant to our existence; thus, we prefer to first cope with phenomena that are closer to us in terms of time and space. Our brain is recruited more vigorously to the processing of various phenomena that are perceived as more relevant to us. We come up, therefore, with mental scenarios of near-future events at a higher concrete level compared to remote-future events.

A common belief is that "rolling about" in a language, by assimilation in the environment in which everybody talks the language on a daily basis, speeds its comprehension and improves our mastery of it. A common explanation for this is that, in such a situation, the individual "must" gain rapid mastery of the language in order to function in the environment, and this "selective pressure" is a very powerful mental catalyst. The relevance component of the new language is greatly intensified in such a situation, and the tendency of our brain to prefer information that has immediate practical and relevant implications is "taken advantage of" in order to gain mastery of the new language quickly. We are biased to act according to the relevancy principle with respect to social matters as well. Sometimes this tendency is unfolded in front of our eyes blatantly—when we feel, for example, that a certain person is taking advantage of another person in an "instrumental" manner in order to promote his agenda. We all commit this sin, but we must be aware of this tendency and refine it by constantly reminding ourselves that the rights of others are not inferior to ours.

End Thoughts—in the Shadow of the Finality of Life

Knowing the end is an insight that is both reinforcing and weakening. It derives from the "excess capabilities" of the human brain, which granted it possession of knowledge that is probably hidden from the rest of the residents of the animal kingdom.

In our generation, many believe that the short period of time during which the flame of the candle of our life is burning is all the time we have.

Time and death walk hand in hand and set the ultimate standard of 100-percent success in their intimate relationship.

The moments of our life are like sand grains falling between spread-apart fingers; the finality of our life is the only certain gamble in the casino of life.

Life sometimes seems like a short-term loan on which we pay the interest of constant fear of annihilation.

An approach that is not pleasant to our palate claims that, as our materialism is doomed to go with the wind, our consciousness will evaporate along with it.

Wondering about the final chord of the tune of our life always accompanies our thoughts about the future. Will the end be Shakespearean—the hero's body appears on stage following a sudden stream of daringness in a fighting retreat against the capricious monsters of life? Or will it be Chekhov-like—the hero accepts his destiny in quiet despair until the end?

At both ends of our life we are totally lonely; we enter the world and leave it in splendid isolation, even if other people are present at these events. Some may say that we are lonely, also, through most of the way between those ends.

Acknowledging the temporal nature of our life might bring along an "awakening experience" that recalibrates our life values and changes our conduct. Sensing that the arms of the clock accelerate their running, and that time is pressing, is a common trigger for the formation of that experience.

"The privilege of knowing the end" is a unique gift that was given to the human brain. We feel we know how our journey on the face of the earth will end, although the specific circumstances are hidden from us. Knowing the end provides an appropriate perspective to our plans and actions.

The insights regarding our last day on planet Earth, our ending thoughts, just before the journey ends; imagining our final moments, looking with a summarizing glance at our life journey, at the good times and less good times… Did we win in bringing the good in us into expression?

The wish of knowing the "knowledge of the end" in advance is reflected in a cartoon scene. Snoopy, the famous cartoon dog, exchanges toward evening philosophical thoughts with a fruit fly, whose life expectancy is a single day. The fly tells Snoopy candidly that it regrets only one thing: "I wish I knew at nine in the morning what I know now." The lost wish of the fruit fly sometimes visits our mind as well.

The insights accumulated by our brain throughout the years enable us to write more and more chapters (and to rewrite some of them) in our personal version of the guide "Life—A Brief Course for Beginners." The possibility of endowing insights that are viewed from the proximity to the finish line by the elderly, to the youngsters who are far away from it, is rather comforting. This educational task is intended to improve the youngsters' future coping with "universal life crises" that they will have to face by virtue of being human beings who sail on the river of life.

Facing the End—Coping with Finality

When dealing with the finality of our life, we often find ourselves in the midst of a cognitive-dissonance storm that destroys our rational arguments like a house of cards. The fear of human logic concerning the "end of logic," which involves the finality of life, is understandable;

thus we find it difficult to look straight at the abyss of our finality and tend to avert our gaze from it.

This is the main reason human cultures are saturated in death-denying concepts intended to sweeten the horrible bitterness of finality. Believing in these concepts is like wearing metaphorical sunglasses—their aim is to ease a sober view on our finality, in the sense of a conceptual infrastructure that relieves the pain of meeting the gloomy reality.

In many religions, personal immortality has turned into a sort of reward given to devout believers as a means of encouraging them to adhere to the dictates of the superior meme of the religion.

The preservation of the soul after the destruction of the body is considered the longed-for reward for adhering to the way. The book of life ends with the tragedy of death, but the epilogue that follows allows for a happy end to the story.

On the other hand, total extinction is, in the opinion of many others, the period at the end of the sentence of our life. In this spirit, once our brain dies, our spirit is gone and our thoughts vanish. In less-softened words, all of our insights and memories become worms' food.

François de La Rochefoucauld once said that "On neither the sun, nor death, can a man look fixedly."

The Roman emperor Marcus Aurelius gave us royal advice: "Do every act of your life as if it were your last." This ethos is also reflected in the "carpe diem" advice: seize the day.

In an earlier era, Epicurus, the philosopher, preached in favor of equanimity with respect to death since, in his opinion, we are never in a state of superposition with death: "… as long as we exist, death is not here. And once it does come, we no longer exist."

Some people include in their daily schedule a courageous practice of meditation that focuses on death first thing in the morning, at the beginning of a new day. Some do so based on the view that death is the central marker of the coordinate system of life.

Memes that were intended to ease the pain of finality are embedded in all human cultures as an attempt to protect us from the frost induced by total extinction.

The comforting wings of imagination are able to carry us above the mountaintops of reality, as far as Shangri La—an earthy paradise where death has no hold (after the magical valley, hidden among the mountaintops of the Himalayas, that was described in James Hilton's book *Lost Horizon*).

Some feel as if they manage to partially escape from the tight grip of death through their offspring, in a pattern of genetic heredity, and perceive their children as chariots that carry part of themselves toward the future (and, assuming their children and the children of their children do not abuse the dictates of the genes, part of them will be carried forward toward the next generations).

Others feel comforted by assuming that the memes that constitute the "list of the best," produced by their brain while it existed, will continue to exist in the brains of the people who continue to live after they're gone. In other words, the spirit's footprints of their brain will continue to exist in the sense of endowing the world with ideas born in their brain, or with ideas that were upgraded in it, through bequeathing of the memes.

Some have a gloomier, less romantic view, which they perceive as more realistic. According to their view, when our self eulogizes its destruction, it is actually an illusion eulogizing an illusion, since the self is nothing but an illusionary entity.

Others come to terms with the finality of our consciousness and find comfort in the fact that the molecules that compose our body will return to Mother Nature, assimilate in the mixed multitude of molecules our world is composed of, and might be part of the founding of another living entity.

Some others find peace and acceptance in the saying that the circumstances of our life and their ending are only the way the universe acts.

Despair

Certainty is the end of hope, and in the absence of hope despair usually takes the reins. Under such circumstances, our brain produces a behavioral repertoire that does not exist under different life circumstances.

Despair, as an existential state, is a powerful human ethos facing our ephemeral materialism, or, as other might claim, the casualness of life.

Despair of the Present Detached From the Burden of the Past and the Promise of the Future

When a person is looking from the deck of his life ship and feels as if he were the captain of the *Titanic* right after hitting the iceberg, his brain comes up with action options that could not have come up in any other scenario. In the same spirit, there is a controversial saying, ascribed to a Jesuit monk: "When we live on the verge of despair, we truly live."

An Awakening Experience

At a young age, we feel that the scenario of the end of our material existence is a possibility, and those who internalize the fact that it is not a possibility but, rather, a certainty experience a conscious awakening. This conscious awakening is a "metanoia"—a perceptional change, a breaking of the "glass ceiling" of the perceptional conceptualization ability and acquisition of a more complex perception that substantially changes the perception of life.

Metanoic experiences might take place, inter alia, following brain injury or "revelation experiences"—i.e., following a structural change or perceptual change that naturally induces a structural change. This phenomenon demonstrates, yet again, that insights and thoughts have a structural and functional manifestation, which means they are wired and engraved in a three-dimensional neural structural pattern and a unique electrophysiological action pattern. Many people among those who experienced near-death experiences undergo metanoia, each of them in an individual, unique pattern.

The Silhouette of Extinction

The silhouette of extinction walks stealthily by our side from our very first steps on the path of life. During childhood we come across flowers that bent their heads and whose petals have withered; we come across aquarium fish that float motionless on the surface of water, or a puppy that did not make it to the other side of the road. There is the kind elderly woman, who used to offer cookies to us, who has suddenly disappeared. As years go by, the trail of people whom we used to meet in life's paths and are no longer with us becomes longer and thicker. The fact that life is ephemeral makes it more valuable.

The shadow of death is cast over many of our thoughts. The sorrow of our finality is soaked in us, hides and reappears, alternately, in front of the eye of our consciousness.

Uninvited thoughts regarding the finality of life jump on the stage of our consciousness, especially when the markers of finality reappear: the end of a day, the end of a season, the end of love.

We all resemble impermanent refugees, with respect to our presence in the world, and the entire world resembles one big refugee camp.

When we march along the expected procession of generations, we first part from the parents of our parents' parents (if they were lucky enough to have great-grandchildren), from our parents' parents, from our parents, and then there is nothing more that separates us from death.

Usually, we are remembered for the blink of an eye and then forgotten for good. The ones who are remembered for a longer time enjoy another brief moment. When we look at the "waterfall of generations" in retrospect, we can see that most of us retain the memory of our parents, a little less the memory of our parents' parents, and beyond our great-grandparents a dark hole gapes, in which the characters are faceless and nameless and are referred to only as part of family folklore, if at all.

Some imagine their final departure from life as a prodigious death that weaves heroic tunes in the gloomy requiem, but most of the time we are doomed to fail in our attempt to plan the rules of encounter with the figure in the black cloak that carries the scythe. The circumstances of our departure are unexpected and soaked in thick layers of capricious randomness.

Death is an eminent sociologist that grants its favors equally to all human creatures, and, at the same time, it is amazingly creative. Each exit door from the hall of life is custom made for each person who uses it, as described by the writer Ivan Turgenev: "Death is an old joke, but each individual encounters it anew."

In a case of a person who does not die abruptly, the journey toward death is a totally personal journey characterized by inherent loneliness.

A sense of missed opportunity is mixed in the cocktail of senses related to departing from life. A familiar part of it is regretting not doing something now that it is too late to do it.

The sense of a satisfactory life, as each individual defines it, however, is in correlation with a lesser sense of missed opportunity.

The philosopher Martin Heidegger defined death as no chance of a second chance.

In Greek mythology, the Cyclops got a horrible punishment—it was informed of the day it would die. In reality, when a terminal patient is informed of the estimated death date, this knowledge might increase his suffering to the point that any chance of reasonable living is abolished.

We all live on borrowed time, but in the case of people whose life is in danger, like those who suffer from serious diseases, the overturned hourglass appears on the screen of their consciousness much more frequently as the sense that closing time is near and the sun is lower in the top of the trees and about to set.

In the evolution process, marvelous sand castles, such as the human brain, are created. These castles are probably doomed to be washed away and vanish. The wastefulness of this process is both magnificent and horrible.

The invisible, but highly realistic, sword of Damocles is always hanging above our heads.

As a semi-macabre quip, a website called the "Death Clock" introduces four questions to the surfer and, according to the answers provided, an estimation of death date is made. The date on which my birth certificate expires—prematurely, no doubt—was set by the website as the fourteenth of March, 2040, which means I have sufficient time to revise my will.

Our life expectancy according to the Death Clock website is an interpolation to a value between zero and 122 years and 164 days, which is the longest human life expectancy according to *Guinness Book of Records*. Let us hope that each one of the readers of this book will increase the top limit of this interpolation with a satisfactory quality of living.

Chapter 9
A Beautiful Mind

Human Intelligence and the Intelligence of the Ones that Share the Planet with Us

Some people claim that animals have a basic sensory awareness that experiences world manifestations in their raw version and that they do not have the tools for a more sophisticated processing, which is considered the preeminence of man.

It is also believed with regard to the ability to foresee the future, which is probably a mental skill that is almost unique to human beings. It seems that most animals—and some claim even all animals, except for man—live with the perception of "eternal present," of stimulus and reaction, of the here and now.

It seems that it is more correct, however, to say that many of our mental skills, as human beings, are on the same spectrum as the skills of other animals. These skills reached their peak in the performance of the human brain, but they are not unique to us and should not be viewed as features of "binary manifestation." In other words, one should not conclude that these skills are fully present in humans and totally absent in animals. These mental skills have probably appeared gradually throughout evolution, rather than flickered at once, in the brain of the Homo sapiens. It seems that our species is lucky, since these skills have crossed a certain threshold from which the capabilities constantly advance, as in crossing the threshold of a singularity point.

From a phylogenetic point of view, we are lucky. The Homo erectus brought about the gift of erect body, and the Homo sapiens brought about the gift of improved intelligence. Our exclusive status as the most intelligent creatures in the known universe brings along a heavy burden of responsibility, however.

As a species we tend toward teleological solipsism: the past led to our creation, the future will be created by us—and, according to this spirit, we often judge the intelligence of animals based on anthropocentric features (which derives from perceiving humans as the center of the world).

The writer Isaac Bashevis Singer once made a claim that is like a punch in the soft belly of emotions: "For the animals in the food industry, each day is an eternal Treblinka."

We are obliged to constantly improve our conduct with respect to animals that are used, or whose products are used, in the food industry. We must treat them with ultimate compassion.

Reality perception might differ among different organisms based on the functions of their nervous system, in which the reality-indication mechanism is assimilated.

Intelligence exists in animals in a hierarchical pattern rather than in a binary pattern of either existing or being absent. A practical conclusion might be that we must treat the ones who share this planet with us respectfully, because we are their older siblings (in the sense of having a more "mature" intelligence), and because many species demonstrate intelligence; and it seems that some of them also have awareness, as it is defined in broad circles, and also due to their own value, regardless of any comparison.

Being Dr. Dolittle—Speaking the Animal Language

Creating a communication interface between the human brain and the brain of animals is sometimes possible through using the grammar rules of the "animal language." An event that illustrates this took place in 2005, when the building of the department of brain sciences of the research center in Cambridge, England, became a battlefield for birds. Every day about ten birds flew toward the new, grand glass front of the building and were knocked to death. The reflection of the bushes and trees on the glass front prevented the birds from perceiving it as a lethal barrier. Attempts were made to prevent the phenomenon; a scarecrow was placed in front of the building, but this solution failed. One of the employees adopted a "bird-like" worldview and, based on this perspective, cut cardboard pieces in the shape of a large vulture and attached it to the glass front. The attempt was successful—the cardboard vulture managed to keep the birds away and prevented them from crashing into the glass front. Adopting the grammar rules of the biological language of the birds helped convey the desirable message.

The Wonders of Nonhuman Intelligence

There are numerous examples of astonishing performances in the animal kingdom. The secrets of these performances are partially beyond our grasp.

The bees create the cells of the beehive in a shape of a hexagon as a genetic dictate, which is based on significant energy saving and a cost-benefit effect.

Bees' mastery of the secrets of the dance that is designed to pass information among them regarding the location of available sources of nectar is innate. The "choreography on the way to the flowers" is ascribed to a gene that produces protein—"the dancing protein." Once this protein gets to the bee's brain, it grants it the ability to communicate through a dance, which gives its sisters accurate navigation instructions to the spring of sweetness.

Salmons have extraordinary navigation skills. They swim thousands of kilometers in the ocean and find their way back to the exact location of the stream of their birth.

Another amazing example is the example of the monarch butterflies that embark on a multigenerational journey across thousands of kilometers toward a specific forest in Mexico where they make love.

The butterflies start their journey in spring time, in a grove located in the center of Mexico. At the beginning of the journey, following their mating, the females lay eggs on the Asclepiadaceae's leaves and then start their journey. Their destination is a forest in southeast Canada, at a distance of four thousand kilometers—a place they have never been to. The caterpillars hatch, fed from the leaves that are toxic to other types of insects, and turn into cocoons. Ten days later, a butterfly emerges from the cocoon, and the next generation follows the mothers. Without any tour guide, they find their way to their destination. The female generation mates, lays eggs on the way, and the grandchildren later join the journey. When autumn comes, the offspring of the fourth generation embark on a journey in the opposite direction—from southeast Canada to the grove in Mexico. Although this generation has never encountered the glory of Mexico, the butterflies land in the same forest from which the parents of their parents' parents left on their way to Canada. The amazing, multigenerational navigation is ascribed by scientists to an innate skill that is concealed in the brain of the butterflies, which is as tiny as a pin. This organ serves as a compass, and it seems that it enables navigation while relying on the outline of the magnetic field of Earth.

People who keep pets are sometimes amazed by the amusing behaviors of their nonhuman friends that seem to reflect thinking. A friend of mine told me that when his dog wishes to go out, it peeks through the main door; if it sees that it is raining heavily, it immediately runs to the back door and peeks outside, as if checking whether it is also raining in the backyard, or, perhaps, whether this door leads to a drier reality. Change of place does not necessarily mean change of luck…

The neurons that build our brain are not substantially different from the ones that build the brain of the chimpanzees or the brain of sea snails. The difference is found especially in the function as a whole. One of the differences between a human brain and the brain of the chimpanzee, which is considered to be the most intelligent animal, derives from the activity of the stopper gene. This gene determines the number of divisions of neurons and, by that, their overall amount in the brain. In humans, the stopper determines about a hundred divisions, which results in an amount of about ninety billion neurons in a human brain on average. The stopper gene in chimpanzees dictates an earlier stop of division, which results in the fact that the size of a chimpanzee's brain is about one-third of an average human brain. The quantitative aspect is also translated into a qualitative aspect. The difference is intensified in the sense that it results in an exponential gap in the number of links between the neurons.

The presence of "soul sight" among the hominina (the super-apes) is controversial, although there is evidence for the presence of some of these skills among the high perimeters.

Being a Person—What Makes Us Human?

Out of the three billion signals that compose the human genome book, less than 1 percent (about fifteen million signals) is different from the chimpanzees' genome book. It all started when our evolutionary pathways were split about six million years ago. The recipe that generates a human being is more than 99 percent identical to the recipe that generates a chimpanzee. The tiny, 1-percent difference is responsible for the "preeminence of man."

With regard to volume: One kilogram of brain tissue distinguishes the human brain, which weighs, on average, 1400 grams, from the chimpanzee's brain, which weighs 400 grams on average. The preeminence of man is embedded in this kilogram and its unique wiring.

Generosity is not necessarily in correlation with the brain's volume. The bonobo, a subspecies of chimpanzee, has a remarkable friendly and pleasant temper, from which humans can also learn a thing or two.

As for the range of our behaviors as humans, we are at limbo (a place between heaven and hell, according to Christian tradition)—in the midst between animal-like and divine.

In this sense, it seems that we are "bipolar" creatures. Our personality and the memes that settle in our brain as a result of environmental impact will determine the direction we take—thus the importance of transforming the "correct" memes, which will make the pendulum of our behavior, as human beings, turn toward the pole of superior characteristics and, as a result, deter us from offensive behavior tendencies.

Human Brain and Digital Computer

The view of the brain as a machine first appeared in the seventeenth century, when many intellectuals tended to explain various world phenomena in mechanistic terms.

The contemporary brain version derives from the view of the brain as a machine, and the terms are taken from a very complex machine (according to the contemporary view): the modern computer.

Silicon intelligence: The computer is a machine intended for formal operation of symbols that are reflected in two digits: zero and one (the appearance of the quantum computer causes a "quantum leap" with respect to the difficulty in explaining its modus operandi). Information processing in temporary calculation machines is done at the configurational layer rather than the content layer of information. The ability to process information according to contents is the "promised land" of artificial intelligence, and it is still a vision for the future.

Complexity is a field in computer science that assesses the effectiveness of solution manners to a certain problem. The coping approaches are assessed primarily according to the time period necessary to solve the problem, in accordance with the required processing resources. A similar approach is often adopted when assessing the coping of our brain with mental challenges.

A neuron that transmits a bioelectrical signal requires five milliseconds for reorganization before it is ready for the next transmission. In other words, the top limit for the pace signals' creation in the neurons is about two hundred hertz (two hundred signals per second). Modern computers are capable of carrying out billions of actions per second, which means that their speed is at least five million times higher than the speed of the neurons in our brain.

The "hundred-steps rule" demonstrates the possibility of the existence of a threshold that must be reached with regard to the speed of information processing in protoplasm brains, in silicon brains, or in any other type of processing and calculation apparatus.

The rule can be explained by the following metaphor: a person is required to carry a hundred sacks, each of which requires the person's full carrying capacity. The sacks are to be carried beyond a mountain range. Crossing the mountain range requires one million steps. If this person hires a hundred people who carry the sacks, the overall time of performance will be shortened by a hundred. The hiring of extra workers, however, will not make the hands of the clock move faster until the finish line, since the simultaneous performance potential has already been utilized, and the average length of a human's step is not affected by the number of people who take that step.

The validity of this rule is controversial. One can claim that an alternate relay run of a thousand people who share the carrying of the hundred sacks in a serial pattern might shorten the duration of the task, since a "fresh" carrier takes the sack from the tired carrier that preceded him. Yet, many others claim that the rule is valid, and that it dictates a rough top limit with regard to the speed of performance based on the type of information that is being processed.

Artificial neural networks are a calculatory mathematic model that resembles a natural neural network in certain aspects. This model still has difficulties with clarifying certain central aspects of brain activity. On the other hand, this model shares surprising similarities with certain aspects of brain activity patterns, such as the fact that most of the information that is stored in our brain is not located in a specific location but scattered throughout the network.

Models of neural networks (neuroid networks—in the shape of neurons) "invented" new capabilities and skills that were not assimilated in them by their creators in the first place. These features, which showed up spontaneously, are remarkably similar to skills that "derive" from the brain, especially in the brain areas that operate according to an "open code" information processing pattern (i.e., the area of the neo- cortex).

The brain and the digital computer are able to process information in a serial pattern and in a parallel pattern, according to the nature of the task.

On the other hand, computers lack personal-experience reality. A computer is not an entity that is aware of the contents of the information it mediates. It is not equipped with meta-memory (the capacity to pass from one memory to the next according to the person's desire), etc.

Some physicists, such as Roger Penrose, claim that there are similarities between the human brain and the quantum computer, with respect to the manner of processing information, in the sense that the brain carries out processing in superposition, which is intensified exponentially.

The Turing test for determining intelligence, proposed by mathematician Alan Turing in 1950, suggests that a machine will be considered intelligent if its output misleads a human being to believe it is human. This threshold is yet to be crossed, and some people doubt it will ever be.

A large, global research project called "the blue brain" is based on the structure and function of the human brain. The researchers who work on this project are developing a conceptual core model of brain function in an attempt to grant the designated computer the skills of coping with vague input, responding in context, and demonstrating independent learning curve, similar to the features of the human brain.

Limited Experts

Expertise (in the sense of extraordinary, unique skills) among people who are limited in their skills in other fields is called "savantism."

The skill of a savant is an extraordinary talent in a person who is extremely limited in other cognitive domains.

Savant syndromes are characterized by extreme height differences in the topographic map of mental capabilities—as if Everest and the Dead Sea were located at the same region. So, for example, autistic people who suffer from severe impairment in language proficiency and basic life skills might demonstrate unique talents related to musical memory or painting talents.

As with a raw diamond compared to a polished one, some claim that savants have more access to raw data coming from the sensory organs, which are normally unavailable to ordinary people since they undergo processing, filtering, and polishing before they are available for conscious processing. As we have been climbing the phylogenetic ladder (the ladder of evolutionary development), our perception mechanisms have become a sophisticated polishing workshop that operates on the raw diamonds (as a metaphor to raw sensory information) that come across it instantaneously. It seems that this perceptual mechanism improves our daily functioning, and was created as a survival adjustment, but, in a sense, it detaches us from the authentic perception of raw information. An unpolished diamond is highly valued in our brain's stock exchange.

An enchanting story by Jorge Borges—*Funes the Memorious*—describes Funes, who, until the age of 19, used to be extremely forgetful. Then, after he fell from a horse and injured his head, he became a grand mnemonist of details who remembered every leaf of every tree, but not the trees as a whole, or, alternatively, remembered all the trees but not the forest. Funes had a rare talent of remembering details but failed in seeing the global picture, and, even when he tried wholeheartedly to forget, he could not escape from the arms of his memory.

This characteristic of remembering details in an amazingly accurate and detailed manner is unique to some of the people who suffer from autism. It is known, however, that autistic people who have this extraordinary skill, like Funes, fail in seeing the "global picture."

There is a claim regarding the manifestations of exceptional excellence among autistic people that their excellence is reflected in meticulous reconstruction of reality, but not in creation of a new reality, which is an important measurement of creativity. They might excel in their ability to come up with a "mental photograph" of a scene and draw it with photographic accuracy, but they will find it difficult to draw a nonconcrete figure. Alternatively, they might excel in playing a musical piece after listening to it once, but it will be difficult for them to compose a new tune.

The savant's phenomenon is highly common among people who suffer from autism; the rate is one savant out of ten autistic people. The phenomenon is six times greater among autistic boys in comparison to autistic girls (the rate of autism, in general, is four to seven times higher among boys in comparison to girls). A similar disproportion between boys and girls exists not only with respect to autism but, also, with respect to other neural impairments such as dyslexia, speech delay, stuttering, and hyperactivity.

In the case of savants, there is exceptional intensification of skills that survived the basic brain impairment. Most of their extraordinary skills are related to nonverbal fields such as music, arithmetic skills, and spatial distinction ability.

Some ascribe the amazing abilities of the savants to expertise differences between the two halves of the brain.

Most thinking-function impairments among autistic people are related to functions of the left hemisphere.

A talent similar to savant ability might appear at a later stage of life, in particular following brain injuries such as stroke or brain tumors. The rather sudden appearance of a talent "out of the blue" (contrary to innate savant syndromes, which usually appear after a long, continuous effort of the savant) suggests that a hidden ability is put from theory into practice. Some ascribe this phenomenon to the release of right-hemisphere ability from the inhibitions and barriers imposed by the left hemisphere in ordinary action pattern. It is primarily ascribed to the left temporal lobe, which is thought to be the residence of the "thought police."

Among people who suffered injury in the left temporal-frontal area, the ability to understand words was damaged, but hidden artistic talents, such as playing and drawing, which characterize the functions of the right temporal-parietal area, appeared. Such is the case of brain researcher Jill Taylor. Taylor had a stroke in 1996 following a hemorrhage in the left hemisphere, which was reflected in the loss of function in the right side of her body and in speech damage. Taylor described the experiences she had after the stroke and claimed that the impairment of the left hemisphere "released" right-hemisphere capabilities that were not expressed earlier.

The ones who are of the opinion that the impairment in the function of the left hemisphere releases the hidden skills of the right hemisphere from their mental handcuffs mostly refer to spatial orientation, and painting and playing talents. The right hemisphere sometimes acts as an iconoclast—shattering conventions—and in such a climate creativity might also spread its wings.

It seems that the engine of creativity is fed by energy that derives mostly from the right hemisphere. In this context, there is an anecdote regarding Mozart, who asked his wife to read him stories while he was composing. A possible explanation for his request might be that the contents of the stories occupied his left hemisphere (which focuses on processing familiar linguistic aspects), "swallowed its mental resources," and prevented it from imposing censorship on the creativity of the right hemisphere.

In the spirit of this explanation, it seems that the scissors of the censor, which were operated by the left temporal and frontal lobes and concealed the artistic talents, vanished and, by doing so, liberated the hidden talents. The artist who was hidden in the right hemisphere no longer bows his head in front of the strict analyst from the left hemisphere.

This supposition is related to an additional assumption, according to which the "genius within" is concealed in each of us—in other words, that hidden savant talents are hidden between the folds of our brain and are not expressed due to constant hindering.

An example of acquired acquisition of extraordinary skills is the case of Rüdiger Gamm, a German young man of average skills, according to his description, who developed exceptional mathematic calculation capabilities that enabled him to come up with the ninth power, or the fifth root, of a number within a very short period of time. When a PET (functional imaging) brain scan was performed, it was found that his brain recruited brain areas that are about five times bigger than the ones used by average people for making calculations. Experts believe that people like him, who developed extraordinary skills, rely on long-term memory, with the help of which they retrieve core facts and information processing methods that make shortcuts, while ordinary people mostly rely on short-term memory.

A Brain in the Dark

The Argentinean writer Jorge Luis Borges lost his eyesight in 1955. He became totally blind in one eye and partially blind in the other one. It happened at the time he was appointed manager of the national library in Buenos Aires; then he said that he received two conflicting gifts at the same time: numerous books and night—words and the lack of ability to read them. He wrote, "Let neither tear nor reproach besmirch the declaration of the mastery of God who, with magnificent irony, granted me both the gift of books and the night."[35]

An example of a sensory, neural injury that leads to the intensification of creativity can be found in the life story of Sargy Mann, a painter who gained his reputation in Britain. During the course of his life, he became blind but continued to draw out of the darkness, and the drawings he drew as a blind painter became very successful. Mann tried to explain his success and said that, when he still had his eyesight, he followed the style of admirable painters, but once he became blind the spring of his inspiration changed and streamed from his own brain. The footprint of "the paintbrush of his brain" became a colorful light coming out of the darkness.

Cracks in the Crystal Ball

In 1900, British experts anticipated that, by 1950, the large number of coaches and horses in London would make the city swamped with horse droppings. They did not imagine the level of technological developments and assumed, wrongly, that technology and transportation would develop in a predictable, linear manner. Often we find it difficult to look beyond the conceptual horizon that is seen from the window of our era.

It should also serve as a lesson with respect to attempts to foresee the future in general, and the future of intelligence in particular.

Brain-Made Intelligence

In the spirit of the words of the computer scientist Alan Kay, who said that the best way to foresee the future is to invent it, the discipline that focuses on the interface between brain and machine yields many inventions.

A neural memory chip, which combines silicone and live neurons from a culture, has managed to retain a shooting pattern that indicates a memory was imprinted on it as the birth moment of a long-term memory, with an artificial-natural chimera as an infrastructure. The imprinting took place in a pattern similar to the pattern in which the neurons "chat" with each other: pulses that are based on changes of voltage, which create a shooting pattern. This shooting pattern is a sequence of signals, a sort of a Morse code, that is also used by memory processes.

In order to make the stiff features of the contemporary silicone-chip-based computer more flexible, it will be possible to interface it with neuron chips and get a sort of a cyborg brain, which combines the artificial and the natural and creates dialogue between the silicone and the neurons. The silicone will be able to function as intensifying thinking skills—a sort of cognitive prosthesis that hastens the running of the legs' thought.

A chip in intracranial fitting: Recruiting the power of an artificial information processing device for reinforcing the capabilities of the human brain through a direct, physical interface between them is an achievable objective which, in fact, has already been partially achieved.

A neural prosthesis is a common term describing an artificial support system that helps intensify the capabilities of a live neural tissue. Thus, for example, there are brain-machine interfaces that enable paralyzed individuals to operate a computer by means of brain waves. Alternatively, programmed brain pacing programmed in electrodes that are inserted into the brain tissue might significantly alleviate various aspects of Parkinson's syndromes. This is an applied method called "deep brain stimulation."

The possibility of a cyborg interface was demonstrated in an experiment that was conducted on monkeys (again, I am not referring to the complex ethical aspects of such problematic experiments), during which the monkeys managed to control a robotic arm, which was attached to their brain by hair's-breadth thin wires that were inserted in the motor cortex area. They operated the artificial arm very skillfully, based on the function of their brain alone, and used this arm to feed themselves marshmallow and fruits.

It is possible to transfer information directly to the brain, as signals across time and space, without the mediation of a certain sense, by directly stimulating input areas in the brain. Thus, it seems that artificial devices that create a direct interface with the brain will be able to replace impaired sensory organs.

Mechanized sight interfaces ("visual prosthesis") and interfaces related to other senses are being developed rapidly.

Telekinesis in practice: There are interfaces of brain waves—machines that enable us to move objects we do not have a direct physical contact with. The brain waves are translated into wireless signals, and they mediate the information to the object that is being moved. In addition to helping invalids, show business is also interested in it: for the computer- games companies, brain waves are the ultimate joystick.

The interface with artificial intelligence might stimulate the more human, emotional side in us, as is reflected in the following humoristic confession: "My computer once beat me at chess, but immediately afterward I beat it in Karate!"

Emotional adjustment in a human-machine interface was demonstrated in a study that found that "talking" GPS systems that are installed in cars succeed more in promoting safe driving if the prosody correlates with the driver's mood.

Each evolutionary cycle probably has its expiry date, and it is possible that this insight also applies to humans. During the evolutionary journey, which lasted for ages—from the amoeba to the primate and then to the Homo sapiens—we left the aquatic living environment, dropped off the fur, chopped the tail, and our brain swelled and swelled… but from the Age of Enlightenment, the spiritual creatures of the human brain have been creating a new evolutionary course that rushes itself in an exponential pattern and is reflected, inter alia, in the creation of supercomputers that perform highly complex calculations as quick as lightning. Will the creatures of artificial intelligence form the next stage of evolution?

The term "singularity" is taken from the world of physics, where it means a point in space in which density is endless. The futurist Ray Kurzweil borrowed this term and gave it the meaning of a point in time in which scientific development progresses so rapidly that the graph's line that describes its progress is almost vertical.[36]

An interesting optional future awaits the human brain if Kurzweil's prediction, according to which in the future "we will merge with our technology," is fulfilled.

Such a cyborg brain is about to lead to a far-reaching evolutionary leap. In such a case, it seems that the nonbiological component of our cognitive skills will better itself in very brief intervals.

Is it possible to mechanize the mind, the emotions, and the consciousness?

Creating consciousness in an artificial brain that is built by human brains will be a substantial breakthrough that will divide human history into "before" and "after." The birth of a Robo sapiens (the thinking robot) will constitute a new stage in the evolution of intelligence.

The ultimate outsourcing of human knowledge might be the delegation of knowledge preservation, and the research of the universe, to artificial-intelligence machines.

Will we be able to accept that there are some riddles of the universe that no human brain can solve? Perhaps artificial intelligence will be more successful in that?

In this context, we fondly remember the computer that was intended to provide an answer to "the ultimate question of life, the universe and everything" in the book T*he Hitchhiker's Guide to the Galaxy.*

A physical merger between the brain and its spiritual creatures, such as supporting computers, is a common futuristic forecast.

The Frankensteins and the Luddites are on both ends of the spectrum. On one end, there are the supporters of the interface between the brain and technology devices that will intensify its capabilities, for instance, by creating a direct interface between brain cells and silicone chips. Their opponents call them "Frankensteins." On the other hand, there are those who are concerned about the attempt to connect the brain tissue to an artificial tissue directly in order to promote cognitive and functional reinforcement. Their opponents refer to them as "Luddites," after the people who opposed the industrial revolution in England in the nineteenth century.

Cybernetic scientists are working on a new alliance between humans and machines—an alliance that, in a known nightmarish scenario, is broken by the machines.

In this context, some warn that we tend to focus on the progress vision that is reflected in Moore's law while quoting the axiom that says that the processing capacity of computers will be doubled every two years, as a measure of the duration of a technological generation, but tend to underestimate the validity of Murphy's Law, according to which "anything that can go wrong, will go wrong." Under such a scenario, the technological dummy might rise against its maker. It also refers to brain auxiliary devices that might have an effect on our human essence, and not necessarily a positive effect.

Extending the time frame during which the brain is exposed to information includes options of cognitive reinforcement.

In this context, a fascinating field with a dreamlike potential of development is "hypnopedia"—sleep-learning while being very cautious not to bring about brain failure as a result of overload.

Perhaps we will also be able to improve our capabilities by rewriting the genetic code that encodes us.

For the first time in history, we are able to review the spiral volumes of our DNA.

Climbing the spiral staircase of the DNA makes it clear that life is information, but it seems that the coded information is multilayered, and, at this stage, only a part of its complexity is known to us.

This hunch was expressed at one of the latest Ig Nobel Prize ceremonies by geneticist Eric Lander, who said, "Genome. Bought the book. Hard to read."

Nevertheless, the climbing of the spiral staircase will continue, and the future will enable us to reprogram the DNA. Later on, according to the boldest among the thinkers, we will be able to upgrade our milieu into a totally new and improved one. Perhaps a thorough understanding of what is written in the book of genome, alongside progress made in other scientific fields, will enable us to translate soul functions into information. It is possible that such an information device will be able to contain our "soul," and our human essence will then wear a totally different garment.

It seems that no medication will be able to totally prevent the aging of the human body and brain. We are postmodern but not posthuman. Science will probably not be able to prevent the degeneration of our body, but it might find a way to perpetuate the products of our brain; thus, after our death, the ghost of our soul will hover around the virtual Internet space instead of ancient, moss-filled castles.

Aspects of Thinking Sceneries that Characterize Various Brain Syndromes

The drums beat in different rhythms in the brains of people who have various innate syndromes. They seem different to us, since a different type of "inner music" creates the choreography of their conduct.

The "who" that reflects the personal uniqueness of the person who has the syndrome comes out of the "what"—the stereotypical pattern of characteristics of the people who suffer from the syndrome.

The "genius within"—it seems that savant skills are not that rare, and it is possible that they are concealed in every one of us. A unique intracranial climate is required in order to bring them into practice, however.

It seems that there are regulation mechanisms inside our brain that activate feedback mechanisms of inhibition release. Thus, hearing impairment that creates forced unemployment in the auditory cortex, where the initial audio input is centered, causes, through a "release" mechanism, complex sound hallucinations. It seems that this is how synesthetic skills appear after a person becomes blind, similar to savant skills that appear after damage to the left hemisphere.

There are cases in which a functional impairment in the left temporal lobe brought along a stream of artistic talents that seemed to be hidden inside the brain all along.

Autism is mostly a multifrontal development disorder: impairment in aspects of perception, thinking functions, emotional and social skills, and language skills. In most cases, the formal IQ of autists is lower than 70.

The genetic aspect of autism is the dominant one: in identical twins, if one twin suffers from autism, the chances of the other twin to suffer from it are 80–90 percent.

One of the definitions of autism is "human intelligence that has difficulty with acknowledging other intelligences." In other words, a difficulty in experiencing the world as inhabited by other intelligent entities.

Looking at a person's face activates an area of designated cells in our brain called the fusiform gyrus, which is in charge of recognizing human faces. On the other hand, looking at a standstill object, such as a table, activates a brain area called the inferior temporal gyrus. It was found that when people who suffer from autism are looking at a person's face, no nerve activity takes place in the fusiform gyrus, but, at the same time, intense activity is shown in the brain area that is in charge of recognizing standstill objects.

Many of the people who suffer from autism suffer from sensory flooding and oversensitivity to sounds and touch.

An interesting supposition that has factual support is that the brain maps of autistic people are not classified in a way that is sufficiently differentiating and that they are expanded and networked in a scattered manner.

The result of this expanded networking (which derives from the lack of sufficient differentiation) is sensory flooding, which makes it hard for the individual to act in our world, in which we are constantly swamped by multisensory stimulations.

It seems that a paralyzing murmur of "white noise" covers the brain of those who suffer from autism in a mixture of stimuli that has no end and no beginning.

People who suffer from autism are known for their ability to identify a hidden pattern in a "camouflaging picture" (a picture in which the desirable print appears next to similar patterns, and so they "camouflage" it), a task that is difficult for people who do not suffer from autism. A plausible explanation for that might be related to the basic method of mining information. Ordinary people usually look for a match to a familiar pattern as the first stage of information processing (examining the level of "magnetizing" of the new information to the familiar patterns that are mostly encoded in the left hemisphere). The tendency is toward seeing the whole picture at the expense of putting less focus on the patterns of the various details. Thus, they immediately see the forest that "shows up" from the trees. On the other hand, people who suffer from autism tend to focus on the small details and have difficulty in seeing the "general picture." In other words, they see the various leaves' foliage but not the whole forest. The principle of "controlled blindness" is built into the "preferring-the-global-picture" approach typical to most of us. This is a selective view that "refuses" to see all the options and places some of them at the cognitive "blind spot" area.

Indifference to the moving melody that is played on the harp of emotions is one of the possible characteristics of Asperger's syndrome—"soft autism" that might induce emotional numbness.

Klüver-Bucy syndrome, which arises from damage to the temporal lobes, and the amygdala in particular, is expressed, inter alia, in a highly reduced emotional response to events in the surroundings and in a decreased, or even lack of, sense of fear in dangerous situations. The resulting misjudgment leads to impaired interpersonal communication and might be life-threatening at times. This behavioral pattern shares similarities with phenomena observed among people who suffer from autism. The sense of fear, in accordance with the circumstances, is either blunted or missing among autistics as well. In this context, the abnormal response is ascribed to malfunction of the amygdala, which is in charge of triggering fear in situations that require caution and discretion.

Williams syndrome is a genetic syndrome that derives from the deletion of a cluster of genes (about twenty-six on average) from the long arm of chromosome number 7. It is reflected by a unique mix of strengths and weaknesses.

People who suffer from this syndrome are characterized by increased empathy and great friendliness, even toward strangers, to the point that they tend to trust strangers without any reservations. On the other hand, they suffer from severe limitations with regard to spatial, visual functions and have severe difficulties with the practical aspects of living.

From the structural-anatomic aspect, the brains of the ones who suffer from Williams syndrome have unique features; the volume of the average brain is 20 percent smaller than the brains of people who do not suffer from this syndrome. The decrease in volume is ascribed to the smaller size of the structures at the back of the brain—the parietal and occipital lobes. On the other hand, the size of the temporal lobes is normal, and might be even larger than average. This variation is probably the cause of the severe impairment with respect to visual, spatial functions among those who suffer from the syndrome. On the other hand, the larger-than-normal temporal lobes, in certain cases, are probably at the basis of a great love of music and high verbal skills among people who suffer from this syndrome.

Their brains, which are different in terms of structure and function, produce a unique personality profile and behavior patterns.

Patients who suffer from Williams syndrome, similar to people who suffer from autism, are good at seeing the details but find it difficult to see the global picture. They are good at noticing the trees but have difficulties in seeing the entire forest.

After meeting the friendly people who suffer from Williams syndrome, one might raise the question of whether human society would have been friendlier and more peace loving if all people had had Williams syndrome. The answer to this question might be positive; however, we should also ask ourselves whether the severe impairments related to the syndrome would not have stood in the way of other aspects of human civilization progression.

Females who suffer from Turner's syndrome, a genetic disorder that affects only females and is caused by complete deletion (monosomy) or partial deletion of the two X chromosomes, tend to suffer from dyscalculia or even acalculia (major difficulty in performing arithmetic calculations).

In a test called "counting Stroop," two figures of different physical sizes are presented. For example, figure 6 is presented in a large size while figure 8 is presented in a smaller size. In an experiment in which the participants were asked to indicate the figure whose physical size was bigger, while ignoring the arithmetic values of the numbers, people who suffer from dyscalculia, as in Turner's syndrome, achieved better results. It is assumed that their success is due to the fact that they suffer from various levels of "blindness" with respect to the nominal value of numbers, and so the contradiction that is formed between the higher nominal value and the smaller physical size does not slow down their thinking pace significantly.

An interesting finding was found in an experiment in which temporary impairment in brain activity was induced by means of transcranial magnetic stimulation, which causes weak electrical currents in the brain tissue. When the disturbance was induced at the area of the right intra-parietal sulcus of people who do not suffer from dyscalculia, the quantitative-numeral information processing in their brain became, for a short period of time, similar to those who suffer from dyscalculia. This finding raises the suspicion that this brain area is dysfunctional among females who suffer from Turner's syndrome.

A unique reality scenery characterizes the common brain of the twins Krista and Tatiana Hogan, who were born in Canada in 2006. Their case is a unique, extreme example of inter-brain sharing. The twins share a brain, which joins at the core area of the thalamus, and it seems that whatever is seen by one pair of eyes is also perceived in the brain of the other twin, and vice versa.

It seems that some of the sensory impressions that are perceived in the sensory organs systems of one twin is transferred to the joining brain and turns into a common psychic experience, or two experiences that are based on the same sensory input.

Einstein's Brain

Right after the death of Einstein at the age of seventy-six, on April 18, 1955, his brain was taken out of his skull. The brain's weight was 1,230 grams. It was dissected into 240 pieces that, as in a saga that fits the mythical owner of the brain, alternately appeared and disappeared at various laboratories around the globe.

No special particular findings were found in the analysis of the morphological aspects of Einstein's brain. Among the findings that were indicated in the various studies was a slightly higher-than-usual ratio between the glia cells and the neurons at the lower parietal lobe, to which mathematic and spatial information processing is usually ascribed. In addition, an unusual winding pathway of the back side of the groove that separates the parietal lobe from the temporal lobe from the left was found, as well as other morphological findings that did not demonstrate significant differences from an average human brain of a corresponding age.

Numerous researchers criticized these findings and the possibility of drawing significant conclusions based on them. Some even called the researchers who attempted to do so as dealing with neophrenology. (According to this theory, there is correlation between the skull configuration and the skills of the brain that resides within it. The theory was refuted, and ever since it has represented "unreal science" or pseudo-science.)

Features of Intelligence

Cognitive skills are the pillars in the hall of wisdom.

The various skills that are included in the term "intelligence" are highly diversified. The exact outline of the term, in the terminological and semantic senses, is a focus of a long-lasting dispute.

In a paraphrase of the saying "Some are more beautiful than her, but no one is as beautiful as her," which was written by the poet Nathan Alterman, we might say that some brains are more gifted with respect to certain skills, but the map of skills—the strengths and the weaknesses—is unique to the brain of every human being. Like a fingerprint, a person's brain is one of a kind. With respect to certain skills, there are brains that are more gifted than it, but there isn't one like it in its uniqueness.

It is possible to manipulate biology. Biology determines approximate end points, and our location on the spectrum can be changed as a result of our actions. Intensified use of certain brain skills reinforces them in the biological-structural sense as well.

Becoming more proficient in the secrets of the "brain manual" might better its products.

Intensified cognition among human beings does not take place in a linear pattern. Thus, for example, acceleration of cognitive skills takes place in elementary school approximately up to the age of eleven or twelve, just prior to adolescence. Afterward, the progress is significantly slowed down. As a mirror image, the weakening of mental skills at an old age also does not form in a linear pattern.

The psychometry discipline that attempts to quantify the qualitative, evasive traits of human skills usually suffers from partial mapping of the strengths and weaknesses related to the thinking skills of an individual. Psychologist Howard Gardner expressed this approach in his theory of multiple intelligences.

Haier and Jung found in 2005 that there is no difference in the averaging of general intelligence tests between men and women, though there are clear structural differences, such as the fact that women's brains contain relatively more white matter and, on the other hand, men's brains contain more gray matter.[37] Thus, it was concluded that a structural difference does not necessarily mean a difference in quality.

Chapter 10
Stricken Brain—Dealing with Brain Injuries

Descriptions of cases of brain and soul traumas sometimes have a dark attraction. Each of us might enter this frightening kingdom someday.

At the neuroanatomy room in medical schools, which is packed with jars containing human brains in preservation liquid, one might feel a sense of awe combined with a sense of gloom, as if it were a graveyard of lost souls. It seems that vanished worlds reside inside the jars, along with dissolved memories and shattered hopes and, also, spectacular buds of thought that sprouted within these brains—stairs of insight that were created by the pick-axes of thought originated in these brains, that contributed to raise the tower of human knowledge. These jars are far from containing mere preserved flesh.

Brain functions exist out of a confederation of thoughts, memes, emotions, and perception impressions. In cases of failure in brain functions, various components of this cooperative array might be damaged while others continue to function faultlessly. Sometimes, withdrawal of certain components from the confederation might take place while the brain owner is unaware of it, as a procedure that is concealed from the person's consciousness but is noticeable by others.

When people are unaware of their existence, the failures related to brain activity are sometimes perceived as "wrinkles in the texture of reality" that cast a shadow and make it difficult to see the horizon beyond them—as mountains that block their sight line.

"Losing Myself"—Alzheimer's Disease

On November 25, 1901, a fifty-one-year-old patient, Auguste Deter, was examined by Dr. Alois Alzheimer at a psychiatric hospital in Frankfurt. The words of the first documented Alzheimer's patient are moving due to their frankness and the porthole they open to the essence of the inner being of the ones who suffer from the disease. The patient said, "I lost myself…." The sense of loss of the self, which gradually evaporates, turns the patient into a spectator who is compelled to watch a horror play that takes place at the theater located inside his skull.

A person who loses his grip on his brain is in a situation in which the organ that is in charge of dealing with the challenges of life gradually retires from coping.

The outline of Alzheimer's disease sometimes resembles a nightmare without awakening. The sick person is dealing with an obstinate process that brings about the loss of lifetime memories.

The shades of the mental view that is reflected in the eyes of the brain fade away. The view that is reflected from the rearview mirror of the life coach is grim, and the front lights are unable to penetrate the fog.

Dementia is like an inverted loom; instead of weaving the threads of wisdom in our brain, it unweaves them.

If we are unlucky and toward the end of our life our brain sinks into the dump of dementia, the end of our life will get a totally different meaning out of the ruins of our brain. Our life will no longer be full of bright consciousness facing the forthcoming eternal silence, but a life of fragments of shattered consciousness that is looking at reality through turbid lenses, to a point that the magic touch of wisdom is lost.

When the brain fails and abandons its owner when he is still alive, and when the lens of consciousness becomes turbid and the brain becomes a wilderness, the brains' lucidity vanishes and is replaced by welcomed and unwelcomed visitors from the land of hallucinations. In such a case we ask ourselves a blunt question: is this a life worth living? If the road to destruction winds through this twilight zone, the transition from life to death is no longer sharp.

When the number of people who suffer from dementia is always on the rise, foggy consciousness—a not-rare mode of existence on the spectrum between bright consciousness and turbid, shattered consciousness—becomes a common mode of human existence.

The destruction topography of Alzheimer's disease, which is the most common source of dementia, is characterized by rapid degeneracy in certain brain areas. In Alzheimer's disease, the function of the hippocampus and the temporal and occipital cortex is damaged at a more rapid pace compared to the function of the frontal lobes. This is the inverted version of the degeneracy pattern that takes place in a slow, gradual manner in cases of normal aging.

In structural brain scans of Alzheimer's patients can be observed, at an early stage, a degeneracy of tissue (atrophy) at the medial, temporal areas (the areas of the inner temporal lobe including the hippocampus). In functional imaging scans that reflect the activity at various brain areas, we witness a decrease in metabolism (hypometabolism) at parieto-temporal areas.

In normal cases, our brain first identifies the essential aspects of a stimulation, prior to complete identification, and responds to them. For instance, a long, cylindrical shape will trigger an emotional response even before our brain identifies it as a snake, a pipe, or some other entity. Later on, brain processing moves from general identification to specific identification of the phenomenon, according to the circumstances.

Atavism (backwards evolution) of brain skills characterizes the pattern of cognitive withdrawal among those who suffer from Alzheimer's disease.

One of the functional withdrawal signs among Alzheimer's patients at an early stage is the tendency to define phenomena in general terms and not to continue the processing that will enable them to identify the unique features of a phenomenon. For example, they might refer to all the animals that walk on four limbs as dogs, even when they meet a deer. It seems that brain processing gets "stuck" at the general phase, which defines fundamental aspects of a phenomenon but does not go any deeper. Alzheimer's patients withdraw toward generalization.

The atavism in the conceptual capabilities and information retrieval resembles an inverted mirror image of the development of such capabilities in the brain of a human infant. Remaining at the general definition stage is typical of infants as well when they get to know the world and might refer to all animals with wings as birds. Later on, however, a process of information conceptualization and retrieval is formed among infants; it's in an inverted pattern compared to the process that takes place in the brain of Alzheimer's patients: infants develop the ability to distinguish and differentiate between various details, which enables them to identify one flying entity as a stork and the other as a raven. This ability, on the other hand, is lost forever in the case of Alzheimer's patients, who withdraw toward the phase in which all will be defined only as birds.

The Pharaonic Pyramid of Memory

Our brain generalizes and differentiates, during childhood, the layers of the pyramid of memory, which are built from the bottom upwards—from the pattern of generalization toward the details. The same process takes place in the brain of an expert. On the other hand, in the case of dementia, a gradual destruction of the layers takes place, and the pattern of wear starts from the high downward toward the low—from the level of conceptualizing nuances toward the level of raw conceptualization. The number of remaining layers is in correlation with the severity of dementia.

In a normal, young brain, the memory pyramid is built from the generalizing basis toward the layers that are built upward on top of each other, in which unique details are embedded. The level of memory uniqueness rises gradually until it reaches its exclusive features. It seems that, at the time of assimilating new information in the brain of Alzheimer's patients, the upper layers of the specific memory that encode its unique features are not built, and only the basic layers that include the very general features of the event are built. It is also true with respect to information retrieval—its availability among Alzheimer's patients is limited to the basic layers that store general information. They seem to lose the ability to differentiate information at the time of assimilation and to retrieve unique information.

Love at First Sight—at a Second Sight

A known convention is that our first love is always the last. It is so, in the sense of a virginal experience that we will not experience again as its definition. But, due to the (too-) short refresh time of the screen of consciousness among those who suffer from advanced Alzheimer's disease, which erases the traces of perception recordings prior to their preservation, Alzheimer's patients live in a sort of perpetual present. In such a situation, there is a tendency to be captured by the first love for the second and third time, and the patient enjoys the heady scent of a flower he comes across again and again as if it were the first time. The perception recordings exist within a brief present that vanishes without leaving a trace.

The pleasant scent and the rich shades of the flower the patient comes across numerous times wash over the field of consciousness as if it were the first encounter. The intensity of its presence does not dim, even when it becomes familiar according to standard definitions, but this supposedly romantic aspect cannot dim the tragic consequences of this disease.

"Upside-Down Childhood"

When the children of Alzheimer's patients, with children of their own, have to take care of their parents they face a new responsibility. Contrary to a child that acquires capabilities, the patient loses capabilities. The imaginary encounter between the capabilities of the child, who develops rapidly, on the one hand, and the parent who sinks in a swamp of dementia, on the other hand, takes place when each of them comes from an opposite direction on the spectrum of abilities.

Alzheimer's disease sometimes seems like "reversed childhood," since, in many senses, it is like a mirror image of the stages of child development. The understandable emotional difficulty of the patient's family members sometimes derives from comparing their elderly loved one to a child, but sometimes it cannot be avoided if we want to create reality-compatible expectations among family members who take care of the patients.

The Safe Harbor of Mother Tongue

Studies that were conducted in countries with multiple immigrants show that in people who immigrated as adults to a country where the common language is not their mother tongue and who have become proficient in that language, and whose brain is affected by dementia in old age, the wall of second-language proficiency becomes cracked; these people retreat to the haven of their mother tongue. This language constitutes a more solid defense line against the attack of linguistic fading.

First language, like first love, is usually unforgettable. We use our first language to create most linguistic patterns. Thus, our scope of expression is wider in this language. With respect to a second language, we are usually equipped with a more limited repertoire of patterns, and using this language requires more mental energy and effort compared to our first language.

It might constitute a plausible explanation for the fact that people who speak two languages and suffer from dementia first lose their mastery of the second language and preserve their mother tongue better.

Insulin: Not Just for Treating Diabetes?

Insulin plays a role in memory and learning. A study shows that taking insulin improves memory performance soon after the intake. The number of insulin receptors in the brain areas that have an important role in memory and learning functions is significantly higher in the brains of healthy people compared to the brains of Alzheimer's patients. Diabetes patients who are not balanced might be at a higher risk of Alzheimer's disease compared to people who do not suffer from diabetes. That is another reason to keep a strict balance of glucose levels in the blood. It was found that inhaling insulin out of an inhaler improved memory performance among healthy subjects, and it might lead to developing such a treatment with respect to memory disorders.

So What Shall We Do with It?

Some medical practitioners doubt the efficiency of early detection of the process of dementia due to the unfortunate fact that the reversibility component, with regard to these syndromes, is relatively minor. A common assumption is that 10 percent of the disorders generating dementia are curable nowadays, such as lack of vitamins, malfunction of the thyroid gland, and problems in the liver and kidneys. Although there is no cure, treatment that might improve the quality of life of the patients, their family members, and the people who take care of them is of great importance, however.

Disruption of Internal Insinuation

The faces of people who suffer from the Parkinson's disease wear an involuntary mask due to physiological disruption in innervations. The lack of dopamine causes weakness in the muscles of facial expression.

With regard to people who suffer from Parkinson's disease, the hands of the time clock slow down in several senses; in addition to the slow movements, called "bradykinesia," and the slow pace of thinking, called "bradyphrenia," the personal sense of time slows down as if enchanted, in accordance with the slowing down of movement and thought. Often the patient does not experience his actions and movements as slower than usual, and when he measures himself against an external time pacer—another person or a standard clock—it often seems to him that they act more quickly than usual.

Self-regulation directed by the "internal insinuation" mechanism is disrupted among Parkinson's patients. In order for their speech to be clearer, they are required to recalibrate the volume of their voice and the fluency of their speech not according to their inner sense but, rather, in accordance with external measurements. It is similar to the recommendation given to pilots at a state of vertigo—to change from flight mode based on senses and intuition to flight mode based on devices and external information.

Behavioral Effects of Frontal Lobe Injury

Injury at the lateral areas of the frontal lobes induces a state of mind that is characterized by apathy and lack of initiation. Injury at the central or orbito-frontal areas reduces or even eliminates inhibitions with regard to primeval urges and raw thoughts that have not been processed through the filter of discretion. In such a case, the injured person tends to exhibit judgeless, impulsive and shameless behavior that is inconsiderate toward others.

Single Impressions That Do not Become Unified

A stroke might induce, as a rare phenomenon, a condition called simultanagnosia—difficulty in perceiving various aspects of sensory experience simultaneously. Thus, for example, vocal simultanagnosia might be formed as a result of damage to the centers of auditory-input processing, an injury that does not affect the ability to perceive the pitch and the rhythm separately but affects the ability to perceive them as combined into a melody.

Visual simultanagnosia might be formed as a result of an injury at the meeting areas between the parietal lobes and the occipital lobes on both sides of the brain. Its outcome is Balint's syndrome, which is expressed, inter alia, in difficulty in seeing the overall visual picture, and seeing the objects that compose the global picture as single entities detached from the surrounding context.

The Profile of Stroke

Stroke at the area called fusiform gyrus might severely damage the ability to recognize human faces. A person might not recognize his parents, his children, and even his own reflection in the mirror.

From a Lover to an Impostor

Tragic failure in the perceptual field of vision combined with a normal sensory field of vision is reflected in the rare Capgras delusion, which results from brain injury. In this syndrome, the patient believes that a family member or a loved acquaintance was replaced by an impostor. The sensory input is identical, but its interpretation changes. The person who has the syndrome sees the face of his acquaintances impeccably but is convinced that a talented impostor, and not the real person, stands in front of him.

When Consciousness Roars and the Voice is Mute

A painful and moving illustration of a rare syndrome that might be caused by stroke was described in the French movie *The Diving Bell and the Butterfly*. The movie tells the story of Jean-Dominique Bauby, the editor of a fashion magazine, who had a stroke in December 1995. Due to the unique pattern of brain damage at the brain stem caused by the stroke, Bauby was unable to talk or to move his body except for his eyes—although the sharpness of his brain and his comprehension were not damaged at all.

He had a syndrome called "Locked-in Syndrome." People who suffer from this syndrome live a full life with regard to their soul, but they are locked inside a paralyzed body that cannot communicate with the world (except for moving the eyes). Bauby passed away in 1997.

The Trees That Conceal the Forest

Brain syndromes such as Balint's syndrome, Williams syndrome, and autism have unique characteristics. Their causes vary, and the way they are reflected also depends on the unique personality of the person who suffers from them. Nevertheless, they are all characterized by a difficulty in seeing the overall picture, as if the trees conceal the forest—the essence is drowned in the sea of details.

Hyperphysiological Conditions of the Brain

A brain tissue whose function was disrupted might turn into a source of arbitrary signals that might jam the lines of communication in the brain and prevent well-timed signals of a healthy brain tissue from reaching their destination.

Patterns of overarousal of certain brain areas, which disrupt normal functioning, can be found in conditions such as epilepsy, which is caused by an electrical storm that takes place in the brain and might induce involuntary movements and disruption of consciousness. Other examples are the mania that induces hyperactivity, emotional tide, and disturbance in the course of thinking; and dyskinesia—movement spasm as a reflection of motor automation, which mostly affects people who suffer from Parkinson's as a side effect of the drugs they are treated with.

To the Depth of the Matters

The depth dimension in eyesight (the stereoscopic dimension) relies, as an important, though not exclusive, source of information, on a spatial-horizontal gap that is customarily measured between the two pupils and whose average length is six centimeters.

Even in the case of monovision, the brain manages to create a sense of depth, though not as clearly as in two-eyed vision. It does so by relying on environmental clues, such as movement shifts (changing of the input during movement and changing of the angle of view that derives from it), perspective, and concealing.

In this case, brain processing "adds a dimension" to the visual perception recordings of each of the eyes separately. The carpet of the retina is almost two-dimensional (although there are several cell layers). It is very thin, and it sends raw, two-dimensional impressions of world manifestations from each of the eyes to the brain. Brain processing, which weaves together the input coming from each of the eyes, is the main contributor to experiencing the depth dimension.

Voluntary neck movements of low amplitude might assist people who rely on monovision to sample the vision input from angles that are a bit different, which will improve the ability of the brain to produce the depth dimension out of this information. Neck movement can also help people with one-ear hearing to sample the sounds from directions that are a bit different and, by that, to add the depth dimension.

With regard to spotting the source of sounds, the "stereophonic gap" is a time gap of several thousandths of a second in the echoing of sound waves between the two ears.

People with one-ear hearing might rely on changes in the intensity of sound waves when they are echoed from the surface and from various textures in the environment to help them find the direction of the source of sound.

Loss of the depth dimension might take place at the time of perceiving a visual input—which will cause the loss of stereoscopic ability—and with regard to perceiving an auditory input—the loss of stereophonic ability. The depth dimension enriches the ability of experiencing world phenomena, and, on the practical aspect, it serves as an efficient tool for assessing the distance from the viewed, or heard, phenomenon. Some claim that the omission of the depth dimension from the auditory or visual input makes the perception of the experience duller also, in the sense of the emotional echo it triggers.

The Anarchic Hand Phenomenon

The alien or anarchic hand phenomenon, which appears in the degenerative brain disease called "corticobasal degeneration (CBD)," is reflected in a condition in which one of the hands acts in an independent manner, without the patient's voluntary sense of control. A CBD patient described that while one of his hands was buttoning up his shirt, the other hand was involuntarily unbuttoning them. A process of "alienation" takes place. The hand becomes a stranger to its owner with respect to his sense of control over it.

Aspects of Brain Rehabilitation Following an Injury

Aspects of Rehabilitation of Motor Function

When the brain suffers from an injury, it tries to improvise a response in an attempt to compensate for the damage.

The brain's compensation ability depends on the severity and duration of the damage and also on the networking patterns that characterize the specific brain.

When the traffic of signals in the neural pathway is disrupted, the brain tries to compensate for that by exposing bypass routes that are intended to preserve the neural network in a functional mode and to enable the signals to be transferred from the original location to their destination. The brain attempts to do so even if it is done not through the main road, and even through pathways that resemble winding, mountainous courses. As these pathways are used time and time again, they become wider and more user-friendly.

In the hope of rehabilitating their brain, people with nervous system injuries often walk down arduous paths of treatment.

A key question with regard to rehabilitation of patients who suffer from neurological damage focuses on how to induce inactivation of the prior action pattern, which is no longer applicable after the injury, and, at the same time, promote new learning in the shortest way, in an attempt to avoid the arduous process of prolonged practice that deters many of the patients.

Among the principles of optimization of rehabilitating function, as shown in various studies, there is the focused, short practice, which shows more concrete improvement compared to practice that spreads over a longer period of time with lower frequency of practice. The patient should be faced with challenges with a rising level of difficulty. The practice should focus on skills that are similar to those required for daily functioning.

In brain injuries that involve disruption of the natural skill of walking, sometimes it is worthwhile to learn how to walk from the beginning while being provided tools to cope with the deficiencies that caused the disruption. Such learning creates a new networking of the brain map, which encodes walking, instead of the old map, which is worn and disrupted, and skills that compensate for the brain damage are imbedded in the new map.

It is somewhat similar to reacquiring "gravitation feet" among astronauts who return from missions in space that involve conditions of almost zero gravity. The astronauts need a few days of practicing walking—like infants who take their first steps—until their "earthy feet" come back to them.

In a case of a broken limb that does not function, as in a case of a prolonged period of wearing a cast on a joint of a limb that prevents normal movement, the limb loses its "representation areas" in the representation maps of the body in the brain for the sake of functional areas in the body, which deprives it of its territory. If the limb's function is fully or partially retained, however, it is forced to reconquer the territory of its brain representation. The rehabilitation of the limb itself is not sufficient. In order for it to function appropriately, the rehabilitation of its neural representation in the brain is also needed—in the absence of which, the function of the limb will remain faulty, even after the limb has fully recuperated.

The brain map representing the injured limb unweaves and dissolves due to limited use or lack of use. This process is called "acquired lack of use." As if in a vicious circle, the tendency of acquired lack of use is reinforced, and the disability becomes more and more fixed. Studies show that, among people who suffer extreme weakness of the arm following a stroke, the motor representation map of the arm shrinks up to half of the representation territory it had prior to the damage. Reversing this trend by forced activation of the limb and reviving the neural network—i.e., the brain map that encodes the function of the limb—sugarcoats the pill and often greatly improves function of the damaged limb. In the case of limb weakness due to brain damage, it was found that intentionally restricting the movement of the opposite, healthy limb forces the brain to use the damaged limb, and its function is improved in the long term. A similar principle is used in treating a "lazy eye" (an eye that does not function well at the level of the cortex that processes the visual information coming from it): an eye patch that is worn over the healthy eye forces the brain to use the lazy eye, and the performance of the brain areas that process the information deriving from it improves as time goes by. The treatment is successful, especially during the time window of early childhood. The brain creates new networks of function maps based on the new artificial constraints. It seems that many rehabilitation programs for improving brain functions can benefit from these findings.

Tourette's syndrome is characterized by meaningless movements (ticks), which the person who suffers from the syndrome feels an irrepressible urge to perform. One possible treatment of the syndrome is to try to recruit these forced spouts of movement for the sake of focused, target-oriented actions that result in purposeful behavior.

Life as a song: People who suffer from brain injuries, particularly at the frontal lobes, which cause difficulty in performing sequences of daily activities, might find it helpful to use functional directions for performing a sequence of actions transcribed as a rhymed song with a familiar melody.

Aspects of Rehabilitating Speech

Aphasia is damage to the skills of producing or understanding speech.

When we produce speech, our brain exchanges a sequence of symbols, in the figures of words that conceptualize thinking, for a sequence of movements in the muscles of the pharynx, the tongue, the palate, and the lips. It seems that the brain area that orchestrates these movements into a single melodic sequence is the premotor cortex at the left frontal lobe. An injury at this area is commonly reflected in difficulty with producing syllables (phonemes)—the basic units of speech.

Semantic aphasia is the inability to process representations, symbolization objects, and abstract concepts, and it mostly derives from injury at a particular area in the left, temporal lobe, which is in charge of granting meaning to the contents of speech. In accordance with the severity of injury, people who suffer from such injury have difficulties in understanding the contents of speech when other people speak; on the other hand, the content of the sentences they produce is usually incomprehensible (gibberish), or embedded with meaningful islands surrounded by a sea of meaningless wording.

According to common estimation, about 40 percent of people who have had a stroke in the left hemisphere suffer from aphasia. In those cases, it was also found that an intense practice program improves the condition more significantly compared to prolonged practice programs with long intervals between one training period to the other.

It was found that the people who suffer from aphasia, who are unable to talk, are sometimes able to sing. The fact that the stream of words has not dried up completely might serve as an encouragement for them. It seems that a familiar composed transcription, in which threads of emotion are weaved and whose content is "ready-made" and does not require "syntactic engineering," often enables them to overcome the wall of aphasia that blocks purposeful speech, and it is based on self-creation of contents and syntax.

A concrete example of rapid acquisition of an acquired skill is acquiring a new language while staying in a foreign country. The constraint of communicating in the local language; the intense, frequent exposure; the daily implementation—all of these accelerate the pace of language acquisition and deepen its traces in our brain. Similarly, intense and frequent exposure to practicing of language skills usually accelerates speech rehabilitation among those who suffer from aphasia.

Aspects of Coping with Degenerative Brain Diseases

Nowadays, treatment of degenerative brain diseases might seem like a sequence of rearguard battles intended to slow down degeneration.

Sometimes the shreds of functioning that survived in the shattered brain cannot be patched up—brain plasticity also has unfortunate limitations that cannot be overcome.

Brain diseases, especially those that affect the frontal lobes, are able to reshape personality and calibrate beliefs, habits, and typical behaviors in a new pattern.

Dementia deprives the brain of its most magnificent assets. The consequences of dementia are destructive. A brain of a wise person who suffers from dementia resembles an architectural masterpiece that turned into ruins following an earthquake.

Patients who suffer from dementia, which envelops their consciousness in loneliness, are like a wanderer in the wilderness of the cold tundra; they are accompanied only by a gloomy silence and sometimes warm themselves up in the light of music, which is capable of wakening dormant emotions.

Physical Exercise and the Brain—the Brain in Sports Shoes

Physical exercise intensifies the creation of new neurons in the brain.

An experiment that was conducted on mice showed that urging them to perform aerobic physical exercise doubled the pace of the production of new neurons in the brain area called the dentate gyrus, which is located in the hippocampus. Other studies also found that aerobic physical exercise is a habit that has maximal positive effect on the brain.

Brain researcher Art Kramer and his associates found that among people at the age of sixty and above who regularly performed aerobic exercise for a period of six months, there was an increase in the volume of the gray matter in the frontal lobes, which are considered the most vulnerable to aging. In addition, they found thickening of the volume of the white matter in the corpus callosum, an area in which there are numerous axons that constitute a "latitude road" that mediates the streaming of information between the two hemispheres. At the practical level, the active group demonstrated improved performance with respect to focusing attention and the ability to ignore distracting information (functions that are ascribed to the frontal lobes).[38]

Physical exercise also triggers the production of brain-derived neurotrophic factor (BDNF), which also has an important role with respect to the structural and functional flexibility of the brain.

As we further expand our education and become more active physically and more socially involved, we improve our coping capabilities, and our potential compensation repertoire, related to coping with the process of dementia in the unfortunate case it affects us in the future.

Aspects of Rehabilitation of Emotional Regulation

Long periods of stress or deep depression might cause a reduction in the size of the hippocampus. On the other hand, it was found that antidepressant drugs that increase the level of the neurotransmitter serotonin at the neural junctions brings about an increase in the amount of stem cells that turn into neurons in the hippocampus.

Usually it takes three weeks or more until the effect of such an antidepressant is noticeable. It is also the duration of the period required for newly created neurons to network themselves in the hippocampus and create new neural pathways. Thus, some of the antidepressant drugs might have a neurotropic effect—one that encourages the thriving of neurons and their links. It might be wise to prescribe them as part of the routine therapeutic process in cases of other degenerative brain diseases like Parkinson's and Alzheimer's.

Doctors of the soul have noticed that therapy that includes talks improves patients' mood, and the ability to perform cognitive tasks, among those who suffer from depression. It might be that a therapeutic talk can change the biochemistry in the brain to such an extent that creation of new neurons in the hippocampus is encouraged and networking among them is intensified. According to this supposition, we might be able to claim that words are formed into new links between neurons.

In the case of adults who suffer from depression and also experienced a traumatic event during childhood, it was found that the volume of the hippocampus is about 20 percent smaller compared to adults who suffer from depression but did not experience significant trauma during childhood. When stress does not last long, there is a component of reversibility, and the original size of the hippocampus is restored. It seems that beyond a certain period of time, however, the damage is for good.

How the Brain Copes with Injury

Injuries in the structure of the brain, such as in areas that remain scarred after strokes, usually do not have a linear connection to functional loss. The correlation between them is usually related to a certain threshold. In other words, once the structural injuries cross a certain threshold, which derives from their accumulated effect, it results in a significant functional impairment that was not noticed at the subthreshold level.

Shades of gray: The effects of brain injury can be noticed on a spectrum. Damage to the gray matter and the white matter in the brain is often manifested in the gray area, rather than in absolute black or white terms.

The scope of flexibility in rehabilitating neural wear also derives from the ability, though limited, of creating neurons in the brain in adulthood. Studies show that the support cells in the brain, which constitute the logistic infrastructure that supports the activity of neurons, have a potential ability to exchange their identity and turn into neurons due to the influence of certain regulation proteins that induce this transformation.

Most brain cells belong to the glia type of support cells. Some of them have the shape of a star, which is the source of their name—astroglia. At the early stages of brain development, the astroglia cells constitute a kind of reserve, and some of them become functional neurons. At later stages of development, the functional and structural transformation capability is lost. Researchers have found a way to reintroduce this magnificent past ability of structural and functional transformation, however, by clicking the buttons of the right genes.

It has far-reaching implications with respect to a future option of rehabilitating the neurons' tissue following neural wear caused by various factors.

In 1868, Jules Cotard studied children who suffered from a serious brain disease that caused severe injury in the left hemisphere. Despite the injury, these children spoke normally. Hemispherectomy is a term that describes a surgery during which a whole brain hemisphere is removed. This extreme surgery is performed only as a last resort, mostly in people who suffer from brain injury in which one of the hemispheres generates repetitive electrical storms (epileptogenic zone) that severely impair life quality, and when all other treatment alternatives fail to bring about significant improvement.

Some children who underwent hemispherectomy, mostly due to repetitive convulsion episodes that could not be controlled by medications, managed to acquire almost full language skills. One of the famous cases in medical literature is of a girl who underwent surgery at the age of eight. Her left hemisphere was removed, but she managed to acquire good mastery of both Turkish and German. It seems that the significant brain flexibility during childhood enabled language skills to skip to the right hemisphere and find in it a neural infrastructure in which to settle.

One of the main causes of the gap between the high rehabilitative abilities of a child's brain and the more limited abilities of an adult's brain (the "recuperation gap") derives from the fact that children's brains are more flexible and plastic and have not yet accumulated core insights that are hostile toward incompatible new information. On the other hand, adults' brains are, in this sense, "expert brains" with regard to world phenomena. They have reduced flexibility, and when they are damaged their rehabilitative potential is also reduced.

A Shining Path to Treatment—Electricity-Aided Therapeutic Approaches

Transcranial magnetic stimulation (TMS) is a therapeutic method designed to trigger a different action pattern in brain areas that suffer from malfunctioning; it's delivered by means of magnetic fields. Through a ring of copper coil placed as a crown on the patient's head, electrical current is transferred and induces the creation of a changing magnetic field. The changing magnetic field induces electrical current around it. This is a noninvasive method of triggering electrical currents in the neurons. This method is used to activate certain brain areas or, on the other hand, to block their activity. The triggering of neurons creates an "echoing cycle," which continues to operate for a while, even after the triggering through the device has ended.

This method proved to be effective with respect to treating depression, and it is being examined as a treatment for degenerative brain diseases such as Alzheimer's and Parkinson's, as well as for bipolar disorder.

It was also found that therapy based on transcranial direct-current stimulation, directed at the cortex at the area of the frontal lobe from the left, might improve the ability to name different objects among people who suffered a stroke that was expressed as aphasia.

Deep brain stimulation (DBS) is a therapeutic method in which direct stimulation of deep brain areas is made through an electrode that is placed in them. This method has significantly improved the condition of some of Parkinson's patients. The parameters of stimulation are controllable and can be recalibrated by a device that is external to the patient's body. This therapy mostly relieves the severe motor symptoms of the disease. Among brain areas that are electrically stimulated this way, there are the subthalamic nucleus, the inner part of a brain structure called the globus pallidus, and the ventral area of the thalamus.

In the spirit of the saying *"similia similibus curentur"* ("like is cured by like" in Latin,) which is taken from homeopathy, electrical stimulations of the brain are also intended to relieve the electrical storm that takes place in the brain of those who suffer from epilepsy when it is difficult to control by means of medications.

Chapter 11
Memetics—When the Brain Meets an Idea

Memes are ideas that duplicate themselves by "skipping" from one brain to another by means of transferring information from one person to the next.

As units of self-duplication, memes share similarities with our genes, the components of genetic heredity. Like them, they are entities that duplicate themselves when transferring from one person to another.

Memetics is the discipline that focuses on the nature and the manner of transfer of memes from one brain to another.

Survival—The Neuron Version of the Reality Show

Survival of memes is a process related to Darwinism of ideas.

The brain is an ecological system in which networks of neurons struggle for their survival. At another, parallel layer, a survival battle of ideas takes place. The brain is a battlefield in which an evolutionary survival war takes place fiercely, both at the physical front (the survival of the fittest neuron networks) and at the ideation front (the survival of insights and beliefs, in which there is also an aspect of physical infrastructure).

An idea that is encoded in a network that is richer in neurons has a better chance of surviving. A multitude of synapses (i.e., connecting junctions between the neurons) is a powerful prediction factor with respect to survivability of ideas. A "multisynaptic idea" usually survives longer.

Neural Darwinism sometimes seems like a Hobbesian jungle in which, in the spirit of Hobbes's philosophical theory, an idea is a wolf to another idea. Memes are units of information that are forced by their nature to duplicate themselves in a "selfish" manner. In the functional sense, a neural network that retains a certain idea attempts to ascribe to itself energy resources and cause the neurons that take part in encoding a competitive idea to desert that idea and join its ranks.

In a somewhat simplistic tone, we might say that evolution is a combination of two processes, one of which is random and the second of which is not; the creation of mutations is a random process. The selection process is mostly a nonrandom process that is dictated by the environment. And, as an analogy to memes, a subconscious, creative process that causes the creation of an innovative meme mostly includes many features of a "random" process. On the other hand, success in spreading the meme to other brains includes many features of a nonrandom process that is "environment-dependent."

Merging of Memes

A success in the implementation of an abstract idea mostly constitutes the climax of a collective learning process in which multiple memes were fused together and numerous upgrades deriving from multiple brains merged into an impressive product.

Conceptual hybridization—i.e., the combination of similar concepts—often brings about a whole that is bigger than its parts. In other words, the resulting understanding is better than the understanding that results from each of the separate explanations. It can be seen as "the different mathematics of thoughts."

As a grain of salt or spray of lemon, as a small quantity that betters the taste of the whole, so is a word in the right place; an additional little "meme" can significantly improve a whole paragraph or the entire set of memes.

Engraved in Neuron—the Structural Infrastructure of Memes

The abstract conceptual infrastructure is fixated as a neural, structural infrastructure inside our brain.

Perceptions and ideas change the wiring of our brain. Different memes wire the brain in different manners, and different wiring even causes various mutations among the memes.

A possible practical implementation is adapting higher selectivity with regard to memes that visit our brain (some of which will become permanent residents) through selective exposure to them.

Intergenerational Confrontation—Senior Memes and Novice Memes

We tend to prefer information that fits our beliefs. Senior memes magnetize memes that are compatible with them and reject the ones that contradict them. There is a threshold, however, and once this threshold is crossed, our consciousness, with the senior memes it contains, confronts the novice memes and challenges them.

In other words, in order to facilitate the assimilation process of new memes, backward compatibility is required. Memes that contain information that contradicts the senior memes, which are established as patterns, will be met with suspicion.

Creativity and Memes

The Fusion Kitchen of Memes

The dishes of our brain casserole, which merge different insights, are at the basis of creativity. A new conceptual combination made by creating new links between memes is the essence of creativity.

Sometimes a comprehensive meme is created, which groups together memes that were not combined earlier. Sometimes the cluster of memes is dismantled into separate components, which are regrouped in a different pattern.

Important facts, when they are out of context, might constitute "white noise" that makes the clarification and understanding of a phenomenon more difficult. When these facts are connected in the right context, however, a conceptual structure that was previously invisible is built out of them.

Measured doses of "loose association" and "flight of ideas" (when the dosage is too high, we might find ourselves in the isolation ward of a psychiatric institute) are needed in order to match conceptual creatures that come from brain continents that are located far away from each other. As our mental grip on a certain thread of thought becomes tighter, our grip on other threads of thoughts becomes looser. Yet, in order to bring together numerous threads of thoughts, such a loose grip is sometimes required.

At times, a meme that sprouted in our brain wins our appreciation only after another brain brings it to our mental attention, although the meme was there, inside our brain, all along. It is similar to a saying according to which the insight that the pupil heard from the wise man following an arduous journey to the top of a mountain is sometimes the same insight the pupil himself brought with him.

Sometimes the ingredients of a marvelous conceptual recipe float in the space of interbrain discourse (i.e., information shared by numerous brains), and once one of the brains turns them into a unified whole, a winning recipe is created in the cookbook of ideas—in other words, a super-meme is born.

An infectious conscious metamorphosis is, for example, the "illogical" spatial shapes that were invented by the British mathematician Roger Penrose, which served as inspiration for the drawings of Maurits Cornelis Escher, such as the famous painting of the ascending-descending stairs.

The Birth of a Supermeme—the Fractal Insight Example

A super-meme is a captivating idea that shakes the foundations of our senior memes and gets to infect numerous brains at once.

An example of the creation of a super-meme is the fractal insight—a broken dimension that exists in between a line and a surface. Once this insight was thought of, it was rapidly assimilated in many disciplines.

Configurational hybrids with fractal features are widespread in our world: the veins of the falling leaves, the pyrotechnic show of lightning, and the pathways of brain axons. The fingerprint of fractal architecture can also be found in the structure of the brain, visible to the naked eye, which resembles a walnut or a cauliflower.

Fractals have configurational features that exist in between the dimensions—between a line (one dimension) to a surface (two dimensions).

In 1975 the mathematician Benoit Mandelbrot coined the term "fractal," which means "fracture," as a reflection of the fact that figures, structures, and lines whose dimensions are in between one dimension and another can be expressed as a number plus a fracture of a number. Mandelbrot stirred up a conceptual revolution with respect to the perception of natural shapes and, in the linguistic sense, exposed to the light of verbal consciousness a perceptual layer that was concealed from the linguistic memes' consciousness—as long as there had not been coined a particular linguistic term that defined it. The birth of the fractal is a wonderful example of a phenomenon that existed before our material eyes but was, in many senses, concealed from our perceptional eyes. Some might claim that it illustrates the fact that language limitations, which were broken by Mandelbrot once he defined the phenomenon and gave it a name, might also be the limitations of our conceptual world. In fact, Mandelbrot gave birth to a conceptual meme that had been there all along but which our perception had tended to skip before it was linguistically defined and explained.

Yellow Submarine at the Dock of Consciousness (Inspired by the Beatles' Song)

Sometimes cultural icons turn into super-memes.

Human culture tends to glorify cultural icons and create a memetic heritage around them. At times, the traits that are ascribed to the cultural icon exceed the limits of the person at its basis. In this context, there is the anecdote about Charlie Chaplin, who participated in a Charlie Chaplin imitators' contest and only won third place.

The Hemispheric Identity of the Memes

The hemispheric identity of memes is determined by the pattern of their birth and development. Many of them are the result of a mix that reflects the contribution of both hemispheres.

The seeds of the memes that were born in our brain, fed on the condition of the ecological niche that is unique to our brain and developed in a unique pattern, will be prone to innovation and will be characterized as right-hemisphere oriented. On the other hand, the memes that we receive in one piece and that conveniently integrate with senior memes in our brain will be characterized as left-hemisphere oriented.

We mostly store the memes whose perceptional representation is "common insights regarding our world" in the left hemisphere.

The Actuarial Science of Memes

A meme that has a survival ability is mostly one that is characterized by a relatively low level of randomality noise compared to other memes. With respect to ideas that manage to survive crises over time, it seems that the origin of their preferred resistance derives from being less infected by randomality in the database on which they rely.

The creation of echoing memes (which, metaphorically speaking, triggers continuous vibrations in the communication circles in the brain and involve numerous neuron networks) reinforces their assimilation and impact. From a pedagogical point of view, it is important to pump into the brain conceptual mint ideas (memes) with an educational message so that they will echo in the space of thought for a long period of time and project on many thoughts.

Environment and Genetics as Memes' Designers

Our brain is a machine that absorbs, preserves, creates, and distributes memes. Some of the memes are genetically imbedded in the brain, while others are absorbed from the surroundings. The unique contribution of the brain is in creating new memes.

The interrelations between genetics and environment are prominent in the brain. Conceptual feeding and environmental stimulation make the brain adapt itself to them (a different cultural environment creates a different brain).

Memes' mutation takes place in the brain. Brains are like "parallel universes" in which the physics laws are similar but not identical (the general anatomy represents the similar aspects, and the different wiring represents the variance). The various brains are like greenhouses of different climates, which are unique to each brain and in which there is a different evolutionary course to each of the memes. Thus, in different brains, different versions of a meme that is considered an ancestor—a common meme in the community in which the brain owner was raised—are created.

Epigenetics and Memes

Epigenetics ("epi" means "above" in Greek; as a whole, beyond genetics) refers to the entire processes that influence heredity; they are not mediated directly by the genes but rather by different conditions that affect the expression of the genes' products.

An example of the workings of epigenetics is proteins that activate or silence expression of particular genes according to environmental conditions. The principle of epigenetics also works for memes ("epimemetics"). Different memes might remain dormant for a long time and be resurrected when the memes' climate around them encourages it.

There are types of animals (like the blind mole) that do not carry out memes' heredity from one generation to the next, since their lifestyle does not include "parental guidance"; they are abandoned a short time after birth and live their life as lone individuals. They are forced to develop their understanding of the world by themselves almost from the beginning (almost, since they are equipped with changing doses of innate wisdom). Each of these animals resembles a cognitive Robinson Crusoe, who independently develops a worldview that fits his personal island; in the mole's case, the ecological niche it lives in. From the animal's perspective, this niche is the entire world. In the absence of accessible accumulated wisdom, which stores the knowledge of previous generations and transfers it forward to the next generations (by training and imitation, for example), the animals are forced to create insights regarding the world, almost from scratch, in a manner of experiencing sequential hardships, and the price is sometimes a death toll.

As a result, the depth and scope of the insights of these animals are humble. These kinds of animals resemble lonesome horsemen who ride on the paths of life without receiving past insights from other members of their species and do not pass on accumulated insights to the next generation.

The Memes as Designers of the Pattern of Information Passing

There are at least four ways of transferring information between generations:

Genetic transfer in the language of DNA.

Transfer by means of feedback mechanisms that respond to the environment and turn the genes on/off in accordance to environmental conditions; this is transfer in an epigenetic pattern.

Transfer by means of direct behavioral imitation.

Transfer in a way that is unique to human beings—conceptualization by means of symbols, such as written language—i.e., memetics.

It seems that the memes have an impact on the pattern of information transfer, at least in the case of the last three sections.

The Genes–Memes Interface

There are two suns in human skies, in the sense that humans exist in the shade of two evolutionary processes. The one—biological—duplicates the genes and is epigenetic (depends upon proteins), and the second—cultural—duplicates memes. In the clash between the decree of the gene and protein and the decree of the meme, the meme often wins. For instance, the meme that brings about the lifestyle that contradicts genes' duplication—monasticism.

In this sense, our brain serves as a storage container, but it is not a passive agent of memes' duplication. Memes' mutations are created in every human brain as a result of recombination and hybridizations between memes or parts of memes that have already acquired a foothold in the brain of that person. If the person shares them with others, the newly born memes' creatures wander to humanity's world of ideas.

In the information age, the memes have become more viral and infect populations more rapidly, and in a much more parallel manner, compared to genes. The meme does not require a courtship period and an additional nine months in order to be transferred to another body. In order to win massive, immediate exposure, all it takes is uploading to the Internet.

In many senses, the memes' evolution reinforces the human brain's capabilities even more than the genetic-biological evolution.

The memes—the various ideas and skills encoded in our brain, which are the basis of human culture—change our brain in the anatomic and physiological sense (i.e., they have a clear material signature). In a sense, a new course that "bypasses traditional evolution" is created here, since, through epigenetic means, the memes press the activation buttons and, alternatively, inactivation buttons of genes. They press the keys of protein production and change the structure of neural communication networks. In other words, they influence the gene and, accordingly, the protein encoded by it and, as a result, the wiring of neurons.

Memes can change the expression of the genes in an epigenetic manner. They do not change the composition of the genes directly, but they do change the regulation proteins that regulate activation or inactivation of genes, which means they affect the start/stop switch of the genes. Sometimes they encourage activation, and, by that, lead to reproduction of the genes to the proteins that are the genes' executive arm; at other times they promote inactivation, which results in lack of reproduction. In fact, they create a balance of terror against the genes and constitute counterweight to genetic influence.

That being the case, thoughts that are expressed as memes are capable of changing the reproduction pattern and, as a result, to narrow or expand the expression of genes.

The pressing of the memes on the genes' keys leads to the creation of sounds that are similar, in terms of their material manifestation, to proteins that create new music, which is not composed by the laws of the old composition of evolution.

The opposite is also true. It seems that a certain genetic array is more prone to absorbing certain memes, as in the common supposition according to which the human brain, out of its genetic conditioning, which determines its structure and wiring, is prone to "connect" to the meme of believing in *force majeure*; and it seems that there is a difference among people in terms of the willingness of their brain to fixate the concept.

We might say that success from the point of view of the gene, which is reflected in increasing its expression in the genes' reservoir of the next generations, shares similarities with the success of memes. Thus, for example, with a meme of religious belief, the brains that contain it mostly aspire to "infect" other brains with it.

More Similarities

The nature of genes and memes is embedded in them, and they lack a pattern of free choice. In other words, the ideas do not aspire to "distribute" themselves, and the genes are not equipped with free will. Nevertheless, there are memes that acquire a foothold in our brain, which they use to distribute themselves. They have a hold on us more than we have a hold on them. They use us more than we use them (exactly like the genes..).

Another similarity is the existence of a multigene phenomenon and the multimeme phenomenon. Various physical conditions reflect the involvement of several genes (a multigene consequence), similar to complex ideas whose creation involves different memes (a multimeme consequence).

The Memes as Lamarckian Evolution

As aforementioned, epigenetics deals with heredity processes that are not performed according to the classic Mendelian pattern of change in the order of the nucleotides in the DNA—which means heredity not through intergenerational transfer of genes' duplicates. Some refer to epigenetic heredity as "soft heredity," which is not stiff like the etching of information in the DNA strands.

In assimilation of "memes" that affect youngsters' behavior, a type of epigenetic heredity can be noticed.

Jean-Baptiste Lamarck, the eighteenth-century French biologist, assumed that traits that are acquired throughout a person's life might be transferred to his offspring. At first, this assumption had numerous supporters (including Darwin!), but it was later excluded from the heredity discipline. Recently, however, several mechanisms were found through which information trickles from one generation to the next not by means of "blind" mutations in the DNA nucleotides. These mechanisms have an important part in directing the arrow of species' development.

Thus, for example, environmental conditions and nutrition might significantly affect the pattern of development. The bee's maggot, which is hidden in a beehive's cell, will either turn into a queen bee or an ordinary working bee in accordance with the nutrition it gets. The queen bee is fed royal jelly throughout its development, from the maggot stage until it is crowned, which takes sixteen days, on average. On the other hand, the beehive's proletariat is fed royal jelly only through the first three days of development and is later fed poorly on honey mixed with pollen. The different nutrition turns on and turns off different genes; thus, although the maggot has the same array of genes (identical genotype), the activation–inactivation pattern of the genes will determine the course of its development and its different final shape (different phenotype).

It is known that when humans move from a certain food culture to another food culture, such as when immigrating to another country, their morbidity epidemiology also changes.

Along the genetic dictate, our development is environmental oriented, and, among humans, it is also culture oriented. The environment supplements genes-oriented heredity and, in certain aspects, is even more important than it.

Human culture exists in a Lamarckian heredity pattern. The memes are assimilated by adults in the brain of youngsters from generation to generation. It seems that it is not less important than "classic" genetic heredity and is often even more important than it.

One might claim that, due to constant improvement in our ability to conceptualize the world (discoveries expose to our conceptual eyes nuances that were not known to past generations, and new lingual concepts that conceptualize these new distinctions are constantly born in increasing rate)—improved capabilities that are also reflected in language—the transferring of the terms of "improved language" to the next generation is, in a way, a "perceptional Lamarckism"—an intergenerational leap in conceptualization capabilities and understanding of the world.

The memes allow for Lamarckian evolution, which, within one generation, or even within the same generation, brings about changes that, for the first time since life appeared on Earth, enable us to bypass the arduous course of Darwin's evolution.

Memetic Bang

A memetic bang might be defined as a burst idea that casts a shadow over numerous conceptual systems; under its shadow, they can be understood, anew, in a more "correct" manner. The meme at the basis of the explosion serves as an intellectual godfather for many secondary memes that derive from it. Its appearance resembles an explosive bang that stirs things up, in terms of the most basic concepts of our culture, and reorganizes them afterward. This is a type of super-meme of the highest level. An example of such a meme is the meme of the evolution concept.

The big explosion of memes that took place in 1859 when Darwin's book *The Origin of Species* was first published and the dawn of the meme of evolution in the sky of ideas make us sometimes refer to the people who lived before 1859, whose brains did not encounter or assimilate the evolution super-insight—gullible by force of circumstances whose understanding of humanity was missing a highly important piece of the puzzle.

Although we are, in distance, a tiny grain of time in the hourglass of evolution away from the people who lived on Earth in 1859, before

Darwin's book was published, in terms of evolutionary perception, it seems that our understanding of the world is far away from theirs. It seems that they come from a totally different conceptual era, which is ancient and deprived. The conceptual bang of the super-meme, with respect to the natural selection, has brought to us, for the first time, a natural explanation regarding world phenomena, which we were almost unable to evaluate beforehand, except through mystic, religious terms. The year 1859 is a borderline between two eras of memes' evolution, since the concept of evolution has turned into a huge cascade from which springs of sub-memes derive with respect to all areas of life.

We are not more intelligent than the people who lived in 1859, but it seems our conceptual toolbox contains key tools that allow for understanding world manifestations in a manner that is closer to truth.

Does "memes' impairment" turn the people who lived before the evolution concept was discovered into an archaic version of a lower value? Can we say the same thing about the brains of people who lived prior to 1905, when Einstein talked about the meme bang based on the equivalence principle of mass and energy? On a similar note, an idea that is concealed from our collective consciousness at the moment, and that might cause a "conceptual bang" in the future, will turn our brain into a "memetic- deprived" brain in the eyes of generations to come.

The Memes as a New Type of Evolution Outliners

The Homo memeticus is the upgraded Homo sapiens.

From an evolutionary point of view, our brain, with all its various functions, was not "intended" to enable us to understand the world but, rather, to survive in it—to "hang in there" for a sufficient period of time that will allow us to pass our genes forward. The additional tasks and requirements we load on our brain have nothing to do with evolution; thus, we need to come up with unique strategies in order to improve them. The excess capabilities of our brain are a type of extra advantage that was not preplanned by evolution (but, on the other hand, evolution did not "oppose" them). Another possible conclusion is that we are sailing along a cruise route that deviates, for the first time on our planet, from ordinary evolution routes. Moreover, we are creating such a "different evolution" that some of its rules are not written in the book of laws of ordinary evolution. We invent and enact these rules by ourselves and for ourselves.

The outcome of the encounter between the human brain and the memes is a combination that bypasses Darwinian evolution.

The human brain is becoming a tool that bypasses evolution through the memes in the sense that the result of their joint activity is a process of evolutionary Lamarckism. Within a single generation (and sometimes within an intragenerational period of time), the memes are improved so that the insight of the next generation sometimes performs a "quantic insight leap." The biological evolution plods on, while intellectual evolution moves forward in an exponential pattern.

Human beings and lab mice: The Internet and other technological progress devices, creatures of the human brain, change the pattern of our relationships and, at a most basic level, the pattern of our brain. We all take part in a huge experiment in which we cross the boundaries of ordinary biological evolution. This is an evolutionary leap that reinforces, as if in a magic cycle, the gadgets' repertoire of our brain. These are the "luxuries"—excessive qualifications that were acquired by our brain, beyond the scope of capabilities that were granted to it by evolution.

Memes' Heredity and Genetic Heredity

In humans, as in any other organism that reproduces through sexual reproduction, there is a 50-percent chance, on average, to pass a certain gene (out of the two gene manifestations—the alleles) to each of the offspring. The parallel gene of the partner has an identical chance of being passed to the offspring. The gene array of the offspring is weaved as if according to binomial probability, which is composed of experiments that do not depend on each other's results and, in each of the experiments, there are two possibilities of even chances: success and failure. On the other hand, the memes, once memorized by the next generation, have a 100-percent chance of meeting the neurons in the offspring's brain but will not necessarily acquire foothold in it.

"Memes' penetrability" is a term that is taken from genetics, in which the term "genes' penetrability" reflects the percentage of carriers of a certain array of genes (genotype) who physically reflect the phenomena deriving from the specific array (phenotype). These terms are usually used in relation to impaired genes' array on the genotype side and to the level of their reflection in terms of morbidity on the phenotype side.

According to an interpretative view that uses genetics' terms, one might claim that memes of ideas that acquired foothold in a specific brain can be referred to as "memetic genotype." Their influence on the thoughts and actions of the brain owner equals the "memetic phenotype," and the extent to which these ideas cause a typical behavioral, conceptual/practical outcome might be considered "memetic penetrability." On the other hand, there might be other terminological interpretations, such as the definition of "memetic penetrability" as the level of the ideas' (memes') tendency to set foot in other brains (how infectious they are), and the definition of the "memetic phenotype" as the level according to which the ideas are put into practice (i.e., their implementation in the practical world by the brain owners who host them).

A danger of "random eradication" hovers over unique memes, as it hovers over unique genes, if the person whose brain invented the unique meme (or the carrier of the unique gene) passes away in random circumstances before the meme or the gene is transferred to other brains, or is perpetuated in some other manner. There is no doubt that numerous potential founders of conceptual revolutions were lost to humanity as victims who died in vain during battles, and we all pay the price for it, mostly without even being aware of it.

Our behavior is the common phenotype of our genes and the memes that reside in our brains, both of which, in various proportions, contribute to our different behavioral patterns.

We are like dancers who follow the choreography created by both the genes and the memes.

Co-Evolution of Genes and Culture

The genetic outline of our brain leaves sufficient room for the influence of culture and environment.

The capabilities we acquire from birth and on are added to the innate skills, expand their owner's options, and reinforce their competence and survivability. It can be seen as the blessing of the acquired course. Sometimes the intensity of acquired memes surpasses the intensity of innate memes with respect to dictating behavior. An example of that can be found in people's willingness to risk their life in the name of the concept of brotherhood, for the sake of saving others, although the genetic dictate perceives the survivability of the body (and, more accurately, the survivability of the genes that the body carries and "distributes") as the supreme value.

The acquired course is at the foundation of the development of human culture, and some might say it constitutes a new phase of evolution also, in the sense that it enables a transition from passive adaptation of man to the surrounding to adaptation of the environments to man's needs. The acquired course and the culture that derives from it reinforce, in an exponential manner, the capabilities of human kind. It enables expansion of the narrow slit through which our senses have sampled the world to a wide opening that ranges from the microscopic dimensions of the subatomic world to the dimensions of the galaxies that exist at a distance of billions of light years in the depths of the universe. The adaptation of the environment to man's needs, as has been and is promoted by mankind, brings thriving and welfare to humanity but, at the same time, serves as a two-edged sword, since accelerated, unplanned change involves a breach of the ecological balance, which puts the future of Earth at risk.

In accordance with the degrees of freedom in behavior that are at the disposal of man, man is not a passive victim of his own capabilities; he has the power to reverse the reversible damage done to the environment and re-achieve ecological balance.

Faith is not just an idea you grasp, but an idea that grasps you. There are super-memes that encircle our brain in an invisible manner, which is hard to get rid of.

Monogamous adherence to a meme might be part of an extreme worldview that does not allow the facts to confuse it. Although there are memes that we should to be married to in monogamous matrimony, there are some that require bigamy, polygamy, or even swingerism. It is all in the name of pragmatism and being loyal to the truth, whose deeper understanding dynamically changes our memetic perception of the world. There is also the serial-monogamy approach, according to which, at every given moment, loyalty is devoted to a single idea, but the idea changes.

Rigid social topography, such as social classes, and rigid geographical topography, such as mountain ridges, prevented, in the past, the social and geographical mobility of people.

Many of the geographical and social barriers have been removed, and the physical and memetic mobility has improved dramatically. People who were born in different societies meet one another, and, at the same time, memes created in different cultures meet, clash, or assimilate with each other and bring closer the fulfilling of the vision of a memetic, universal codex with respect to core issues related to the future of mankind (such as the treaty that prohibits child labor).

The Creation of Memes' Clusters (Meme-Complex)

Just as genes appear in clusters, which improves the complexity of the organism and increases the chances of duplication and transfer of themselves to the next generation, memes also appear in clusters (i.e., memes that are bounded, support, and have been supported by each other). The inter-memetic-collaboration strategy often intensifies memes' survivability and their chance of being distributed to other brains.

"Sociable memes," which have a friendly interface for connecting to other memes, are more prone to integrating into clusters of memes and survive better—like genes that tend to "collaborate" with other genes. A collaboration in this sense is not a "zero-sum game" (a competition in which the success of one party means the failure of the other)—it might benefit all the participants (again, without attributing "will" or intention, in the teleological sense, to the memes or the genes).

A cluster of memes can also create a common memetic core that serves as an anchor point for the various memes that are included in the conceptual cluster. Such a pattern might exist in an inclusive pattern that groups a group of submemes and encodes the common between them.

The paths of history are embedded with warning signs—summoned by circumstances under which, in different societies and different times, brains of an entire generation (and sometimes entire generations) were captured by the dark spell of a memetic complex—that many of the memes that compose it are morally impaired. A typical case is that the outline of the memetic complex is based on a super idea, which turns the ones that carry it into marionettes that move along the sounds of the conceptual collectivism; their lives run in its shadow, and they are a "recruited" generation in the service of its distribution. Such examples can also be found nowadays.

A memetic complex (i.e., a mix of memes) is determined by averaging the shades of the memes that compose it and not according to the shades of single memes. The zealous supporters of the Soviet empire might illustrate that.

USSR—the red empire—following the 1917 revolution, claimed to lead humanity to the end of history. Its founders assumed that Marxist-Leninist philosophy was the ultimate answer to all that was wrong in human societies. The equality meme that enchanted with its clear shades (and, literally, on the moral level, it does have a high moral value) was mixed with numerous turbid memes that turned its shade into turbid red from the blood of the "victims of the system."

Naive people who not infrequently believed they had good intentions became "color-blind" servants of the complex of memes dictated by the commissars of the communist USSR and turned, according to Lenin, into "useful idiots." The ones who served the regime caused great suffering to numerous people in the name of a "noble ideal."

A gloomy but evident conclusion is that those who choose to focus on a single meme out of a conceptual structure that is a mix of memes, and at the same time sweepingly adapt the multimemes' inclusive concept without criticizing the rest of the components, voluntarily choose "color blindness," and their discretion is highly questionable.

Real-Estate Battles of the Inner Skull

The generator of memes, the workshop in which new memes are formed, is mostly based on old models that undergo change and adaptation to other memes that reside within the same brain. There is co-evolution among the memes as well. Sometimes a certain meme reinforces another meme and, at other times, it weakens it. The main structural meaning of reinforcing or weakening the meme's grip on our brain is increasing or decreasing, respectively, the number of synapses devoted to the encoding of a specific meme. The memes are territorial creatures that compete over living area within the limited space inside the skull.

The Meme of the Self

The experience of the brain user creates the term "self" as a super-meme.

Some claim that the "self," which is by definition an autonomous entity that exists on a sequence of time, from past to present, is a delusional meme that has a conceptual bond with many other memes in a pattern of comprehensive conceptual cluster (memeplex).

Means of Transferring Memes and Infecting with Memes

Once the mythological order "Beam me up Scotty" is given in the science fiction series *Star Trek*, the physical body immediately shifts its position to a new destination in space by means of teleportation. The body of the time traveler is scanned, turns into information, and is recreated, according to this information, at a different site in space. We might say, metaphorically, that teleportation of memes is, in fact, what memetic evolution does. The information travels from one brain to another without any physical transfer of the objects that carry it.

Our brain is designed to process information and spread it in a narrative pattern. Despite the suspicious attitude "hard-science" people sometime have toward "soft-sciences" (such as humanities) people, due to the narrative pattern combined in their studies, there is no doubt that their role is highly important with respect to promoting culture, since they are like "agents" who spread memes of great importance in the conceptual war of all against all that takes place in the battlefields of human brains.

Senior memes tend to embrace memes that are close to them "in spirit" and reject the ones that are conceptually remote in a pattern that might be called "meme brings meme," as in "friend brings friend." Due to the array of memes networked in our brain, which creates a bias to adapt memes that will reinforce their predecessors' grip and support them, it seems our brain tends to act as a lawyer who selects evidence according to its compatibility with his clients' arguments more than as an objective judge in pursuit of the truth.

It seems that memes that find an anchor point to the memes that exist in our brain (i.e., patterns that reinforce or share similarities with the patterns that already exist in our brain) bring about intellectual pleasure, which is probably manifested in a certain level of physical pleasure, and results in an increase in dopamine production in the reward cycle.

On the other hand, memes that contradict other memes that reside in our brain create basic resistance, since they induce a state of concern and insecurity. It partially explains the success of self-help books, which group together ideas intended to improve our life quality. The truths listed in them are, in fact, mostly truths we are all familiar with. Their seeds, as beneficial ideas, exist in our brain, and the echoing of these seeds of memes in our brain, combined with external memes of similar spirit that sprouted into mature ideas and which are presented in such books, reinforce familiar memes and create the same pleasurable echoing in our brain in the sense of the "truth speaks for itself."

"Skeuomorphism" is a term that is used, inter alia, for describing the assimilation of an old product's features in a new product. Combining of familiar features with new ones leads to an encounter that is characterized by psychological calmness, due to the fact that the familiar features, which are reflected as senior "memes," improve the interface between the new product (and its reflection as a new meme) and the consuming brain areas.

This marketing insight, which relies on the "familiar within the old," is applied—for example, in the field of product design, in which senior configurational traits are assimilated in a new product and facilitate the rapid anchoring of the new product to old memes that represent a familiar product and, by that, increase the chance of the new product befriending the customer's brain. In this spirit, the first cars that were produced were similar in shape to the coaches that preceded them.

In order for a memetic seed to sprout into a mature insight, it would be very supportive for it to be planted in a sympathetic, memetic bedding that will constitute a growth platform.

A "natural meme" is a meme that seems to have connected to a brain structure that fits it conveniently, as if it were made for it. Such a meme is highly infectious and has a great chance of surviving.

Memetic Gerontocracy

The older memes are deeply rooted in our brain, since their seniority grants them a special status. It is most difficult for memes that do not match the old memes and do not interface with them to set foot in our brain. In this context, we can relate to the wonderful story "Nightfall," written by Isaac Asimov. The story illustrates the impact of a highly exceptional event on the brain of people who are not ready for it in the conceptual (memetic) sense. The conceptual core change enforced by the event leads to complete chaos.[39] When the most basic memes of our reality perception are challenged, we enter a state of mental chaos.

Memes and Technology

Technology changes the conceptual contents of memes and the way we infect and are infected by them, as well as the way we perceive and process them. The navigation maps created by technology enable us to navigate in the oceans of information that rage around us and to find the shortest route between our brain and the desired information.

In the era that preceded television and the Internet, bookstores were the ones that sold ideas to which one could be exposed in a selective pattern.

In the age of the Internet, our species can be referred to as "Homo interneticus." The brain-Internet interface is a type of a new phase in human evolution.

The utopians might say that the availability of information, which increases under the auspices of technology, enables us to experience an enriched reality. The ones who disagree might say that the attack on our senses uncontrollably gulps our brain's processing resources. It seems that, in light of the information flooding nowadays, we must be more selective, in relation to the memes to which we expose our brain, in order to protect its resources. The Internet became an infectious vector of memes, and it constitutes a founder phase in the evolution of human memes.

Thus, for example, it is possible to conceptualize in it the global human mix of emotions, as in a pixilated color balloon. It can be done with software that scans tens of thousands of blogs from around the world, seeks words that describe emotions, and afterward translates them into colorful dots according to a predefined color-emotion key. The resulting color mix reflects, at any given moment, the temporary state of mind that is common in the blogosphere, and some might see it as the reflection of the global, human state of mind—as a summation that reflects an emotional super consciousness.

Memes at the Right Time—in the Splint of the Era

The "right time" is fateful timing for the memes. A meme (in the shape of an idea) that is too foreign, in terms of the conceptual world of the people who live in a specific era (i.e., a meme that does not find an anchorage point in the memes that reside in people's brain,) is likely to be rejected and considered to be faulty or imaginary.

In the satirical novella by Edwin Abbott from the nineteenth century, *Flatland, a Romance of Many Dimensions*, the two-dimensional square from Flatland is exposed to different dimensions and different geometry from those it was used to within the boundaries of its flat world. It visits a land with a single dimension—Lineland—and a land of no dimensions—Pointland—where the monarch who rules it lacks dimensions, with regard to consciousness as well, and believes that he is surrounded only by creatures of his own imagination, in a sort of thinking pattern that resembles solipsism. On the other hand, the square is also exposed to the third dimension, where he is thrown to the depths of Spaceland under the patronage of a ball. The cross-dimensional journeys engulf his conceptual world, reshuffle his system of concepts and recalibrate his beliefs. The mindset change he undergoes leads him to share his insights with his subjects in Flatland, but the priests consider him a heretic who jeopardizes the serenity of Flatland's citizens and deprive him of his freedom. That is how the wandering square is materialistically, though not spiritually, confined to a land with only two dimensions.[40]

The Roman emperor Hadrian said, "Being right too early is just as bad as being wrong."

A genius idea that is ahead of its time will have difficulty becoming an infectious meme that will conquer the brains of the people. It does not drain into the common drainage basins that are wired in the brains of the people who are exposed to it. In other words, there are "deviational" patterns of knowledge our brain refuses to assimilate, regardless of their correctness.

In this context, we might say that one should communicate with people of different ages in accordance with the level of their brain maturity—i.e., in accordance with the repertoire of knowledge patterns (memes' representations) stored in their brain. For example, when we talk to children, we should use "their own language" and pour into their brain ideas that will be able to drain into the basins that exist in their brain. It is also true for adolescents, or when we talk to adults or elderly people, taking into consideration their fields of knowledge and the topographic map of their various skills. No one knows everything about everything.

Our brain is unable to assimilate every piece of knowledge, but it has the ability to assimilate pieces of knowledge whose way was paved by previous memes and that contain a certain level of similarity, which allows for an anchoring point to a representation of a specific meme that is already stored in the brain.

"No army can withstand the strength of an idea whose time has come" (Victor Hugo). History is rich with examples of stubborn geniuses, such as Copernicus, Galileo, and others, who were outcast in their time and whose insights were not acknowledged appropriately. Nevertheless, with time, these geniuses made the world see things as they saw them in the first place in their minds' eye.

The Terms of Conceptualization in the Mirror of the Era

In each era, humans dig with a metaphorical spade that conceptualizes concepts in terms of the systems that the people of the era are familiar with.

We conceptualize complex ideas by creating analogies to a system of complex concepts we are familiar with and which serve as a sort of a prototype of complex systems in our time.

Each era has its own "model system," which is often a system that represents the technological forefront of that era, in relation to which analogies are created. Thus, for example, the ancient Greeks saw musical scales and geometric structures as a model for understanding the laws of physics. In Europe, during the Age of

Enlightenment, the clock served as a conceptualization of reality. During the industrial revolution, the universe was compared to a steam machine that produces entropy. Even today, we cannot avoid the cultural bias that is part of human conceptualization. We tend to explain the dynamics of the operation of complex systems, such as the human brain, through analogy to the digital computer.

The patterns—the memes we are familiar with—serve as an anchor point for conceptualizing ideas (memes) that we were not familiar with before they became familiar patterns.

A Trendy Meme

Trendy memes sometimes do not pass the test of time, and their enchantment expires as time goes by.

An example of that is the meme of the "science" of phrenology. Phrenology blossomed during the first half of the nineteenth century, mostly in Europe. The theory of phrenology divides the surface of the skull into thirty-five areas, each of which represents the person's skills in a certain field. According to the topography of the skull, which is deduced from a thorough palpation of the skull, the skills of a person are deduced. Phrenology is an example of a trendy meme that had infected numerous brains and was admired mostly in the lounges of European aristocrats, but, due to its lack of factual basis, its spell broke, and the term "neophrenology" has become synonymous with pseudo-science, which is based on false beliefs that lack factual basis.

Illusory, trendy memes are often characterized by a simplistic guiding principle, which supposedly stores hidden wisdom that is known only to the most expert ones and is spread out in an efficient distribution pattern that is inexpensive and available to numerous brains—the "secret of the world on a T-shirt."

"Memetic Libido"

The desire to match the thoughts that were produced in our brain with thoughts produced in other brains, to "communicate" and talk, can be defined as the "libido of the brain."

Some might claim that, unlike the libido in its usual sense—the sexual libido—which seemed to be more intense among men, women's "memetic libido" is stronger compared to men's. Women tend to discuss memes, especially the ones who have a significant emotional component, with other women more commonly than men do with other men (or women).

Brainstorming is, in a sense, an "orgy of memes"; thoughts deriving from different brains "mate" with each other and create new thinking offspring, whose components sprouted in the wombs of different brains. Some of these multibrain thinking creatures are assimilated in our brain and guide our behavior. In this sense, the psychologist Lev Vygotsky was right when he said "Through others we become ourselves."

Emotion as a Memetic Duplication Agent

Sometimes, when we are exposed to the visible behavioral motivations of others and to their feelings, we might become infected with them as a sort of memetic infecting of emotions.

Ideas that are particularly powerful often accumulate their power through the lower lane, due to their proximity to strong feelings. These ideas might someday be understood as reflecting mirror neurons in action.

Memes that sprout on an emotional bedding of a trust relationship grow stronger roots, and, at the other end, their top is taller. Positive feelings for the person who infected us with the meme will probably intensify the echoing of this meme in our brain; thus the importance of establishing a therapeutic alliance between therapist and patient.

Trust relationship has a crucial effect on the "conceptual infecting" and on the ability of "therapeutic memes" to become a living insight.

The Memes as the Basis of Human Culture

The Memes in the Mirror of Evolution

We are the offspring of the Hominidae, who were victors with respect to genetics, and our brain is full of memes that are the offspring of the successful memes of the past.

Culture grants us the right to drink from the pool of human experience.

From the dawn of the appearance of the Homo sapiens on the surface of Earth, brains similar to ours have experienced a world that is similar to ours in many senses. Out of their insights and their memes, we often enjoy shortcuts when we march in the tracks that were paved by our ancestors.

The memes are like sticks in the intergenerational relay race. In this race, there are phases of life during which our hands are stretched out to get, as in childhood, and, alternately, in which we are willing to give, such as during the second part of our life.

During the Renaissance, it was said that "a person moves forward by looking back." In a sense, we can interpret it as a conditioning of the progress in the future by relying on insights that were acquired in the past (senior memes in the cultural or personal sense).

Thus, for example, regarding a cliché. A cliché is allegedly a worn-out meme that was overly used and that is mostly transferred intergenerationally, but it mostly relies on true foundations. Its banality derives from the fact that it is perceived as obvious, not from its being untrue, and, most of the time, it constitutes a meme that, in numerous brains, is considered a truth that contains concentrated wisdom—a fruit of the experience of multiple generations.

Some find a teleological pattern of development in the development of human cultures, and a view called historicism suggests that the events that take place throughout history are developed with relation to a certain guiding principle, a sort of a supermeme that directs their course.

Culture on the Couch

Freud coined conceptual terms (memes) for describing the main structures of the soul, such as the id, which is at the core of the soul and which is the creator of raw urges. On top of it, there is the ego, which is in charge of awareness and is rational; its role is to moderate and refine the id and channel its primitive urges to channels that are socially accepted.

Freud saw the evolution of human culture as a collective reflection of the constant tug of war between the id and the ego. A society is more advanced when the individuals in it are willing to sacrifice the desires created by their personal id on behalf of the public welfare.[41] The memes that describe the soul, according to Freud, became super-memes, but their glow faded down the road.

The Wisdom of Generations

The wisdom of our ancient mothers, who were most successful, in terms of survival, is wired into the brains of women, and, similarly, the wisdom of our most successful ancient fathers is wired into the brain of men.

Some might claim that the amygdala is, in a sense, the prehistoric man, who still lives within us. In this spirit, the frontal lobes are the "load" the culture puts on the shoulders of the prehistoric brain.

Memes and Science

Science, being created by human beings, is not unbiased. The memes that reside in the brains of scientists determine the starting point of studying a certain phenomenon—which research questions will be asked, and which interpretation of the findings will be selected.

Subjectivity is inherent to science as well. Research methods are a derivative of a scientific worldview.

The way to objective memes ("the facts of the world") relies, at its basis, on subjective memes ("scientific worldview").

It seems that objective knowledge, of any kind, ultimately relies on a collection of subjective knowledge, and some claim that the term "objective knowledge" is an oxymoron in itself.

Scientists hold memetic axioms in light of which they operate. Thus, for example, the English scientist Isaac Newton, and other scientists of his time, worked according to the supposition that a creator—God—enforced a cosmic order on nature, and their role was to discover this order.

Science that seeks the alternative skies, which are not inhibited by angels and demons, is a reflection of a subjective meme of scientists who do not believe in common theological explanations that guide thinking.

Modern science tends to look for causality rather than purposefulness in nature. Its explanations are mostly causal rather than teleological.

The Secret of Foresight

The sentence "If I have seen further, it is by standing on the shoulders of giants" is ascribed to Isaac Newton, who is responsible for revolutionary scientific insights.

Newton was poetically commemorated in the words of the poet Alexander Pope: "Nature and nature's laws lay hid in night, God said: Let Newton be! And all was light."

Newton merged his glowing insights with those of other brains and, as part of the dialectics of science development, improved them with the signature of his unique brain.

The crystallographic data regarding the structure of DNA, shown in Rosalind Franklin's studies, served as crucial information at the basis of the revolutionary insight of Francis Crick and James Watson regarding the DNA structure. There is no doubt that observing the horizon of the language of genetics from the top of the double helix was made possible also by standing on Franklin's shoulders—i.e., by relying on her insights.[42]

The saying "Great minds think alike" is sometimes referred to jokingly when two people (e.g., you and your friend) think the same thought at the same time. It can be seen as support to the argument that the similarity among our brains (and among humans in general) is greater than the difference. On the basis of a similar cultural climate, complex insights might sprout in different brains simultaneously. A famous historical example of that are the insights of differential calculus that were raised simultaneously by Isaac Newton and the German philosopher-mathematician Wilhelm Leibniz. Another example is the natural selection hypothesis that appeared at the same time in the brain of Charles Darwin and the brain of (the less-famous) Alfred Wallace.

It seems that there are universal memes that represent what can be referred to as "substantial insights" related to the nature of the world, whose secrets will be revealed by various inquisitive brains. These brains rely on a similar memetic load that enables them to cross the threshold to understanding the revolutionary meme formed in their brain, by means of interaction between the memetic load stored in their brain and the world of phenomena.

According to a similar view called "cultural determinism," the spirit of the time, with regard to the cultural climate, and the insights available to the brains of people who live in that era are the wheels of the coach of progress rather than the genius of individual ones. According to this view, a pioneering insight resembles a discovery of a new continent that awaits the random discoverer; the wind of circumstances brought his ship to its shores, and if he fails the mission another discoverer will arrive and win the experience of the discovery.

There is probably an "essential and sufficient" memetic mountain, and the person who manages to reach its peak and, on the way up, is exposed to the memes that are piled on top of each other and constitute the mountain, has the chance of looking beyond the mountain to the new horizon of the promised land. The one who manages to do that is able to create a revolutionary insight that contains the past insights, which compose the memetic mountain, but soars higher and takes the horizon forward to a great degree.

Alternately, it seems that the brain owner who infused his brain with a critical mass of memes that were born in many other brains beforehand might start within his brain a chain reaction that results in the glare of the "eureka."

Memes that represent the forefront of human knowledge in a specific discipline are born with a built-in expiration date, though it is not a specific date. The hourglass that times their life expectancy rolls over and trickles its grains from the birth of the renewing insight, when the last grain announces the appearance of the newer insight, which is the "new truth." It can be defined as the "half-life-time of truth" (in this context, we may ask the question, is it possible at all for our mortal brain to produce immortal insights?).

Prior to the Second World War, Albert Einstein visited Britain and, at an event that was prepared for him, a famous playwright, Bernard Shaw, proposed a toast. He said that Ptolemy's theory lasted for two thousand years, Newton's theory lasted for two hundred years, and he preferred not to guess how long Einstein's theory would last at the forefront of human knowledge in its field. Einstein himself considered his insights to be an "interim station" and tried until the end of his life to deepen the insights beyond the outline he had sketched for the theory of relativity.

The Gift of Culture

Culture is the information highway that connects past generations to future generations. World insights, which are, in fact, mental patterns that were acquired by humanity collectively, are passed through culture. The entire human knowledge relies on the incorporation of human brains—i.e., it is networked and spread across the brains of living human beings as collective memory instruments.

The magnitude of the rooting of memes that can echo throughout a generation and more, while skipping from brain to brain intergenerationally, is reflected in the saying of the great educator Janusz Korczak: "The one concerned with days, plants wheat; with years, plants trees; with generations, educates people."

Man was mostly spared the *via dolorosa*, through which various animals acquire basic insights about the nature of the world, due to human culture that uses the various symbolization languages and mediates a large treasure of accumulated past knowledge to present generations. Thus, humans become partners to the secrets of the collective wisdom. Access to this huge insights' library greatly improves the cognitive strength of each individual in human society.

Culture can be seen as a collection of the best memes—in other words, the collection of the most successful memes that were ever born in the brains of human beings (successful in the sense that they managed to survive the passage of generations and the leaps from brain to brain). This is an agnostic claim, in terms of morality, since some might claim that some of the memetic hits that reside in the brains of people of our time do not pass the test of ethics and are not "moral hits."

According to the instrumental view, the quality of an idea is measured according to its practical value—in other words, according to its usability and applicability in terms of solving problems in life. There are memes, such as memes that are connected to the world of art, whose level of instrumentalism is sometimes in question but, nevertheless, managed to survive the test of generations and ages of memetic evolution.

Along its visible hands, cultural heritage also has invisible hands whose fingers shape the brains of multiple consecutive generations and create a path for our tendencies, which is sometimes hidden from the eyes of consciousness. "The language of culture" we were raised with, in terms of its common world interpretation and the stimulation-response dictionary that is used by it in common circumstances in life, is mostly the language that, according to its grammar rules, we "speak ourselves," in the sense of behavior.

The memes that create the core of a cultural heritage routinely infect all brain owners who are raised with the same heritage. Assimilation of the core memes of a certain culture is the main training process the individual goes through on his way to becoming a member of that culture. The memes are the ones that weave the threads of linkage between the individual and the society he belongs to.

The ability to wipe the traces of these memes is mostly limited and involves agonizing "delearning" efforts.

Cultural shaping: The values that are assimilated in us, as intergenerational assimilation—the people we have contact with, the experiences we choose to experience (such as which book to read)—are all factors that shape our brain and, sometimes, our fate.

The shadow of the culture we are raised in is cast over our life. The way of life that is dictated by this culture directly affects our health and thinking patterns.

In the spirit of Plato's "Allegory of the Cave," culture at its best resembles a continuous journey of exiting the shadows' cave in an attempt to capture the true light of reality.

Sometimes, however, the "blessing of culture" is, at least partially, a curse. Certain societies and cultures are, in fact, modern shadows' caves; they pass the "knowledge of the world" through a prism that reflects their values, and the conceptual core they assimilate into the brains of their people is fabricated and unrealistic, or it distorts reality completely. The reality perception of people who live in such societies resembles a "shadows' show" that reflects the nature of the conceptual, cultural prism that mediates the information more than it reflects reality manifestations as they are.

The Ages of Wisdom?

Sources from various times contributed to the wisdom available to our brain:

The wisdom of climbing hoards DNA-encoded behavioral tendencies that were accumulated throughout man's evolutionary development and is millions of years old.

The wisdom of culture hoards the treasure of humankind's knowledge of the last thousands of years, from the time when abstract symbols, and particularly language and script, became widespread as a means of storing information.

The wisdom of the group hoards unique knowledge about specific fields, such as in an association of professionals; its age is usually a few dozen years.

The wisdom of the individual is as old as the person who hoards it.

To the Personal and Beyond

Expanding the boundaries of wisdom beyond the personal is based on learning from the experience of others while using the unique tools invented by human culture for passing information.

In the stories of the sages of various ethnic groups in our world, tradition has a voice. The stories of the sages around the bonfires at night in the African savannas are often told in a way that pleases the ears of various crowds: the crowd of the spirits of the deceased men and women of the clan that hover around at night and, on the other hand, the crowd made of the children of the clan who shudder joyfully.

"Leave the dead some room to dance," said Wole Soyinka, a Nigerian writer who received a Nobel Prize in 1986. He referred to the sense of belonging that binds us by means of memory of the past, the past generation, and the present generation. The memes flow in the waterfall of the generations.

The Contents of the Safe at the Wisdom Bank

The number of information pieces that constitute an infrastructure for a complex cognitive skill or a high level skill that reflects expertise (a layer in which a person is considered a master), such as complete mastery of languages or mastery of chess, based on indirect findings, is assessed as approximately fifty thousand pieces of information that relate to the skill and includes both theoretical and practical pieces of information.

The Steps of Time to the Top of the Tower of Wisdom

We have innate skills to perceive basic perception impressions, but developing an expert brain that allows us to perceive and differentiate between complexities and nuances in our world requires time and experiences.

Experts assess that acquisition of a complex skill with a high level of mastery (i.e., acquiring expert skills in a certain field) requires about ten years of concentrated effort, during which a person focuses on the skill for three or four hours a day.

From another point of view, the amount of practice time needed to become the owner of an "expert brain," with respect to a complex skill, is estimated as ten thousand hours. It seems that continuous practice that lasts for ten thousand hours creates a neural infrastructure that relies on columns of neurons that serve as a trampoline for insight jumps, which enable the performance of the complex skill at a high and thorough level within a short period of time.

Memes at the Right Time—the Memetic Gradient

Multistage memetic selectivity—i.e., graded exposure to insights about reality—is a desirable pattern for matching the information to the level of maturity of the exposed brain and for assimilating correct core values in it, which will grant the brain and the environment in which it develops a satisfactory and fruitful life.

The memes to which young brains are exposed should match their mental age. The unsightly facts of life should be filtered and softened for the young brain.

For instance, exposing the brain of a young child to information about abnormal human behavior—such as a horrific affair of a serial killer, which, due to its sensational nature, is reported massively by all media channels—exposes the young child to an extremely distorted value system that might assimilate in his brain an attitude of lack of trust and of suspicion as a prominent personality pattern.

The above recommendation does not apply to an adult brain. Adults are supposed to encounter the various manifestations in life, including the tragic, unsightly aspects. Indeed, truth should always be our guiding principle, but, exactly as occupational information is fitted to the professional level of an individual, it is not recommended to mix insights of novices, young children, or adolescents with insights of an expert in life—a grownup individual, since, in that case, there is great danger of shaping a young brain in a problematic manner.

Memes that contain the unsavory facts of life should encounter a young brain in a pattern of a truth gradient that is based on mental age. In other words, insights for kindergarten are different from insights for school years, and from youth movements' insights, parenting insights or golden age insights.

It seems that there is a need for a universal memetic pedagogy determined by an international body (such as the UN organization for education) to struggle with the problem and determine core memes that should be assimilated in the human's brain in accordance with its location in the ascent of the mountain of the years.

There are memes that can be defined as "mediating memes" whose purpose is to mediate reality manifestations without compromising their basic nature but, rather, through the necessary adaptations to the containment capability and the mental scenery that is reflected from the eyes of the human being who is exposed to the information.

The desirable "age of truth," in which a person should be exposed to all available world knowledge without any mediating memes that "soften reality" is, as some people believe, when a person reaches mental maturity; this is not a specific age but can change from one individual to the next.

Instilling the core of moral memes, whose purpose is to divert people's hearts toward generous, humanistic way of thinking, should include a graded pattern of exposing the truth that is easy to digest and assimilate. The magnitude of the bitter taste of life, which hits more violently as people grow older, is an insight that is suitable for an expert's brain and not for a novice's brain.

The Importance of Instilling Morally Correct Memes to Young Brains

In an era in which the glow of various religions fades away, an organized system of moral memes is a must.

Also, a person who does not believe in a divine entity can relate to a secular, spiritual testament, such as that heaven might not exist, but we should aspire to be worthy of it. In the absence of an obligatory moral doctrine, local versions of the Marquis de Sade and Dracula could appear in every town or village.

Eugenics is the betterment of genes by selective hybridization. We must apply memetic eugenics ("betterment of the genes" of the memetic ideas stored in our brains, and in the brains of all human beings that share the human race, which might be defined as "eumetics") so that a destructive memetic phenotype (such as war) will disappear from the stage of our life.

If we do not succeed in annihilating a memetic genotype that preaches violence and injustice toward others, we can at least dull its sharp points by attempting to decrease its penetration and, thus, its expression in our world as a memetic phenotype.

The spiritual value of human beings should be determined according to the "quality of memes" that reside in their brain and in their actions. The brain is, in the wider sense, a creature of our experiences, the culture we live in, and the memes we are exposed to. It resembles an agricultural cycle of seeding, sprouting, growing, and reaping. The seeds of the memes that are seeded in young brains sprout and grow inside these brains and become core insights that constitute an infrastructure for their mental life.

The Tabula Rasa of the Brain—Morally Agnostic?

The spouting source of the ethical codes that guide our behavior and our attitude toward life has been controversial for many years.

There are contradicting views. One claims that morality is a strict language based on universal internal grammar, while according to the opposite view, the language of morality is encoded in an open code that constantly changes in accordance with the circumstances. The first view is called the deontological approach, and the other is the teleological approach.

According to the deontological approach, certain moralistic positions have positive or negative ethical value that does not depend on the circumstances or the results of moral choice. Kant's categorical imperative is an example of a deontological perception of morality.

On the other hand, there is the teleological view of morality, which considers the results of the moral decisions and the circumstantial context in which they are made as the fundamental element.

Another aspect of the controversy is reflected in the ancient question, is man's heart good or evil from his youth? It seems that, even in the shadows of these controversies, there is a consensus, according to which culture and the moral load it pours into our brain in the form of memes has a central role—some might even claim an exclusive role—with regard to shaping our core moral values. The brain of a newborn is a "natural nihilist," in the sense that it lacks beliefs and values—these will be poured into it by the memes he will be exposed to at a later stage.

Thus we see the importance of education and exposure to the "right memes," which will instill the appropriate significance to values such as humanism, human dignity, compassion toward animals, ecological awareness, and so on, at a young age, before memes with different contents have a chance to shape the "moral personality."

Initiated annihilation (delearning) of a pattern of behavior, in the form of a habit or a core thought, by the dismantling of neural networks that constitute the basis of a certain behavioral habit is more difficult than establishing an infrastructure of neural networking for the representation of a new acquired skill (learning). The more senior skill enjoys a competitive advantage in controlling brain resources that the new skill is still "trying hard" to acquire.

For these reasons, instilling memes is easier at a young age and more difficult at an older age; thus, extreme ideological movements that are aware of this fact choose to focus on the younger generation, knowing that it is easier to influence it ideologically. Extreme ideological movements prefer to hunt for their prey among youngsters, who are like "captivated babies." A young brain is an ideal seedbed for the purpose of planting an idea that will strike deep roots.

This is why it is extremely important to transfer "good, moral core values" to the young generation that will constitute the basis of its future world of thought. Faulty habits and false beliefs ("bad memes") that conquer representation territory, before the brain has a chance to be exposed to the "good memes," will enjoy a competitive advantage that might help the bad memes strike deeper roots.

Representation of an idea in the brain at the structural level is the neural networking pattern that encodes the meme.

Ideological exposure that is not selective might network, in young brains, unessential maps that will get priority over maps that are more essential and important to their future as individuals and the future of all humanity.

The senior memes we have stored in our brain become the reference point for all the memes that come afterward—thus the great importance of early exposure to positive "supermemes"—basic insights that will serve as the foundations for all the structures of ideas that will be built afterward. The conceptual, moral world of a person is likely to be based on these super-memes that will constitute the moral core.

The ideas that become core ideas (super-memes) and that shape our attitude in life leave their sign, to a great extent, on the nature of our experience throughout our life on the face of Earth. These super-memes become the historians of our past, the spokesmen of our present, and, to a great extent, the prophets of our future.

Memetics and Meme Ethics

Just as the development of genetic engineering requires ethical consideration (the gen-ethics of genetics), the accelerated development of memetics (the means of transferring information among humans' brains) requires such consideration as well.

The memes acquired from our life environment constitute, along innate tendencies, the moral menu our brain is fed on; according to it our moral view is formed and our behavior is shaped. There is an Indian fable in which a father tells his son that in every person's soul there is a struggle between the dark wolf, who is fed on hatred, rage, intolerance, and other dark sides of human personality, and the illuminated wolf, who is fed on compassion, kindliness, tolerance, and all other positive aspects of human personality. When the son asks, worriedly, which wolf will win the struggle within his soul, the father says, "The one you feed more…"—and, in a broader context, most of the times-it's the one that your environment guides you to feed more.

In order to choose a moral path for our life we need a compass that outlines a reliable, replicable directionality and not a compass-rose that changes its direction according to each trendy gust of wind, i.e. we need a preserved representation at the ideological level as well.

Feeding a brain on ideological memes is a vision we should aspire to while keeping in mind that defining an ideological meme is not that simple and will probably raise controversy. On the other hand, it might be that through defining by means of contradiction—i.e., by stating the things that are not moral in the universal sense—such as violence and racist hatred—we will be able to move forward and overcome these obstacles.

The pyramid of human knowledge becomes higher in an exponential pattern. Some claim it will reach a hyperexponential singular point where knowledge crosses a certain critical threshold. On the other hand, our life expectancy and the processing capabilities of our brain do not develop at a synchronized matching rate. Thus, we might anticipate an "information crisis" that requires overselectivity with respect to the core information we wish to assimilate in the brains of young people in an attempt to better their life and their understanding.

The brain, especially a young one, is hungry for guiding insights (show me the way!). In the absence of independent navigating attempts, man is carried on the waves of the ideas of people who are braver than him.

Emanuel Kant provided a definition for enlightenment when he wrote:

"Enlightenment is man's emergence from his self-incurred immaturity. Immaturity is man's inability to make use of his intellect without guidance from another. Have courage to use your own reason! That is the motto of enlightenment."

Kant—one of the most disciplined soldiers in the army of the intellect—took his courage into both hands, as he previously recommended, and forced upon himself a daily dose of strict "contemplation discipline," which yielded an impressive intellectual yield.

In praise of the originality regarding our mental products, it was said, "You are born an original, don't die a copy" (attributed to John Mason).

Our brains make us unique, and we should express this uniqueness for good by actions that contribute to a better world. No other brain's owner can give our special contribution, and vice versa.

We are in charge of shaping such a moral core that, when the owners of the young brains mature and reach their personal enlightenment stage, their brains will provide them with moral guidelines for respecting the other and humanism:

Always be generous toward those who "walk in the dark with us" (the rest of human beings), remembering that many of our weaknesses are universal, in the sense that they are written in the source code of the "being human" software. Judge people according to merciful justice rather than just justice, in the spirit of Mahatma Gandhi's saying "An eye for an eye will only make the whole world blind." These are values that will contribute to the shaping of a better "brave new world."

Thus, it seems that, in the near future, we will need to determine essential, universal core memes and expose our children's brain to them, in a selective manner, by means of universal core studies, which will be taught to all children of the world.

The Offspring of Thought and Flesh

We tend to get emotionally attached to ideas that we have nurtured by dedicating time and thought resources to them, as we do with our biological offspring, in whom we invest our best efforts. We wish to see the fruit of our investment in the generations to come.

In the codex of parenthood, the definition of "meme" must include instilling memes for life—memes whose usage will improve the offspring's success in the world and also meet the criteria of bettering the status of the people around us.

Reading and writing are young skills in terms of evolution, and, as such, they have not yet been granted brain areas that are committed to the processing of these skills from birth. In other words, it seems that there is not any brain area that is designated by a genetic dictate to serve reading and writing skills, as there is for the skill of speech.

A child who learns how to read goes through a shorter version of the process that human culture has gone through since the development of literacy. This is a central role of transferring core memes: shortcuts that will facilitate life's basic training for every young human being and instill life skills that match the era they live in in an enjoyable, fascinating, and effective way.

Words and memes have a chemical manifestation. They change the brain in the structural and functional sense.

Fervently, we might claim that every lesson at school is a minor brain surgery for the pupil's brain. It changes pupils' brain in the structural and functional (biochemical) sense. Thus, we should be very cautious when instilling worthy memes in young brains. For instance, the idea of solving disputes through violence requires initiated, constant extinction from young brains in order to end up with a young generation that is more tolerant and generous compared to former generations.

The neuroplasticity of the human brain, especially a young brain, enables continuous adaptability to changing circumstances. When used intelligently, it can be used for instilling memes that will better the life of individuals, the life of the society this individual belongs to, and the life of all humanity. Cynical use of its features, however, might create an epidemic of defective memes, such as coercion and violence.

A conceptual vaccine against distorted ideas is mandatory with respect to educating the young generation, and, similar to ordinary vaccines against infectious diseases, it requires a "booster" from time to time that will renew its vaccination-conceptual potency.

Just as it's appropriate to be concerned about the quality of our atmosphere, it's appropriate, also, to be concerned about, and act accordingly toward, the betterment, and the prevention of deterioration, of the internal atmosphere in the brains of the people who reside on our planet. As aforementioned, it is advisable that an organization, such as the UN, be in charge of selecting a "universal core of memes" based on respect toward human beings, animals, and the environment in a humanistic spirit, and that it also enforce its implementation. The internal climate in people's brains will determine the fate of our planet, no less (and perhaps even more) the environmental climate and the atmosphere. A society that brainwashes its youngsters with destructive memes, such as xenophobia based on background, nationality, religion, race, etc., must be forced to stop instilling these memes and start instilling humanistic, universal core memes, even if the only way to do it is by means of sanctions, since poisonous memes in the brains of youngsters will poison the general brain atmosphere in the only world we have.

The Ethics of the Heart

Emotions have a major role in making our ethical decisions, and their contribution often surpasses the contribution of logic.

Plato's utopian heart's desire was a world ruled by logic and, at its summit, a regime, motivated by pure wisdom, under whose wing the subjects would enjoy prosperity and justice.

Emanuel Kant coined the term "the categorical imperative," which refers to a moral decision as a natural inference of rational thinking.[43] It seems, however, that emotions, most of which exist in the unconscious processing layer, are the ones that guide most of the decisions that relate to morality, and the rational brain produces reasoning in retrospect for a decision that was made without its involvement, or only with a minor contribution on its part.

Various studies have shown that in scenarios in which participants are presented with the option of saving several people at the expense of the life of a single person, an interesting discovery is made. Most of the participants (in a cross-cultural pattern) prefer to save as many people as possible when they are able to act according to a pattern of non action or indirect action in order to promote such an outcome. In a scenario in which they are required to directly cause, with their own hands, the death of the person in order to save others, however, many of them will withdraw from their previous preference.[44] In brain-imaging studies that allow us to trace the level of activity in various brain areas at a given moment, it was found that the brain areas whose level of activity was in strong correlation with outlining a moral decision were the brain areas that are in charge of the production of emotions. In the first scenario, the areas in charge of emotions were moderately active and employed "cold," rationalistic considerations to determine the action approach. On the other hand, when the option of direct killing was raised, the emotional flame became very intense, which resulted in avoiding taking any direct action toward killing the person, even when the other people were about to pay the price for it.

It seems that once the moral conflict is resolved, an *a priori* coercion takes place within us; the lower path of the emotional brain makes the decisions and dresses them up in the festive attire of rationalistic explanations in retrospect.

Memes generate and regulate emotions, which constitutes another impact pattern of them on ethical decisions.

Ideas' Engineering

We all have core ideas that modularly compose the towers of our thoughts. The core ideas are the mental building blocks from which the tower of thoughts is built. Thus, it is highly important that we own the blocks that will enable us to modularly assemble a "good" tower of thought, as it is of the utmost importance that the right memes are instilled and that the mental environment delegates these memes. The implications are especially crucial with respect to children, who are in the midst of the process in which core ideas about the world are formed. By instilling an appropriate conceptual arsenal in them, we enable them to become better-ideas engineers. People who instilled in us memes that affected our insights are like living scaffolding in the construction of tower of our thoughts.

All of our brains serve as cement and building blocks in the human tower of insights.

The death of any person resembles a fire in a library—a library of ideas, which is unique and irreplaceable in this sense—thus the importance of transferring our selected insights to other brains, especially young ones, that will walk the paths of life after we are gone.

Memes' Arithmetic

The main part of our role is to instill in the next generations memes that constitute core memes—a concentrated summary of insights (experts' insights) that will make their "basic training" in life easier.

The fate of humanity will be determined by key vectors of the memetic universal parallelogram of forces—i.e., the total of all the memes that reside in the brains of living people at a given time and the direction of development they outline.

The Destructive Power of Morally Distorted Memes

Just as encephalitis (brain infection) might infect others and cause the death of the person who suffers from it, a morally distorted idea with a "dark spell," such as the concept of violent racism, might affect other brains and cause suffering and destruction to the people who adapted the idea and to the people around them (and, metaphorically speaking, cause "memetic encephalitis").

We must beware of being infected with memes that will turn us into "useful idiots" in the service of a "great ideology" based on distorted values.

Auschwitz existed in spite of Goethe, Leibniz, and Kant… Even a cultural climate that has been flourishing for many years is not immune to a cultural "ice age" that freezes basic moral values to death.

Ideas such as extreme nationalism or hatred based on race or religion are memes that induce apoptosis (planned death) of the entire human race, and we must create an effective immune system that will prevent "memetic apoptosis" and neutralize the effect of these toxic memes.

How Do You Say Snow in Eskimo?

You say it in dozens of different ways, according to the unique features of the observed snow.

Human languages are similar to each other due to the very similar physiological and mental traits among the various groups that compose human race. Special populations that live in climate zones that are characterized by "extraordinary" features—such as the Inuit, who live in areas close to the Arctic, and, alternately, the Tuareg clans, the wanderers of the Sahara desert—have developed unique linguistic features that reflect the "ecological niche" they live in.

There are memes that are valid in specific ecological or cultural niches and not in others (sectorial memes) and, on the other hand, universal memes that are valid among all people in the various time and space zones. Civilizations' clashes often derive from clashes of sectorial memes. One possible way to resolve such conflicts is searching for a universal meme that contains components of sectorial memes of both sides of the conflict.

Focusing the investment on core values in the shape of universal memes is the means that paves the way for improvement of humanity. Discrimination based on ethnicity is improper, but, on the other hand, discrimination based on memes is required and essential. Although cultural relativism adds a dimension of complexity to the characterization of the core values of human universal memetics, it must not prevent the distinction between a meme that is dangerous to the future of humanity, such as memes that discriminate on an ethnic basis, and memes that promote improvement and harmony between people. The meme of cultural relativism, which takes into account the uniqueness of the various human cultures, cannot provide an overall excuse for morally faulty behaviors. Examples for that, from the darkest end of the repertoire of behaviors that were accepted (some of which still exist) in certain human societies, are female circumcision and cannibalism. The future of humanity will be determined according to the memes that are common in the brains of human beings in our generation and the next ones. Thus it is critically important to create core memes (universal moral ideas) and instill them in our children. These will be the memes that will promote the future of the human race and will prefer universal considerations over sectorial ones. Beneficial memes might constitute "perpetuum mobile" of grace.

We should instill an anthropocentric scale of values, such that puts man at the center but is also generous and considerate toward the surroundings: animals, plants, and whatever else is around.

It is essential to design a human memetic codex that will be composed of beneficial universal memes whose carrier is not only contemporary, but a person of all times. Perhaps this wish is somewhat utopian and is not realistic enough, and a more realistic aspiration would be to create a core of universal memes that will be updated during each period according to the challenges with which humanity deals, some of which cannot be viewed from our contemporary point of view.

Aspects of Memes' Classification

Heuristics

Generalizations of all types are inclusive memes; although they commit the sin of generalization, they still serve as essential anchoring points for making decisions that grant us a quick leap in terms of our prediction ability regarding specific circumstantial scenarios. Nevertheless, sometimes they tend toward extreme stereotypes, which are overgeneralizations that are not supported by evidence.

Heuristics are a sort of general super-meme that outlines a fast-access lane for "solving problems," such as all sorts of "rules of thumb," but sometimes they contain built-in bias, which makes them less compatible to the world of phenomena around us.

Interface Memes

The replicability of "interface memes"—which restart the screen of consciousness in borderline situations in which the modes of consciousness change, such as the very first moments after we wake up in the morning, when half of our brain is still in the land of sleep and the other half is already awake—enables the replicability of perception and memory. Instead of building "basic world insights" from anew every morning, we can look ahead from the shoulders of all our past days, which provided us with "shortcut" insights about the world. The awakening confusion (the hypnopompic state) in the first minutes after waking up from a night sleep is mostly characterized by numbness, and then the basic perceptional memes (basic activating system) pop up on the screen of consciousness and restart our mental activity in a wakefulness "mode."

Detached Memes

Detached memes are these memes that do not find an anchoring point among the memes that already exist in the brain and do not manage to network themselves within the neural infrastructure that encodes information. In such a state, the thread of their life is short, and they are prone to falling into the abyss of oblivion.

Memes from Nature

Imitable memes do not only constitute some of the most successful tricks of memetics, they might also be a source of inspiration in terms of changing the purpose of a meme. Numerous memes exist around us and around natural appearances that surround us. An invention induced by memes coming from nature is called biomemetics. An invention in the spirit of biomemetics is ascribed, inter alia, to Leonardo da Vinci. The inspiration for the spiral staircase he thought of probably came from observing the internal skeleton of seashells.

The inspiration for the development of Velcro came from sticky plants. Other ideas in the spirit of biomemetics include a desalination method that imitates the way mangrove trees growing in salt water refine the water into fresh water; ideas for sterilization of operating rooms taken from the scales of Galapagos sharks, which are hostile toward all sorts of bacteria that live in the ocean; gathering dew from fog in arid areas by imitating the Namib desert beetle, whose body is a walking dew trap; and "artificial neurons networks": information processing apparatuses that imitate the information processing pattern of human brain cells.

Memes and Language

The treasure of language contains the concentrated wisdom of multiple generations. Language terms are creations of different brains. Language memes (i.e., its various terms) have shown an excelling survivability and competence in terms of patterning the world.

Language and culture in general are clusters of memes that survived multiple "brain generations" and have been upgraded and adapted to the world of phenomena. Language and culture in general are always in the midst of a continuous improvement.

Language, as a collection of memes, constantly gives birth to "mutations," as is the case with memes. A main expression of these mutations is slang, which frequently changes.

The birth of a lingual meme starts with the spring of thought and ends when the words are uttered in the "mouth's estuary."

The ease with which human babies acquire language makes us think, sometimes, that our brain is, among other things, a biological word processor.

The meaning of words mostly constitutes semantic knowledge that lacks personal shade but enables symbolic conceptualization of world entities, which is accessible to all.

The word that represents a certain entity is, most of the time, an arbitrary convention that cannot be inferred logically, except for a few words whose sound is onomatopoeic to the entity they represent. It seems, however, that there is a partial genetic tendency for a correlation between entity and sound, as was shown in the Bouba-Kiki study that was described in the chapter about perception.

The expression "different language, different soul" suggests that each language cherishes cultural memes.

The cultural load that is weaved into each language wires the brain of its speakers in a unique pattern.

Language reflects, in more than one way, the patterns that the human brain tends to produce.

Human culture has developed a number of conceptualization manners through symbols: alongside language, which is the main one, there is also music, mathematics, etc.

Language allows for reality representations that are in changing correlation with the true nature of things. Language creates reality representations, part of which are artificial creatures that are based more on language than on reality.

Our brain prefers narrative information to numeric information. Our brain likes letters more than it likes numbers. Words are usually loyal servants of memes. When they fail to be so, however, it can be frustrating, as reflected in the words of the author Gustave Flaubert: "Language is a cracked kettle on which we beat out tunes for bears to dance to, while all the time we long to move the stars to pity." Some see the limited conceptualization capabilities of language as a "prison of consciousness," as reflected in the resolute words of the writer José Saramago regarding the built-in limitations of words: "Human words are like shadows, and shadows are incapable of explaining light and between shadow and light there is the opaque body from which words are born."

Yet, it seems that we cannot disagree with the fact that innovative memes that are conceptualized through language sometimes open our eyes and expand our mind, in the sense of providing conceptual perception of layers of meaning that were previously concealed from the eye of consciousness.

A super-meme may be formed by a creation of a new entity out of the blue. A meme that is poured into a mold of words may change the way we look at the world. Words have an ability to create new worlds by turning the eyes of our mind toward a direction that was not examined in the past.

On the other hand, due to the importance of language in terms of our reality perception, whatever is not represented by language is prone to being shifted sideways to the outermost part of consciousness or, even, becoming totally concealed from the eyes of perception.

The categorical array of language structures our worldview, to a great extent, in the sense of "the limits of my language are the limits of my soul." Language creates distinction between details within the world image, which are sometimes artificial creatures that are not segregated entities by nature. These details are taken from the environment, selected by a linguistic tweezers, and their selection grants them "classified distinctness" that is sometimes justified, regarding their objective nature, and sometime is not.

In an attempt to bridge the language gaps between people, which make it difficult for the memes to pass from one brain to another, several super-languages were proposed. In the late eighteenth century, Wilhelm Leibniz started to develop a language that he hoped would become a global language. This language was named "Characteristica Universalis," and its script was based on mathematical foundations.

The Austrian philosopher Otto Neurath proposed a picture-language that he hoped would turn into a global language, and he expressed its basic concept as "Words divide us, pictures unite."

Chapter 12
Unity of Contradictions:
Two Minds, One Brain—the Hemispheres

Development of the Hemispheres

The right hemisphere develops more rapidly during infancy. It experiences an accelerated growth period, which starts when the baby is born until about the age of two and enables the brain to establish perception skills from the very first time it encounters the world. The left hemisphere develops more slowly and keeps changing—even long after birth. Some claim that, as our brain gets older, the effect of the left hemisphere on its conceptual and linguistic capabilities grows, and its inhibitive effect on the right hemisphere becomes more intense.

There are cases described in medical literature in which an entire cerebral hemisphere is injured and the affected person keeps on living with only one functioning hemisphere. An example of that is a baby girl whose left hemisphere, due to a problem in the left common carotid artery, which constitutes a central source of blood supply to half of the brain, almost did not develop while she was an embryo, and, consequently, only her right hemisphere was functioning when she was born.

This patient accomplished, through self-training, some impressive achievements with regard to remembering details. Abstract thinking, however, such as interpreting proverbs, was difficult for her. In addition, she suffered from a prominent impairment in her right field of vision and weakness on the right side of her body.

Other examples are known from the cases of patients who underwent "hemispherectomy"—a broad, rare operation in which a whole hemisphere (left or right) is removed. In the extremely rare cases of people who function with a single hemisphere, the questions that are asked include which functions survive when the cerebral living space is reduced by half? Will complex abilities, such as rich verbal expression, distinguishing between nuances of emotions, and the like (in accordance with the injured areas), survive? And "how much brain" is needed in order for a person to "be a person"?

Structural and Functional Characteristics of the Hemispheres

The range of structural and functional differences between the two hemispheres is broad and stretches from the micro level to the macro level.

A central structural difference between the two cerebral hemispheres, which also has functional implications, is embedded in the connectivity pattern, which is different among the different areas within the hemispheres: the neural channels in the right hemisphere are more prone to connecting remote brain areas, while in the left hemisphere they are more prone to connecting adjoining brain areas.

Another important difference derives from information processing processes in the two hemispheres: the right hemisphere is more prone to "multichannel" information processing, while the left one is more prone to "monochannel" information processing. The right hemisphere prefers processing at the areas of the multisensory associative cortex (multichannel), which merges perception impressions from several senses into a single coherent experience. The left hemisphere gives priority to processing at areas of the cortex that are devoted to monosensory input—i.e., the information that is mediated by a single sensory organ, such as the auditory cortex to which the auditory input is channeled from the ears.

It may be that the inclination of the right hemisphere toward multichannel processing gives it an advantage when dealing with new information and of serving as the preferred home port for sailings to the land of imagination.

In the service of the transportation of the signals that sustain the interhemispheric dialogue, there are "latitudinal roads" that are composed of clusters of axons, which mediate transfer of information between the two hemispheres, such as the tracks called the corpus callosum and the anterior and posterior commissure.

Among two-thirds of left-handed people, the hemispheric expertise profile is similar to that of right-handed people. Among one-third of the left-handed people, it seems that the hemispheric expertise profile is inverted, as in a mirror image.

Typical Thinking Skills of the Hemispheres

We have two hemispheres and multiple opinions about them:

From the mid-1970s, the common division of the brain was according to features of hemispheric expertise. Abstract and symbolic skills, such as language and arithmetic, are traditionally ascribed to the left half of the brain. The right half was traditionally viewed as the residence of nonverbal skills, orientation, and imagination. The left hemisphere was characterized as rationalistic, logical, and as the half that prefers verbal conceptualization. On the other hand, the right hemisphere was characterized as processing information in a holistic pattern and as attentive to emotions and spatial information.

Another common assumption is that the right brain excels in parallel processing—i.e., simultaneous processing of pieces of information that are merged into a coherent insight. Serial processing is usually ascribed to the left brain. It seems that both hemispheres have the ability to process information in a serial or parallel pattern, but they have different tendencies.

Another common characterization sees the left hemisphere as the one that contributes the most to self-identification and the hemisphere that generates self-awareness, whereas the right hemisphere is considered to be the main contributor to identification of others.

There is evidence that suggest that, while an experience is being processed, the left prefrontal lobe is in charge of gathering the various details while the right one is in charge of extracting core ideas out of the whole experience (separating the wheat from the chaff) and getting the "point."

The left hemisphere is more oriented to creating compatibility between the perception impressions and the patterns (the memes) that are stored in the hemisphere, and from time to time the "Baron Munchausen" arises within it and fabricates "evidence" in order to create the desired compatibility.

It seems that the left brain has a central role in connecting the patches of our being into a unified quilt, a coherent being.

The right prefrontal cortex is the core of our brain's crystal ball, or the womb from which our foresight is born. This area produces continuous prediction in terms of potential future scenarios. The prediction capacity of people who suffer from impairment in the functioning of the right prefrontal lobe is damaged.

The separated self, in each of the two hemispheres, has a tendency to hope—from the left—and a tendency to color reality in a gloomy shade—from the right. Cynics might say that hope is not a strategy, but it is important in terms of motivating mental processes and in terms of mental resistance. Truth in itself, when it is not diluted by hope, can resemble a bitter medication. The human goblet of the soul is filled with straight interpretation of reality along with a certain amount of hope. The mix between the two varies and depends on the circumstances. Pouring hope into the cocktail of reality perception is sometimes essential in order to create continuous inner mental motivation. The hemisphere that inspires hope—i.e., the left hemisphere—often moderates the tendency to give strict interpretations to most reality manifestations reflected from the eyes of the right hemisphere. Our reality test eventually derives from this inter-hemispheric dialogue.

Reflection of the Hemispheric Mix in our Mental Sphere

It seems that there is no stage in which one of the hemispheres controls our reality perception exclusively. There is always a changing mix of right reality perception and left reality perception, and at any given time one of the hemispheres, and the reality perception it produces, holds the reins, though not exclusively.

Since there is a difference in the rules according to which information is organized and processed in each of the hemispheres, we, in fact, get, at any given moment, two different reality reflections. These reality reflections might sometimes be very different from one another. It seems that our different view of a similar situation, even within short periods of time, is caused by a change in the mix of reality perception in the power game between the two hemispheres. The right hemisphere tends toward a more reliable reality reflection, which is often more pessimistic, and serves as a greenhouse for doubts (often healthy doubts). The left hemisphere weaves an iota (sometimes more) of fiction into the reality reflection and tends toward a more comforting view of the world.

Personal Cognitive Style as a Reflection of a Unique Left-Right Mix

We all tend toward an information processing style that relies on both hemispheres in different mixes. The right hemisphere is more dominant among some people, whereas the left hemisphere is more dominant among others. The relative mix of the usage of both hemispheres is the element that creates the personal cognitive style. As years pile up, it seems that the general tendency tilts more toward the left hemisphere, but even with this general tendency in the background, the unique thinking style is retained and affects the pace of the inter-hemispheric mix changing in the aspect of information processing.

In general, though generalizations are far from being perfect, we might say that the ones who are more prone to taking the right-brain view tend to see the differences, while those who are more prone to taking the left-brain view tend to see the similarities.

There is no doubt that each of the two hemispheres contributes some unique insights, and that they reflect the various faces of reality. The combination between them enriches and deepens world insights.

Areas of Happiness and Sadness—Emotions "Take Sides"

There is a common supposition among brain researchers according to which the left frontal lobe is the main source of the formation of positive emotions, while the right frontal lobe is the main source of the formation of negative emotions along with continuous interaction with the amygdala structures at both sides.

A person who suffers from damage to the left frontal areas is more prone to depression and to a decrease in initiation and inner motivation. This probably derives from the fact that the right frontal hemisphere, which is where the gloomy knight lives, becomes the main generator of this person's emotions. Damage at the right frontal area promotes a tendency to be in a good mood, which might go as far as mania.

This is also the source of the strange expression "happy stroke," which refers to damage to the brain tissue of the right frontal lobe, mostly due to ischemia (damage caused by a reduction in blood supply to the tissue, followed by a lack of sufficient amount of oxygen and glucose). The term "happy stroke," which sounds oxymoronic in this context, derives from the fact that the person who suffered the damage often underestimates the severity of the consequences of the stroke and seems relatively calm.

On the other hand, "sad stroke" refers to ischemic damage in the brain tissue of the left frontal lobe. The word "sad," in this context, refers to the mental expressions of such a stroke, such as depression and lack of hope and purpose with regard to the consequences of the stroke and the chances of recuperation and improvement.

Support for this common hypothesis can be found in the "Wada test," which is a neuropsychological test named after neurologist John Wada, who invented it.

This is a reversible pharmacological process that is used for functional evaluation prior to brain surgery, mostly for those who suffer from epilepsy with seizures that are refractory to medical treatment. During this procedure, one-half of the brain is "deactivated" by anesthesia through means of selective injection of a short-term anesthetic substance (sodium amytal). This substance is channeled through the carotid artery on each side to the target hemisphere, and, later, once the anesthesia fades away on one side, the other half is deactivated. The active half of the brain (the right and the left alternately) is required to perform verbal tasks, mostly in order to assess its compensation potential in case the other hemisphere is operated on and becomes damaged. During the procedure, the patient becomes the owner of a single active hemisphere, a different hemisphere alternately.

When observing the patients who underwent this procedure, it was found that the person in the right-brain version, when the left brain was deactivated, had extreme speech difficulties, to the point of muteness, and his mental state was painted in dark shades. On the other hand, the person in the left-brain version, when the right brain was deactivated, tended to be talkative and reflected a vernal and cheerful inner emotional climate.

Additional support can be found in magnetic stimulation through TMS (transcranial magnetic stimulation). When this method is used to trigger activity at the left frontal lobe, it improves the mood of depressed patients; on the other hand, triggering activity at the right frontal lobe affects the mood of these patients in a negative way.

According to common assumptions, the activity of the left frontal lobe is the main generator of the sense of "feeling good." This feeling is induced by flooding the brain with the nectar of gods—a cocktail that mainly contains the neurotransmitters dopamine, serotonin, and a touch of endorphins.

A step in the shoes of the right hemisphere echoes as a requiem, contrary to a step in the shoes of the left hemisphere, which echoes as a jolly march.

Processing of nonverbal information, mostly in a social context, usually takes place in the right hemisphere. For instance, prosody, the tone that accompanies words (the melody which it sounds the words are said to), is processed at the right hemisphere. Damage to the right hemisphere might lead to impaired social behavior that does not distinguish between nuances, or a tendency to be clumsy and inattentive to shades of emotions. Such behavior is characterized by a lack of social sensitivity (which is also referred to as "social awkwardness").

The right amygdala is in charge of adding intense somber shades to the picture of the soul. It resembles a fountain of grumpiness. Soul doctors as cartoon characters would certainly diagnose Grumpy—the grumpy dwarf in *Snow White*—as suffering from hyperactivity of the right amygdala. This area is like a generator that induces grumpiness and gloom, which cast their long shadow over the inner world of each and every one of us (and, just as in the sleep spell in *Snow White*, a kiss might erase their depressive effect here as well).

On the other hand, the melodies of minor-scale, sorrowful tunes are the emotional soundtracks produced by the harp of the left amygdala.

People who suffer from depression tend to lie less (to themselves and to others). It seems that this reflects the intense activity of the right frontal hemisphere in their brain, which induces sadness but also (over) realism and tends to perceive things as they really are.

Often the reality-as-is potion is too bitter for the ego, and a certain dose of self-deception (induced by the left frontal hemisphere) constitutes an essential sweetening that enables us to sip the potion of reality.

Pouring into the Glass of the Soul

The emotional disposition of the hemispheres also derives from the different mental worlds they create.

It can be assumed that there is a connection between the cognitive-operation profile that characterizes a person and his "emotional personality"—i.e., the manner in which he expresses emotions.

A person's temperament probably derives from different operation patterns of brain areas. The left hemisphere pours sweetness into the glass of the soul, while the right one adds bitterness. Those who are naturally prone to feeling "joie de vivre" are blessed with more intense activity of the left frontal lobe, whereas the ones who are prone to feeling glumness suffer from intense activity of the right frontal lobe. Our "collection of moods" is the mix of the activity in both hemispheres, which results in the "bitter-sweet" taste of life.

The U-Curve of Happiness[45]

In a telephonic survey that was conducted in 2008, thousands of men and women between the ages of eighteen and eighty-five were asked about their feelings regarding different aspects of life.

It was found that a sense of sadness increased in a gradual pattern, from a low level at the age of eighteen to a peak at the age of fifty, then decreased consistently from the age of fifty to about the age of seventy-five, on average, and then increased again, a little more moderately, until the age of eighty-five.

The sense of gratification and happiness was reported as being high at the age of eighteen and, from this point on, decreased consistently and moderately until it reached its lowest point at about the age of fifty. From this point on, the trend changed for the next twenty-five years (until about the age of seventy-five), and the participants reported gratification levels that increased consistently and that, at a certain point, started to decrease again in a moderate manner.

Another study, in which two million people in eighty countries were surveyed, shows a similar pattern that does not depend on gender or geography: we are at the lowest point with regard to satisfaction from life in our midforties, and the levels of gratification are highest close to both edges of life—during adolescence and old age.

It is possible that the U-curve of human happiness, with respect to age, has a universal pattern that is not culture dependent. Nevertheless, of course, it does not mean that this pattern is necessarily true at the individual level. It is a statistical statement whose validity is questioned by some "happiologists" (mental-welfare researchers), who stress that different life circumstances of people prevent us from considering the mere chronological age as the dominant marker of the internal emotional climate.

Taking all reservations into consideration, is it possible that during the fifth decade of our life, the right frontal hemisphere sings its swan song in our emotional lake and then, gradually, the more illuminated shades of the left frontal hemisphere are reflected in our emotional lake as a mental compensation mechanism that is intended to help us cope with the hardships and ailments of old age?

Interaction Between the Hemispheres—Unity of Contradiction

The lasting duet between the right hemisphere and the left hemisphere creates the song of our life. Sometimes there is constant averaging of the two voices of the brain, while at other times one voice surpasses the other. This duet often includes a thread of critical dialogue.

There is considerable similarity between the two hemispheres of our brain, but, still, the different processes of information processing that takes place in each one of them turns brain unity, as a result of the close collaboration between the two hemispheres, into a unity of contradictions. A constant and critical dialogue is conducted between the two hemispheres through axons that transfer information between the hemispheres, mainly through the cluster of axons in the brain core, which is called the corpus callosum, as well as through the anterior commissure tracks and the posterior commissure tracks.

The duopoly of the left and right hemispheres exists in a liquid status quo.

The mutual relations between the two halves of our brain take place as a critical dialogue. The left brain is the generator of theories and one that preserves the "brain constitution," which was written based on familiarity with the world manifestations. The right brain acts as a fighting opposition, which often questions the validity of the "basic laws" embedded in our brain.

The right hemisphere performs constant reality testing and challenges the suppositions of the left hemisphere in cases in which they do not match its perception of reality.

Some might claim that the constant dialectic process, which is at the basis of science and which claims lasting tattooed attempts of the "wall of theory" by means of new hypotheses that wish to be closer to the truth of reality, is a creation of the right hemisphere.

The pattern of intimate relationship between the right brain and the left brain is usually similar to an open marriage. The hemispheres live under one roof, and there is constant interaction between them, but, at the same time, each one of them flirts with reality manifestations in its own unique style.

Sometimes the relationship between the hemispheres suffers from a crisis, in a pattern that might remind us of Jonathan Swift's book *Gulliver's Travels*. This book describes a hybrid brain that was made of two brains of two people in an attempt to resolve a political dispute. One-half of the brain supports a certain political party, while the other half supports a rival party.

A similar idea might be a reciprocal switching of a single hemisphere between two brains of two human beings, so that the skull of each of them hosts half of the original being and half of the "foreign" being, and the two halves try to integrate in the mental sphere within the skull—a sort of mental heterogeneousness that creates a chimeric consciousness. We can also imagine, in our mind's eye, how the creatures of a brain that is composed of Einstein's left hemisphere and van Gogh's right hemisphere would look—and, alternately, the left hemisphere of Marie Curie paired with the right hemisphere of Helen Keller or, in a case of intergender heterogeneousness, the left hemisphere of Agatha Christie with the right hemisphere of Salvador Dali.

The Night Edition of Brain Hemispheres

It seems that during night sleep, the balance of power between the hemispheres changes alternately, which results in changes in the balance of power between their typical realities perceptions.

When we are able to identify the "times of left and right" in our own brain, we have the advantage of understanding the structural bias of our reality perception at a certain moment.

While dolphins sleep, their brain hemispheres take shifts of alternate sleep and wakefulness. Thus, at any given time, one of the hemispheres is awake while the other is asleep, due to the constant need to supervise breathing, which is done by rising above the water surface.

It seems that during the various stages of night sleep, there is an asymmetrical pattern of hemispheric activity in which the hemispheres take turns holding the wand of affecting brain activity (this hypothesis is based on self-observation). There is no such thing as complete shutdown or full activation of the hemispheres, as is the case of the dolphins' hemispheres, but, at each of the stages of sleep, a different pattern of hemispheric control is formed. In other words, at each of the stages there is a different pattern of information processing that tends more toward the right school or the left school alternately, when one of the hemispheres is more active than the other. The switch between the hemispheres, with regard to the firm grip on the wheel of information processing, is based on shifts and timings that are typical of the specific person. Certain hours are characterized by right dominance, while others are characterized by left dominance. I have learned from personal experience that when I wake up, or when I am half asleep at about three a.m., my thoughts are painted in dark shades, which can be defined as hyper-realism that lacks comforting shades, and the contents of my thoughts are autumnal/wintery (an emotional and cognitive climate that is typical of right-hemisphere dominance in which the right amygdala has a central role). Two hours later, my personal mental space is filled with comforting light, and my thoughts are characterized by more positive contents (though often not more "realistic"). The mental spring blossoms in a pattern that matches left-hemisphere dominance.

Thus, I tend to think (in the spirit of induction) that the seasons of the soul exist 24/7, during night or day, in a cycle whose exact timings are probably unique to every person, but it may also be connected to a universal, human pattern.

The Familiar to the Left, the Unfamiliar to the Right Hypothesis

An innovative hypothesis explains the duality in the roles of the hemispheres by assuming that the right hemisphere is more prone to processing new information and the left one is more prone to storing the archive of insights that were already processed and acquired. The right hemisphere tends to serve as the "insights producer," whereas the left one tends to serve as the residence in which the treasures of memories and insights we have collected throughout our lives are stored. In other words, the right hemisphere prefers to process unfamiliar information, whereas the left hemisphere prefers to manage familiar information. This generalization is true for most types of information processed by our brain. This hypothesis was supported by observations that were made among people who suffer from damage in the right hemisphere who demonstrated a desire to avoid new experiences.

The right lobe is the main generator of insights. The left lobe is the chief curator of insights. New reality insights (memes) tend to be created in the right hemisphere. It might be due to its ability to pattern new information into its material infrastructure as "structure of knowledge"—in the shape of a three-dimensional "cloud" pattern of interconnected neurons. On the other hand, the left hemisphere, according to this supposition, is "less fluid" and tends to preserve its "structures of information" and create new structures of information less frequently, compared to the right hemisphere.

The right hemisphere resembles a weariless cartographer who is constantly busy "sketching" new brain maps, while the left hemisphere serves as the archive of maps that have already been validated and found suitable for navigating the paths of life.

The right hemisphere and the left hemisphere can be referred to as "the Magellan hemisphere" and the "there-is-no-place-like-home hemisphere" accordingly. The right hemisphere is the one that processes the new and the unfamiliar. It pursues the different and the exotic, and its horizons wish to reach beyond the present horizon of thought. The Portuguese seaman Ferdinand Magellan, who circled the world, summarizes central components of its nature. On the other hand, the left hemisphere groups a cluster of insights for coping with situations in which we have gained experience, and it usually channels our consciousness toward the familiar and routine. A possible motto for it is "there is no place like home."

There is a dynamic balance between the innovative insights, which the right hemisphere tends to create, and the conservative insights, which the left hemisphere tends to hoard.

The conservatism of the left intends to protect our knowledge of the world, but it sometimes compromises the update of this knowledge.

Language and the Hemispheres

The surface of reality is covered in a fog of uncertainty. Words resemble a retreat toward certainty—conceptualization of terms we became familiar with through our experiences in the world.

Words are also capable of creating reality; this is one of the common interpretations of the magical decree "*abra kadabra*" - in ancient Aramaic "*abra kedabra*" which is: "let it create as I say". Word as working magic that create reality by mere words.

The main, though not exclusive, impresario of the words' show is the left hemisphere, which is the residence of the familiar. However, the saying of the seventeenth-century French neurologist Paul Broca, who said that man speaks through the left hemisphere, is only partially true.

The assumption regarding total hegemony of the left hemisphere over language has been cracked recently.

New findings refute the supposition regarding verbal blindness of the right hemisphere. It turns out that innovative creatures of language, such as metaphors, multiple-meaning words, puns, and layers of meaning, are first processed in the right hemisphere.

Moreover, the right hemisphere has an important role in understanding poetry, which is full of innovative idioms that can be linguistically challenging.

It also seems that language acquisition during childhood follows a rule, according to which new verbal information tends to be processed by the right hemisphere. Since the sounds and words of language are new to a child's brain, the main entrance of language into the halls of the brain is through the right side. On the other hand, once language is acquired, its repertoire of familiar terms is, in fact, a collection of generic patterns that tends to be hoarded, like any other type of familiar information, within the left hemisphere.

The traditional view regarding hemispheric function, which ascribed all language functions to the left hemisphere, is inaccurate, to say the least, and some might even play a requiem for it.

Shoals at the Left Bank of the River of Language

The incidence of dyslexia (a disorder related to reading development among children) among left-handed people is significantly higher than among right-handed people. It might reflect damage to the left hemisphere during early stages of development, which led to selecting the left hand as the dominant hand. The hemisphere that controls the movement of the hand is located opposite side to it, since movement control pathways are crossed.

"Aphasia" (language-function impairment) is also mostly caused as a result of damage to the function of the left hemisphere. The rare condition called "hyperphasia," however, characterized by a tendency to be overtalkative and mechanically pronounce long successions of words, is mostly detected among people who suffer from a rare brain syndrome that is named after Williams and is related to a left hemisphere that is bigger than normal.

It was found that women have greater brain flexibility with respect to preserving language functions, which is also reflected in higher compensation ability at times of injury to the main language areas in the brain. Among women who suffer from damage to the left hemisphere, language functions are usually impaired less severely than those of men who suffer similar damage.

"Morphing" is a term that describes a merger of two distinguished, separate patterns into a single pattern that preserves the characteristics of the two patterns. An experiment that included the use of morphing showed findings that might confirm the hemispheric expertise hypothesis. During the experiment, a gradual merging of the face of the participant with the face of Marilyn Monroe was conducted on a sequence of chimeric portraits on a computer screen. The profiles' contribution to the common portrait varied. Afterward, a supervised, selective anesthesia of the right hemisphere in the patient's brain was performed by injecting short-term anesthetic (sodium amytal) to the right carotid artery, and, as a result, the left hemisphere was entrusted with the processing of visual information. When the new portrait was presented to her, the patient failed to identify her own face within the merged portrait but succeeded in identifying the familiar features of Marilyn Monroe. On the other hand, when the left hemisphere was put under anesthesia, and the wand of visual information processing was transferred to the right hemisphere, the patient identified her own features as well as the features of Marilyn Monroe. Possible interpretation that is compatible with the "familiar and unfamiliar" hypothesis is that it seems that Marilyn Monroe's features constitute familiar knowledge, and our own face, when merged with the features of another human being, constitutes a novelty that requires the involvement of the right hemisphere.

Gender Differences in the Brain-Activation Profile

Although this hypothesis constitutes a generalization, which naturally involves faults, and there is no dispute over the fact that any person, man or woman, in our world is unique, particularly as a result of the uniqueness of their brain, some claim, as an educated guess, that in an average male brain certain behavioral patterns will be more prominent. These patterns are usually ascribed to the right hemisphere—for example, the pursuit of novelties, daring, and the urge to wander. According to this supposition, behavioral patterns deriving from the right amygdala, which mediates urges and impulsive reactions in a more prominent manner, will be found in a male's brain. This hypothesis is controversial, and its evidence base does not suffice in order to see it as an "irrefutable truth."

The Similar and the Different

The reciprocal relations between the two hemispheres—the right, as the one that distinguishes, stresses the different and the new; and the left, as the one that finds similarities to the familiar, includes and tilts toward the drainage basins of familiar insights—are, according to common hypotheses, at the basis of our cognitive products.

Generalization and uniqueness refer to humans, as well. Although all human beings are made according to a similar physiological format, which also leads to a similar psychological format and to numerous similarities in terms of cognitive processes, on the other hand, each person creates a unique emotional and cognitive climate in the planet of his skull, which constitutes a unique ecological niche.

Some might claim that the mix of the right style, which is about finding the difference, and the left style, which is more about detecting the similar, is evident not only with respect to "passive perception of stimulation"—a reflection of the world of phenomena by the sensory organs—but also with respect to producing original thoughts, which are born in our inner world.

Sayings that resemble Koheleth's saying "There is nothing new under the sun" can be seen as reflecting the insight of the left hemisphere; the words were probably said by Koheleth at an old age, at a time when, through the lenses of the left hemisphere, many things seem as variations on the same theme.

The spiritual heritage of Charles Darwin, who was responsible for one of the most shaking revolutions of thought in human history, might serve as an example of the advantages that are part of seeing the similar—i.e., by looking through the left hemisphere's lens.

For years, naturalists have been busy with finding more and more species of animals and plants while emphasizing the difference between them. Semiology became very popular when more and more species and subspecies were added and marked according to the slight differences between them. Right-hemisphere-oriented thinking ruled. On the other hand, Darwin noticed the similar, the similarities between the various forms, which was the source of his insight regarding the organizing principle of the development of life. We might say that Darwin's right hemisphere, the explorer of the unknown, was responsible for his decision to set sail on the *Beagle* for the other part of the world and spend five years on board. In the process of consolidating his insights throughout the years, however, it is the left hemisphere that seems to have had the upper hand. His insight ranged from one end, of the uniqueness of details, to the end of "common ground."

Darwin published his book *On the Origin of Species* at a relatively old age, perhaps as a reflection of the left hemisphere approach that matured in him.

A broom—is it a device for moving dust or for moving witches? An interesting assumption is that creativity is enabled due to the availability of structural infrastructure that is prone to plasticity and the creation of new neural networking, which encodes new information, probably mostly in the right hemisphere. The left hemisphere tends to connect new concepts in a concrete "earthy" manner and usually does not sail to the remote lands of imagination. It seems that it is prone to less flexibility with regard to structural infrastructure that encodes information. The left hemisphere is likely to relate to a broom as a "device for moving dust" from one place to another, whereas the right hemisphere is likely to relate to it as a "device for moving witches from one place to another."

Age-Dependent Changes at the Hemispheric Activation Process

As we become older, our brain goes through reorganization that is characterized by both advantages and disadvantages.

A young brain is more prone to an asymmetrical activity pattern, in which case the right hemisphere is more active. The brain of a middle-aged person is prone to a profile of activity that involves more brain areas compared to a brain of a young person. It contradicts the hypothesis, which was common in the past, according to which a middle-aged brain is less active. The middle-aged brain tends toward a more symmetrical pattern of activity, which involves the left hemisphere more but does not neglect the right hemisphere altogether. It seems that such a pattern of activity makes it easier for emotions and common sense to live under one roof and integrate with one another.

The Hemispheric Pendulum Movement

The view that links the movement of the cognition's center of gravity with the emotions' center of gravity from the right hemisphere to the left hemisphere in the course of our life might serve as an explanation for the changes in the operation pattern of the brain, which seems to match the insight that is common in the field of geometry regarding a necessary and sufficient condition. This view also reflects the spirit of the principle of simplification called "Ockham's Razor," according to which the simple explanation that explains the nature of things appropriately is better than a more sophisticated explanation that explains them.

As aforesaid, as people grow older a more cooperative operation pattern is noticed in the brain, such that involves the two hemispheres. Nevertheless, this pattern reflects more usage of the left hemisphere compared to its use during adolescence. According to this hypothesis, each new acquired pattern is born at the delivery room located in the right hemisphere. When it becomes more mature, as a result of constant encounters with the facts of life and critical dialogue with other insights that exist in the living space of the brain, it moves, with due respect, to the residence of the well-established, stable insights located in the left hemisphere. The left hemisphere is the residence of generic insights—the descriptive and guiding ones.

Defining the hemispheric expertise profile does not involve setting strict limits regarding the contents of the pattern. An information pattern that is considered a novelty today will become a familiar pattern tomorrow. The difference relates to the level of familiarity between the brain and the contents of the information pattern at the time of their encounter.

The movement of the mental pendulum from right to left is probably a phenomenon that is at the core of all learning processes. Nevertheless, it does not mean that as we grow older we abandon the right hemisphere. The operation pattern becomes more cooperative, however, and more room is given to the left hemisphere.

Once the quota of patterns stored in the left hemisphere crosses a certain threshold, and "world knowledge," both descriptive and guiding, has been accumulated sufficiently, the owner of the brain that contains these patterns is able to find his way around, even if the function of the right hemisphere is impaired. He does so based on the reservoir of patterns in the left hemisphere and increasing usage of an approach of patterns identification for problem solving.

Social intelligence prefers the right; various cognitive skills differ from one another in terms of the pace at which the patterns that are related to them wander to the left hemisphere, and there are also some that remain mostly the estate of the right hemisphere.

This is the case with social skills, which are sometimes referred to as "social intelligence." It seems that the spectrum of nuances and variations in the shades of our social world is so broad that they cannot be minimized into a definite number of core patterns. Thus, the main hemisphere that carries the burden of managing our social behavior is the right one. It seems that detecting the "emotional weather" among our fellow humans tends to challenge the familiar patterns, and, similar to real weather, it involves novelties time and time again.

It is generally assumed that damage to the right hemisphere leads to more destructive consequences among children compared to adults, and at the other end of the spectrum of ages the situation is inverted. Among adults, damage to the left hemisphere involves more severe impairments compared to those among children.

Findings that support this hypothesis were found, inter alia, in studies that were conducted among children who underwent hemispherectomy—a broad, radical surgery during which a whole hemisphere, or the lion's share of a hemisphere, is removed and the patient is left with only one functional hemisphere. Usually such an operation is carried out in an attempt to improve the condition of patients who suffer severe epileptic seizures that cannot be controlled by medications, and the focus that generates them is located within one of the hemispheres.

Regression in brain plasticity, which takes place as we become older, might reflect a withdrawal from the right hemisphere, which seems to be the chief generator of brain plasticity to the fortifications of the left hemisphere. As we climb up the mountain of the years and grow older, the center of gravity of our mental life shifts to the left. As years go by, we all tend to become "left-brained."

Constant learning hinders the withdrawal of the right hemisphere from the stage of our cognitive world. Constantly dealing with the new hinders the natural move of the cognitive pendulum, which tends, with the years, to spend less time in the right hemisphere and to prefer brain activity in the left hemisphere.

The Hemispheric Activation Profile as Reflecting a Culture

The findings from a series of experiments suggest that students who were raised in western societies tend to see entities as isolated, which can be seen as preferred observation "through the eyes" of the left hemisphere. On the other hand, students who were raised under the auspices of Japanese culture tend to perceive an object as relating to other entities, a tendency that is characterized as observation "through the eyes" of the right hemisphere.

The tendency toward uniqueness and individualism is probably a cultural suggestion. This tendency is common in the western world. In eastern Asian countries, however, conformism is common, which is probably a cultural suggestion as well. In this spirit, we can interpret the Japanese proverb "The nail that sticks out shall be hammered down" as a recommendation to adapt oneself to the society.

In this context, the traditional hypothesis, which ascribes analytical skills and serial, reductive processing to the left hemisphere and parallel, holistic processing to the right hemisphere, is raised. Comparing the interpretation of world manifestations to common cultural interpretation raises an interesting possibility of cultural preference for a certain hemisphere's skills over the other hemisphere's skills as being culture dependent.

These differences in perception are acquired rather than innate. This assumption is supported, inter alia, by the fact that immigrants from a certain culture to a different culture learn "to perceive things differently" according to the perception pattern that is common in the culture to which they immigrated. It does not mean, however, that a certain perception is better than the other. There are circumstances in life in which a certain perception is more suitable, and other situations in which the other perception is preferable.

Expressions of Impairments in Hemispheric Function

Dr. Roger Sperry, who won the Nobel Prize in Medicine in 1981, based on the research he conducted among people with "split brain" (those who underwent an operation in which their corpus callosum, which connects the two hemispheres, was cut, mostly due to epileptic seizures that could not be controlled by medications), believed, based on his studies, that each hemisphere has a mental sphere of its own. As he once put it, "The joys and senses of the right half of my brain are too great for the left half of my brain to put into words."

In his opinion, the mental sphere of each of the hemispheres is where the perception processes, which are totally different from the perception processes of the other hemisphere, take place. This hypothesis can be defined as "mental duality"—two minds live within one brain or, alternately, two souls share a single boardinghouse.

It seems that the cutting of the corpus callosum leads to flattening of the overall intensity of the experience of perception impressions. This suits the saying "The whole is bigger than its parts." The overall experience is bigger than the sum of its processing in each hemisphere separately. Processing an experience in the "two different minds," from the left and from the right, adds a dimension of depth; on the other hand, cutting of the connecting pathways between the two hemispheres flattens the depth of experience.

Why Did I Laugh?

During an experiment that was conducted on a person whose corpus callosum was cut ("split- brain"), a request, written on a piece of paper, was presented to this person: "Laugh!" The message was presented only to the left field of vision, which channels visual input to the right hemisphere. The patient started laughing in response. When he was asked to state the reason for his laugher, however, he fabricated one. The request to laugh was absent from the visual world of the left hemisphere. Due to the cutting of the main channel of communication between the hemispheres, the corpus callosum, the only addressee of the request, was the right hemisphere. It seems that in the absence of information, the left hemisphere came up with a fabricated story to explain the subject's laughter. Brain researcher Michael Gazzaniga believes that the right hemisphere is the ideal witness; it always tells the truth and nothing but the truth and does not "fill in the gaps" in our perception with false "findings." On the other hand, the left hemisphere tends to ease the discomfort that is involved in partial understanding by creating fictional reality details that never actually happened, although they sound plausible and appropriate.

Fabrication From the Left in Action

In an experiment that was conducted among subjects who were questioned after going through hypnosis, it was found that, when they woke up, they did not remember undergoing conditioning, in which a certain command by the hypnotizer led them to perform a certain action, such as retying their shoe laces. When the sentence was said and the conditioning worked and led them to tie their shoelaces, they explained it by providing reasons that seemed plausible, such as the long walk that awaited them until they got home, and their wish to make sure that their shoes were up to it, by using ad-hoc rationalization. It seems that the left hemisphere came up with an appropriate "cover story," since it was unaware of the hypnosis conditioning.

Imprisoned Melody

The stroke that composer Maurice Ravel suffered in the summer of 1933 took away his god's gift of composing music. He lost the ability to conceptualize his musical ideas in the language of notes. The ability to listen to music and be moved by it was not taken away from him, however, which was comforting in a sense. We might say that his injury was like a mirror image of the deafness of Beethoven, who lost the ability to listen to music but whose composing ability remained intact.

A plausible assumption is that Ravel suffered damage in his left hemisphere, which conceptualizes our thoughts as symbols and turns musical ideas into formal symbols (notes), while the right hemisphere, in which the music experience is perceived as a whole, was probably not injured.

Part B:
The Seasons of the Brain: Changing of the Brain in Different Periods of Life—How Does Our Brain deal with the Different Stages of Life?

Chapter 13
Patterns of Brain Function in Different Periods of Life

Change Fixation

The brain, like Heraclitus's river, is ever changing.

Like Heraclitus' river, each experience that is processed by our brain changes it for good. Reality never encounters an identical brain twice.

The view of our mental world changes constantly. The key concepts in this process are continuous acquisition of world knowledge, and the skills that are necessary for making changes in this knowledge, alongside changes in brain structure and function. The information processing approaches of our brain are prone to change in various stages of our life. This is the "flick-flack" of our learning pendulum.

During early childhood, we tend to study the world by means of overgeneralization. Later on in life, we learn how to notice the nuances and various shades; in old age, it seems that many of us tend to regress to overgeneralization when interpreting world phenomena.

Our brain reacts differently to stimulations in different periods of life.

In each period of life there is a mix of different neurotransmitters that characterizes it and in light of which (or at the shadow of it) our brain functions.

Age-dependent hormonal changes also force upon our brain a new version of its marriage contract with reality perception.

The changing seasons of the soul also color, in their typical color, our inner emotional climate, which changes throughout life and affects our "emotional personality." In this sense, there are different "seasons of the emotions" throughout our life.

Age-Dependent Activation Software

The pattern of activating thinking skills changes over the years. With the climbing of the year's ladder, as we grow older, the mountain of life insights becomes taller, and the method according to which we cope with information relies more and more on pattern identification. On the other hand, thinking functions, such as the ability to split attention, to focus attention for a long period of time (mental resistance), to avoid distractions, and to assimilate new information, become weaker.

Using different brain areas is related to age-dependent characteristics. Brain scans show that between adolescence and young adulthood (ages fourteen to thirty) intense activation of the temporal lobes is detected while performing cognitive tasks. Moreover, a correlation between the level of education and the level of using the temporal lobes was found. On the other hand, among subjects who were older than sixty-five, intense activity was detected at the frontal lobes while performing thinking tasks, and, once more, a correlation between the level of usage of the frontal lobes and the level of education was found.

Brain Seasonality

Our brain operates according to cycles of seasonality on different time axes: throughout the day, throughout the month, and throughout the years (periods of life). The biochemical profile that characterizes this seasonality undergoes changes. Every period has its own "fingerprints," in the sense of the cocktail of neurotransmitters and hormones that create the internal climate, under the auspices of which our brain functions. In each season, our brain is "prone" to producing performances that reflect its internal climate.

Sometimes this weather is capricious and shakes the ship of our brain aggressively. The traditional example of such a situation is premenstrual syndrome. Sometimes its impact is moderate and gradual, as a slow change that takes place throughout all periods of life.

The basic temperament of people seems to derive mostly from the activity profile of the neurotransmitters and their receptors in the various brain pathways and the structure aspect—the communication interface (wiring) between various brain areas.

Familiarity with brain seasonality and our place in it will enable us to assess our advantages and disadvantages at a given time. We should not consider the inner climate as the essence of it all. Our insights related to it might help in creating conscious climate changes, up to a point. Nevertheless, in other senses, we are like biochemical marionettes with a maneuvering range that derives from being aware of our state and from our ability to react out of "understanding the position" in which we live.

At all stages of our life, we will make better decisions if we have a clearer picture of the season that affects our brain, which will grant us better control of our destiny.

Vivaldi—the Brain Version

We can refer to four seasons of the brain—four periods in our mental and emotional life: infancy and childhood, adolescence and adulthood, the "transitional season" of the brain during middle age, and old age.

The various seasons of the brain are characterized by structural, biochemical, and functional changes.

A seemingly natural comparison of spring to youth and winter to old age is not obvious, since threads of spring are woven in old age and threads of winter and fall are woven in adolescence as well.

Brain seasons are daily and monthly and match different periods in life. At least these three layers of time exist in all of the brain seasons among men and women, according to the specific resolution of time that we chose to look at.

Time Changes—Seasons of the Brain

Changes are a matter of routine in our brain, and this is also true for the cycle of our soul life throughout day and night.

Our brain changes also with respects to its approach to information processing, which is carried out according to a daily and monthly cycle and throughout all the seasons of our life.

The daily fluctuation is reversible and cyclic; fluctuation throughout life is less reversible.

Change of Brain Seasons Throughout the Day

As aforementioned, brain seasons also exist according to a twenty-four-hour cycle. The pendulum of hemispheric dominance (dominance in the sense of concentration of the lion's share of brain processing, and not in the sense of "autocracy" of one of the hemispheres) passes from one hemisphere to the other throughout the day.

At any given time, there is a mix of brain activity that is divided between the hemispheres in accordance with the circumstances. At night, when the external sensory input is very limited, it seems that the dominance pattern moves more abruptly between the two hemispheres, and the mix is more prone to reflecting dominance of either the right hemisphere or the left one (in accordance with the stages and times of sleep).

Certain suppositions that are mentioned in this book do not comply with the basic requirements of statistics, since they are based on self-introspection and the sample includes only one subject. Such is the supposition regarding the nightly cycle of the mood pendulum and emotional thinking. Although it is all about self-observation—"individual domain"—it might also be valid with respect to others ("public domain"). According to this observation, at about three a.m., "right dominance" takes place—which induces a realistic point of view that lacks a comforting angle regarding reality. This feeling lasts until about four a.m., the hour of truth of reality testing, just before the comforting morning dew, which takes the form of a neurotransmitters' potion formed under the auspices of "left dominance" and which smoothens the spiky wrinkles of reality and makes sure we are ready to cope with the new day ahead of us.

Chapter 14
How the Brain Deals with the Dimension of Time

Chrono-architecture in the Brain (Synchronism at the Nerves' Pathways)

We run our life according to external time pacers: the clock, whose clicking dictates the pace of our day, and the calendar, whose dates serve as reference points that outline our conduct. Sometimes we forget that internal clocks are embedded within us, time pacers that are built into our body, and we need to listen to the ticking of their hands no less (and perhaps even more) than we listen to the ticking of external time pacers.

"The internal clock" of each and every one of us dictates the rhythms of our life, but it seems that most of us underestimate the speed of its hands.

The Greenwich point of our brain outlines the perceptional reference point with regard to the experience of time (in a humoristic tone, some might say that among married men, this point is often distorted, since, contrary to their perception, they do not live longer; they just see it as a longer time).

The chief candidate to be considered the Greenwich point of the brain, the reference point according to which the "brain time" and a great part of its rhythms are determined, is the cluster of cells in the area called the supra-chiasmatic nucleus (SCN), which is located above the crisscross of optic nerves at the bottom of the frontal lobe. The main role of timing the circadian cycle is preserved for the SCN. In order to calibrate many of our body's rhythms, there are some elements that assist the SCN: the paraventricular nucleus (PVN) at the hypothalamus, the ganglion of the sympathetic system at the top cervical area, and the pineal gland, which is located at the core of the brain.

The SCN has a central role in timing body rhythms—daily, monthly and annually—which affect a variety of physiological derivatives and, particularly, the cycle of sleep-wakefulness.

Our brain operates as a sort of shutter for the framework of perceptual time through which we sample the world.

The hardware we are equipped with—the senses and the perception and processing system in the brain—includes a "time shutter." The impressions our senses mine from the mine of world phenomena are sampled in typical windows of time.

The time resolution of the brain is within the meso range (the middle range between the micro level, which represents time periods that are shorter from those that are "natural" for our consciousness, and the macro level, which represents time periods that are longer than those that are "natural" for our consciousness). Daily world manifestations, as perceived by our senses, exist in accordance with the rhythms of the level of time resolution within the meso range. It is difficult for us to perceive time frames that deviate from the limits of our natural time shutter, such as macro time frames (whose duration is assessed as millions and billions of years) or, at the other end of time amplitude, time frames at the micro level (whose duration is assessed in fractions of a second).

Final forever: A final thought has difficulty in perceiving infinity, so our consciousness has a built-in difficulty in understanding infinite terms. It seems that our consciousness has a "sense of time" that matches human's life expectancy, and not much more beyond it. Our "sense of time" is imperfect, in the sense of evaluating the meaning of long periods of time such as millions or billions of years. It might be the source of our perceptional difficulty of assessing changes along such durations, which are common in descriptions of phenomena related to evolution, geology, and cosmology.

In a paraphrase of a similar claim, we might say that from each point on the face of the earth we are equally remote from infinity. We are ephemeral, and our distance from infinity is infinite.

Time resolution is different with respect to inputs that derive from different senses. Thus, for example, with respect to a visual input, the frames of a film are perceived as successive when they are projected at a rate of twenty-four frames per second and higher, but at a decreased frequency they might be perceived as non successive. These are the "whiteness interruptions" in the early days of cinema. We are not equipped to perceive changes in reality manifestations that occur in time periods of milli (thousandth), micro (millionth), nano (billionth), or pico (trillionth) of a second.

In 1983, brain researchers from the Max Planck Society reported a case of a woman who suffered from movement-perception impairment (called "akinetopsia" in brain research jargon). The patient, referred to as LM, reported that she experienced visual input as discontinuous and nonsuccessive—a sort of a serial collection of separate impressions perceived alternately. She said that when she pours tea from a jar to a cup, the course of the liquid downward is perceived in her brain as a sequence of "frozen situations," which made it difficult for her to assess the amount she pours, and the cup often overflows.[46] This description is a visual picture, which is a sort of mirror image to the successive frames of modern films and a (sad) regression in time to the early days of cinema. The woman who suffered from this disorder was afraid to cross the street, since she perceived the cars as being far away from her and all of the sudden very close to her. Her impaired estimation of distance almost always made her feel as if an accident were about to occur. A CT scan of her brain revealed large lesions around the interface between the parietal lobes and the temporal lobes at both sides of the brain. The injured area in both sides contained the central crest of the temporal lobes, which is in charge of perception of movement.

A material demonstration that is similar to her description can be found in the wonderful work of the Chinese sculptor Tsang Cheung Shing. In *Coffee Kiss*, tea and coffee, poured from different cups, "freeze in time" and become figures that kiss each other.

"The time areas" that are in charge of chrono-architecture in our brain operate in various speeds.

Similar to the time zones on Earth, different brain areas tend to time performance tasks differently. Some of these areas produce output almost in real time, and other areas are characterized by a delayed reaction and contribute to a later chronological layer of the task.

The human male brain and the human female brain dance to the music produced by similar, but different, drums. This results in different synchronization of life rhythms. Thus, for instance, women, on average, tend to fall asleep earlier than men and wake up a little earlier as well. This difference in sleep–wakefulness timing starts in adolescence and lasts until after middle age.

The Brain Copes with the Dimension of Time

Epicurus, the philosopher, raised an interesting argument. According to him, the reference points of our life are like poles of eternity: the time of our birth ends the sequence of eternity that preceded our existence, and our death starts the sequence of eternity in which we do not exist.

The state of not-being after death is similar to the state of not-being before we are born; both last forever, assuming time has no beginning and no end.

In a fable from the Middle Ages, our life is compared to a bird that comes from darkness and frost and flies into a warm, illuminated dance hall full of tables, filled with delicacies, and pleasant sounds of music. After a short time at the dance hall, which is compared to the duration of our life, she flies to a window at the other side of the hall, goes out to the frozen darkness, and has no option of ever coming back.

Our life is immersed in "nothing" on both ends; before our development as an embryo we do not exist, and after our death we do not exist, as well. The boundaries of our existence, at both ends, are nonexistence. The eternity that is ahead of us before we are born, however, is like a deep whale, at the bottom of which there is the water of life and at the other end the eternity that awaits us after our death, like a dark tunnel with no light at the end.

The conceptual evasiveness of time forces us to conceptualize it by means of material objects, like the hands of the clock or the pages of the calendar.

Activities that take a known period of time to perform sometimes serve as time markers, which help us estimate time's passing. Sherlock Holmes once estimated the difficulty of a detective task, in retrospect, according to the number of pipes he smoked from the beginning of the investigation until its end. An example of macabre time estimation is the story of a doctor who was asked by a terminal patient how much time he had left. The doctor said, "I don't recommend that you start reading *War and Peace*."

We rely on the dimension of space as a major time conceptualizer; according to our senses, as they are expressed in language, yesterday stays behind whereas the future is ahead.

The close connection between time and space in our consciousness is reflected in an experiment in which subjects were asked to think about the past and the future. It turned out that the position of the bodies of the ones who thought about the past tilted backward, whereas the ones who were asked to think about the future tended to bend forward.

The metaphor of time as a dimension in space is built in in our perception.

The direction of writing, which has clear spatial rules in the different languages, is also wired in our brain as a marker of the directionality of the arrow of time. So, when asked to describe a sequence of events, traditional Mandarin speakers tend to place past events at the top of a vertical sequence from the right (traditional Chinese is written from top to bottom and from right to left). English speakers will place past events horizontally from the left side, while Arabic and Hebrew speakers will place the events horizontally from the right side.

We are like candles in the wind of time. During childhood we look at the horizon expectantly, and when we grow up we look back from "the residence of the horizon" and often taste the bitter taste of missing. Then we realize more strongly that time is our most precious resource, since we do not know what time will bring in light of our ephemeral materiality. Temporariness is built into our life forever.

Time wears us out unmercifully, and it unbiasedly transfers all of us from this world to the next world while the clock is ticking in the background.

In a spirit of philosophical humor, it was said that time is just God's way of making sure that not everything happens at once (George Carlin). In the language of poets, time pacers were compared to a lover who chose his lover's lips out of all the hands of all the clocks in the world to show him when his time to go came.

Our brain copes with time in its personal aspect and in its public-historical aspect.

The imaginative ability of our brain is incarcerated behind the walls of our world of concepts, which was shaped according to the culture in which we were brought up and confined by the memes we were exposed to. We find it extremely hard to perceive the notion of eternity.

Galileo felt the burning breath of the ecclesiastical establishment on the back of his neck when he contemplated metaphysical issues, but even Galileo, after racking his brain about it in vain, could not decipher one of the perceptional oddities related to the concept of eternity. The infinite series of whole numbers (1, 2, 3, 4…) and the infinite series of whole even numbers (2, 4, 6, 8…)—it seems that both series last endlessly, but isn't the one wholly contained in the other, and isn't its size double?

In studies in which subjects were asked to assess how distant from them were past or future periods of time, it was found that the representation of time in our consciousness is not symmetrical: a period of time in the future is usually perceived as closer than a period of time in the past, which is, in fact, identical to it. A possible explanation for this phenomenon derives from the fact that the time gap between past and present is "full of" memories that "stretch" the psychological time perception, while the time gap between present and future lacks any memories and so "shortens" our psychological time perception.

A "polychronic" (multitemporal) time perception, in which the time frame is not strict, was named in the past, stigmatically, "female time." On the other hand, monochromic (monotemporal) time, experienced as an oriented flow toward future goals, used to be referred to as "male time." These stigmatic suppositions assumed that women are focused on the "existence zones" of the present, and a tendency toward the "promising zones" of the future was ascribed to men.

Many believe that the nullifying of consciousness that involves death becomes a fact that remains valid forever.

The notion of eternity and concepts of nonexistence that involve death are built into "consciousness mines." Human consciousness has a hard time conceptualizing them and dealing with them. It has never encountered them and never will, since, in a state of nonexistence, our consciousness will not exist, and, due to its finality, eternity is far away from it as eternity.

Eternal nonexistence is a conceptual course that is probably inaccessible to the metaphorical feet of our consciousness.

Chrono-architecture

Our brain maps are liquid and ever changing. The constant changing of representation areas of cognitive, behavioral, and motor skills is based on the liquidity in the connections between the neurons as much as the networking pattern between them is at the base of the skill. The changing takes place both in the time dimension and the spatial dimension.

The absolute majority of brain-mapping studies have been conducted so far with respect to the place in which things happen (the spatial component) and not in relation to the timing in which the action pattern exists (the time component).

Tracing the brain activity pattern on the axis of time (chrono-architecture), and not only on the spatial axis (location, structural architecture), is now perceived as essential for a better understanding of the brain's activity pattern.

It seems that each brain area that functions as a functional subsystem is managed on the time axis that is unique to it. In other words, a time-dependent activity pattern is unique to each of the areas (a single area processes information within thousandths of a second; the other within hundredths of a second; and so on).

We can trace the "fingerprint in time" that is unique to the different brain areas by using an imaging method called functional MRI, which enables us to map the brain not only in terms of the place where things are happening but, also, in terms of the timing according to which the activity patterns of the neurons interact.

Thus, for example, in an attempt to follow the procedure of learning a new motor skill, it was found that at first the most active brain areas are those located at the prefrontal cortex. After about five hours, when the skill is already assimilated, it was found that the brain activity related to the task has immigrated backward in the frontal lobe (to a brain area called the premotor cortex) and to the parietal lobe and the cerebellum, where an infrastructure was consolidated for preserving the motor sequence that was learned.

The Where and When in the Brain

While making observations of the brain, the location provides an answer to the "where" question. The order of activation of various brain areas on the time axis provides an answer to the "when" question (structural architecture and chronoarchitecture).

According to Heisenberg's uncertainty principle, it is impossible to determine the location of a particle and the speed of its movement at the same time. It is also true for brain examinations—temporary imaging methods enable us to focus on the location component and the temporal component alternately when accuracy, with respect to evaluating one of the components, requires compromising the level of accuracy with respect to evaluating the other component.

"One Moment, Please"

Delayed feedback creates a delay that gives the input some "time perspective" to assess the input patterns. This is similar to putting our hand in a bucket full of gravel; we will not be able to identify the nature of the element until we feel the different input patterns of the stones' texture on the axis of time. This fact relates to the serial nature of senses' impressions and the threshold that determines what is considered an essential, sufficient input to allow for its reality-compatible conceptualization.

The tempo of thought, in the sense of the pace in which thought is formed, is a unique characteristic of thought according to its nature. The duration of a "thought beat" is an elusive, heterogeneous variable that depends on the nature of the thought and has not been explained sufficiently by science yet.

Chrono-architecture and Consciousness in the Mirror of Time

The system that coordinates attention in our brain seems to work as an oscillator, which becomes a sort of reference clock for multiple internal systems that set their timing according to it. According to one common view, the internal oscillator that allocates attention resources is located at the thalamus nucleuses, but some ascribe this role to the neural network of default.

The reference oscillator, similar to a punctual cesium clock, orchestrates the brain orchestra with various degrees of control. It seems that, at every given moment, there are "rebellious" brain areas that have a schedule and timing of their own. But the more attentive and focused we are, it seems the more brain areas are calibrated and operate compatibly.

During sleep, the uniformity of timing becomes looser. Various brain areas become more autonomous and, as a result, the level of attention decreases.

Descending the stairs of consciousness and the sinking of the consciousness mode decrease both the intensity of illumination and the diameter of the attention beam in accordance with the level of sinking.

In cases of head injuries followed by loss of consciousness, the central regime fades away, and the various brain areas function as a rabble of autonomous entities.

Chrono-architecture—the activity pattern on the time axis is a fundamental component in the existence of consciousness.

The timing of the activity of the bioelectrical action potentials is probably at the basis of consciousness modes, and, according to a central supposition, there is a direct connection between the compatibility of the activity of electrical-action potentials and the level of consciousness. It seems that our consciousness, at its best, resembles a harmonious orchestra.

In our brain there is a large amount of white noise that is, at times, random and chaotic and lacks a clear, compatible pattern as a result of the uncoordinated activity of various brain areas.

Various oscillators (pacers of electrical and biochemical activity) operate in the brain simultaneously, not always in synchronization and harmony, which sometimes results in neuronal cacophony. It seems, however, that there is one superoscillator—"the conductor of the orchestra"—that tries to create harmony between the various oscillators as well. According to a common hypothesis, the compatibility of pacemaking (when multiple brain areas are orchestrated and move to the sounds of the same melody) allows for the production of consciousness.

The pleasant sense of flow, when we feel qualified and succeed in performing a task, might originate from a successful synchronization of brain areas by means of the superoscillator, which allows for the compatibility of the activity in different areas.

According to certain brain researchers, consciousness is, in fact, the preserving of electrical activity between selected brain cycles at forty hertz oscillation. The involved brain cycles preserve the compatibility of their activity by means of feedback. The coordinated fluctuation creates the magic of consciousness.

An intriguing hypothesis called "the forty hertz hypothesis" is based on an observation that revealed that, during a conscious eyesight experience, neurons at the posterior cortex are active (and, in the physiological sense, these neurons "shoot" spurts of bioelectrical/electrochemical charge) at a rate of about forty hertz. According to this hypothesis, each fraction of a second (and, more accurately, the 0.025 part of a second) represents a "unified unit of consciousness." This range of frequencies characterizes gamma-type brain waves.

Other researchers suggest an oscillation rate of sixty hertz as the "consciousness-generator rhythm." In the spirit of these hypotheses, the unity of consciousness is a result of synchronized neural activity that persists on the axis of time. Each second of consciousness is composed of forty to sixty "micro-moments" that create a sense of consistency, similar to movies' frames that, when they are projected at a rate of twenty-four frames or more per second, are perceived as a continuous sight experience. Perhaps this is the source of the saying "Life is like a movie."

Some claim that resonance at a fixed oscillation of forty to sixty hertz, paced by a nucleus at the thalamus, is the generator of compatibility that causes unification of the various fragments of input and it is the "rhythm of perception."

Another common hypothesis among brain researchers is that the neural network of the default system is the reference oscillator (the Greenwich point of frequencies) whose activity synchronizes and modulates the mix of frequencies in our brain.

Expanded Consciousness and Memory

Our consciousness might stretch along the axis of time. "The sense of what is happening" is always affected by the sense of what has happened. The experience of "I am reading this book" carries with it a treasure of enclosed knowledge, derived from a whole life of being you. In this context, it was said that when the painter Pablo Picasso was seventy years old, he was approached by a millionaire who asked him to draw his portrait. Picasso looked at him, sketched for ten minutes, and asked for a quarter of a million dollars for the sketch. The millionaire got upset. "You are asking for a quarter of a million for ten minutes?" Picasso answered, "These are not ten minutes. These are seventy years of Picasso's life experience plus ten minutes."

Chapter 15
Brain Development from Embryo to Adult

A Beginner's Brain

The signature of the electrical activity in the embryonic brain shows that a human embryo's brain is at a state of sleep 95 percent of the time. This continuous state of sleep is divided into two: active sleep, during which there are movements of breathing, eye movement, and sucking, and a quiet sleep, in which tonic muscle activity takes place but there is no eye movement or movement of breathing. These states are similar, respectively, to rapid-eye-movement sleep, which is the main dream stage of sleep in human brains, and the deep sleep stage that is characterized by low-frequency electrical waves. For the rest of the time, the brain is at a state of transition from one mode of sleep to the other.

A baby's brain is cradled in potions that induce sleepiness, like the low pressure of oxygen (similar to the pressure at the top of Everest), the warmth of the amniotic fluid, and the sedative cocktail, produced in the placenta and in the embryo itself, that contains substances that induce relaxation. The brain ripens in Hypnos's cradle.

The wake-up call for a new state of being, which takes place at birth, is mediated by a spurt of noradrenalin in the brain, which starts a new pattern of being.

Pink Brain or Blue Brain

The brains of all embryos look like a female brain until the eighth week—this is the default state determined by nature. A spurt of testosterone in the eighth week initiates the transformation of some of the brains from a generic embryonic brain to a male brain.

Nowadays, it is commonly believed that the exposure of the embryonic brain to testosterone during the development stages in the womb delays the development of the left hemisphere. Since the level of exposure in the brains of male embryos to testosterone is much higher compared to that in the brains of female embryos, the left hemisphere of male babies is less developed than the left hemisphere of female babies. Some relate this to the higher tendency of male babies to suffer from autism and dyslexia and to the high frequency of left-handedness among males.

The Brain During Infancy

During the first years of life, the brain is at the peak of its flexibility. The period of differentiation, during which brain maps are differentiated, takes place within a narrow window of time. The maps can also change in later periods of life, but then there is a higher price to pay in terms of time and energy. Thus, one can learn a foreign language rapidly and elegantly during early childhood, or slowly and in a cumbersome manner during adulthood.

The brain of a newborn weighs 400 grams on average, whereas an adult's brain weighs 1,300 grams on average, which stands for 2 percent of body weight.

A fascinating point of view reveals that the brain of a human child undergoes, in the course of his life, a process of growth that is similar, in measurement, to the growth process of the human brain throughout two million years, from the time of our ancestors until today. Two million years ago, the volume of our ancestors' brain was 500 cubic centimeters and resembled that of contemporary newborns. Since then, in a course, full of hardships, that lasted two million years, the volume of the brain has increased and reached 1,300 cubic centimeters—the volume of an adult brain nowadays.

The right half of the brain is bigger, among both girls and boys, until the end of the second year of life, and it is probably the half that is more dominant in the mental life of infants during the first three years of their life. We can say that infants are mostly "right-brain" creatures. They experience thrills and feelings that are mostly not conceptualized verbally. During the first two years of life, the dialogue between mother and baby takes place primarily as communication between the right half of the mother's brain and the right half of her baby's brain. Facial expressions, nonverbal gestures, and the tone of discourse (the melody of words) are main components in this dialogue.

The age in which a certain mental skill is learned often determines the brain area that will be in charge of encoding this skill. In other words, the same skill learned at different ages will sometimes be encoded in different brain areas. For instance, a language learned in early childhood is encoded in different brain areas from a language learned at a later stage of life.

Freud spoke about windows of psychological development during childhood, which shape our tendencies throughout life, and it seems that the insight regarding the existence of time windows for development is also valid with respect to the complex soul structures that are formed within us.

The ethologist (researcher of animals' behavior) Konrad Lorenz found that newly born goslings ascribe the role of protective mother to the main figure they are exposed to during the time window that ranges from fifteen hours to three days from the moment they are hatched. Such a "maternal imprinting" is a complex mental mechanism that is, of course, not resistant to errors. Lorenz himself became a "mother goose" with respect to newly born goslings that were exposed to his presence within this window of time.[47]

Studies about children's dreams show that they follow, in their content, the children's cognitive and mental development. The reflections in preschool children's dreams are mostly immobile and flat and lack significant mobility components, and just a small part of them reflects emotions and memories. Extrapolation of these findings has led some researchers to conclude that the sleep of a human embryo is practically free of dreams in their customary sense.

The importance of the time-restricted window of development was demonstrated in a famous experiment, though one that involves some problematic ethical aspects, in which it was found that the normal function of the sense of sight depends on exposure to visual stimulation within the narrow window of time of early development. A kitten's eye was sewn shut between the third and eighth week of its life to prevent exposure to visual stimulation. When the stitches were removed and its eye reopened, it was found that this eye was totally blind and that the areas of the visual cortex in the occipital lobe, which would be in charge of processing information from the closed eye, did not develop. Later, these areas took charge of processing visual information from the other eye, together with areas of "normal" visual-processing areas—in the opposite occipital lobe (in accordance with the assumption that real-estate assets in the brain are valuable and there is no such a thing as "deserted property," and that the brain, as a tough employer, does not allow hidden unemployment within its border). The damage was irreversible; as the kitten grew older it remained blind in one eye for the rest of its life. This experiment demonstrated the insight that was also confirmed in other experiments—that life experiences in critical periods of time shape our brain, and in their absence our brain is shaped irreversibly.

It seems that different functional systems in the nervous system have their own typical "critical time windows of development."

Thus, for example, in order to create excellent lingual mastery, the process of language learning should take place by the age of eight, approximately. If learning takes place afterward, the lingual range of sounds is assimilated less accurately, and the language is encoded in a different brain system from the one in which mother tongues are encoded.

A window of opportunity also exists for imprinting a rhythmical preference, as a result of exposure to musical rhythms that are common in the culture in which a person grows up.

Babies are able to notice a wide range of sounds that is composed of all the sounds that exist in human languages. Once the "window of opportunity" is closed, however, with respect to the development of the cortex that processes auditory information, the range of sounds processed by the brain becomes very limited. A brain map that contains the sounds of the language to which the baby is exposed is formed. The number of key sounds that have priority in terms of brain processing is estimated to be about a thousand. As aforementioned, babies who are exposed to Japanese at the age of six months are capable of noticing the difference between "R" and "L," but when they become a year old this distinction between the sounds that are not part of the Japanese language will become very difficult for them.

Our musical taste, which changes throughout the years, reflects, at least to a certain extent, the "brain rhythms" that change in different periods of life. Our brain also has rhythmical seasons; it operates according to different "musical seasons." Perhaps there is a need for compatibility between the rhythms of our brain and musical rhythms in order to experience maximal emotional enjoyment. Since the brain rhythms change in different brain seasons, in different stages of life, it means that our musical taste also changes according to the musical seasons of the brain.

Preferences, which are "acquired tastes" (in the sense that they are formed after birth), are embedded within a neural tissue similar to that of innate preferences, which are genetically dictated. Thus, as adults, we are usually unable to distinguish between acquired preferences and innate ones, since their biological nature becomes identical.

Contribution of the Innate and the Acquired

When a brain is formed, genes and their delegates, the proteins and hormones, sketch the outline of the basic structural infrastructure, to which the reciprocal relations with the environment are added as a shaping element.

Innate information and acquired information outline the communication pathways in the brain together. The language of genetic straitjacket and the language of acquired insights are translated into the same language—the language of wiring neurons in the brain. Our brain is used to shape adaptive reactions to reality tests.

The process of the nullifying of neural junctions in which there is no sufficient activity takes place from the early stages of brain development. We are all born with a brain full of billions of excess neural junctions, which will be reinforced if sufficient activity takes place in them (in the sense of sufficient volume of activity that exceeds the threshold). In case where the activity that takes place in them is not sufficient, however, they will become dismantled and nullified. These junctions of nerves are designed to store neural representations of the world in which we find ourselves. The huge amount of redundant neural junctions (prior to the dilution that takes place in accordance with the activity pattern) has led researchers to claim that they enable representation of almost all possible worlds into which we might be born.

Infectious Behavioral Disorder

We can learn how sensitive our brain tissue is to environmental factors from a surprising finding. It was found that a very small minority of children who were affected with a throat infection generated by Streptococcus developed obsessive-compulsive disorder (OCD) over time.

A structural finding that was found compatible was an increase in volume, at a rate of 25 percent, of a brain structure called the "caudate nucleus." It seems that the correlation is causative. The structural change in the caudate nucleus, which derived from the Streptococcal infection, induces the behavioral disorder. Although it is a very rare condition, it indicates that the landscape of the brain is also prone to undergoing acquired changes that are induced by environmental factors, such as a tiny bacterium, which reshape our brain and result in behavioral changes.

The Gut-Brain Axis

In a broader perspective we can mention the Microbiome—the community of microorganisms that inhabit the niches of the human body's' spaces in health and disease. In this manner, our body is an open ecological system.

According to a well-known demographic assumption, the amount of the microbial population in our body is estimated to outnumber the amount of human cells by ten times. (A funny remark was based on that old estimate: "You are only ten percent of what you think you are…".) Newer estimates assume that the number of microbial cells outnumbering the human cells in our body by a much smaller gap.

We frequently use antibiotics (anti-biotics) that, besides harming the disease –causing microbes, also damage many other microbial cells, including those that work in harmony with our body's cells, and even in synergism. On the other hand, we use too few probiotics (pro-biotics) that enrich the types of microbes that support and improve our health. We should use the "biotics" substances in a more rational way as "eco-biotics"—medication suited to our body as an ecological environment—therefore balancing its ecology to serve our health.

The phrase "gut feeling" is apparently an intuitive expression to a deep truth: the dialog between our body's microbial inhabitants and our emotions. Many substances that are produced by our body's microbiome are absorbed into the blood. Part of them reach our brain and influence its function as well. This complex dialog is termed "the gut-brain axis." It seems that this axis is an important influencing factor on our inner cognitive and emotional climate.

Infancy Memories

Procedural memory is characterized by being nonexplicit and includes information that, after being encoded and consolidated, does not require conceptualization by means of words or projection onto the screen of consciousness. It operates in an automatic pattern, and there is no need to focus our beam of attention on it.

This type of memory is used by babies in their infancy. Today it is believed that only at about the age of two is information encoded in a significant manner as explicit memory, in the sense of conscious memory of information. The conscious memory relies mostly on the crutches of language; once a language is acquired, it is used more often. Thus, our experiences from early infancy are encoded as procedural memories and not explicit memories. Some psychiatrists believe that exposing these unconscious memories to the sun of consciousness might ease the suffering of those who suffer from various mental problems.

Registration of an experience, which includes the time of occurrence, the order of events, and the inner emotional soundtrack that accompanies it (linking self-world), will be remembered consciously only if it is encoded as an episodic memory. In absence of such labeling (along with the physiological sequence typical of encoding such a memory), the event will not be part of the conscious layer, but it might still have an impact on our behavior.

Oblivion—Amnesia of Infancy

Information that does not "flow" to our consciousness, since it is not consciously remembered, is not necessarily inaccessible to pathways that impact our behavior. Its implicit impact, though unconscious, might influence our behavior.

Although there are no explicit memories from the first two years of our life available to our consciousness, there are memories that are immersed in emotions embedded within the limbic system and accompany us for the rest of our life.

During the first two years of life, the hippocampus and the frontal lobes have yet to undergo the structural and functional maturation that will enable them to perform their role appropriately. It might explain the phenomenon of the amnesia of infancy. Memories we encode during these years will not be remembered as episodic memories that we will be able to recall consciously at a later stage. Knowledge, which is important and highly essential for life, is acquired during these years, but it is not consciously available for retrieval; rather, it is embedded in our brain as habits and skills (implicit and semantic knowledge).

Thus, it can be assumed that there is no way of reconstructing conscious experiences that are preserved as episodic memory from approximately the first two years of our life in a reliable manner.

During early childhood, the hippocampus and the frontal lobes are still immature in the structural and functional sense; thus, both the memory-encoding aspect and the memory-retrieval aspect, in their mature version, are still immature as well. At this age, along with the retrieval difficulty there is also difficulty in creating replicable, reality-compatible memory due to the immaturity of the hippocampus (which is a key component in encoding the impressions of the senses as memories).

The immaturity of the structures that are essential to the formation, preserving, and retrieval of episodic memory causes the difficulty in preserving the recordings of events of a personal nature during these years.

When a memory of an episodic nature in someone younger than two years old arises, it is best to refer to it as a memory that was created by means of reconstructing sources of information rather than as a direct episodic memory. It is more likely that it derives from "backward construction" of the components of memory and does not represent a real-time experience from that period.

Episodic "infancy memories" are often a collage of impressions created at a later age out of family documentation, inferences made at a later time, and stories of family members. Sometimes they include a hint of the "sin of backward bias."

Memories from this period are likely to have been created by means of inferences originating in implicit and semantic unconscious memories, or by means of a quilt that was made after listening to parents' stories, reviewing picture albums, reading memoirs, etc.

The cluster of memes and insights that are encoded into patterns that conceptualize the environment during infancy serve as primary mental scaffolding for our perceptual building of the world. Later on, when the tower of insights is already higher or even reconstructed, those primary mental scaffoldings are no longer useful, and so they fade away—thus the assumption that in absence of activity in the neural cycles, which store our infancy insights, they vanish and become nullified. On the other hand, according to an assumption that won great support, these memories remain concealed in the brain but are not available to our consciousness. It may be that the networks between the neurons that were preserved from our first years of life constitute the core of the nerves' networks, which become more expanded and rich in junctions as we grow up.

The Brain During Childhood

Brain-Sculpting Period

During the first years of life, most of the networking cycles in the brain are created in conflicting patterns of connection and disconnection. Along with the process of brain cells grouping into a common cycle of activity, a conflicting process takes place at the same time and sets a threshold of activity as a condition for survival. The synapses in which "traffic load" does not cross the threshold are doomed to gradual fading until they are totally nullified.

From birth until about the age of three, the number of links (synapses) between neurons multiplies (a process called "synaptogenesis") by approximately twenty, so from the age of two or three an expanded process of pruning is formed (called "synaptic pruning"). This process thins out the density of the "synaptic forest." The rate of thinning out is usually estimated at half the number of synapses—thus the number of synapses in different brain areas in the brain of a young adult is estimated as half the number in the brain of a three-year-old child.[48]

Communication junctions (synapses) between the neurons that are selected for networking survive, whereas those that were not selected perish. Thus the number of connections between neurons decreases, but the richness and complexity of our brain cycles increases. The considerable decrease in the number of synapses is formed based on the meritocracy approach (a term taken from social studies and that refers to a society ruled by a select group of the most-talented people). Only the synapses of the neural networks that encode useful, high-quality information survive.

Quite a large number of brain researchers believe that the large-scaled pruning that takes place during the first years of life is, in fact, economical. They believe that the excessive growth of neural networks followed by a controlled pruning process of networks that do not encode information that is substantial to our life is a "cheaper" process, energy-wise, and less time-consuming compared to a process of forming accurate neural networking in the first place.

Childhood is the main period in life in which the saga of eradicating low-activity neural networking junctions takes place. The principles of "whatever is not used is lost" and, on the other hand, "connections that are frequently activated are preserved" exist throughout our life. Paradoxically, the oxymoron "destruction is creation" is valid in this case. The pruning of redundant elements enables brain networking, which encodes complex meanings and insights, similar to creating a sculpture from a raw chunk of stone.

The "Sally–Anne test" is a test in which a sequence of pictures that describe the behavior of two girls, Sally and Anne, is presented. In the first picture, Sally puts her doll in the pram. In the second picture, Sally leaves the room and, after Sally is gone, Anne takes the doll from the pram and puts it in a box. Soon afterward Sally comes back. The children are asked to interpret the sequence of pictures by answering three questions: Where is the doll? Where did Sally put the doll? Where will Sally look for the doll? The age range at which most "small people" manage to predict Sally's thought correctly is three to four years. At the age of three, most children still err and say that Sally will look for the doll in the box. It is difficult for them to put themselves in Sally's mental position and answer according to the knowledge they have about the doll's location. At the age of four, most children give the correct answer: Sally will look for the doll in the pram because this is where she left it before she left the room. This answer reflects their capacity to identify mentally with Sally's thoughts. This is, in fact, a test that examines the skill of "understanding the soul of the other." Success in the Sally–Anne test and the ability to understand the other's soul derive mostly from the operation of the frontal cortex.[49]

Among those who suffer from autism, there is great difficulty with regard to the Sally–Anne test. According to reports, 80 percent of the autists fail the test even at an older age.

Most four-year-olds who do not suffer from neurological disorders do not fail in the test—they have internalized the notion that other people might have false beliefs. In a sense, we might say that the age of four is the age in which we lose our innocence.

Density of Synapses (Connection Junctions Between Neurons) in the Brian At Different Ages

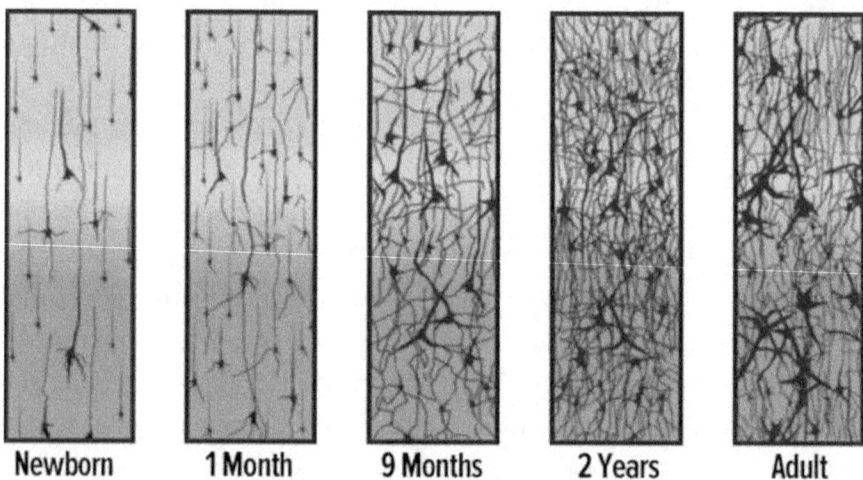

Newborn   1 Month   9 Months   2 Years   Adult

From left to right: a newborn's brain, one month old, nine months old, two years old, an adult brain

The Sally–Anne test shows that, until about the age of four, a child's brain assumes, in an axiomatic manner, that its reality perception is the only one that exists. An important stage in cognitive maturation is the transition from this stage to the stage of understanding that reality has many faces and that there is room for interpretation, which is different from our personal-subjective interpretation. This stage constitutes a milestone in our cognitive development. During adulthood, however, the built-in logical failure reflected in infants' behavior still exists as an initial, raw layer of our perception. Studies show that even as adults, our perception first assumes, for a brief moment, that our raw-subjective perception of reality truly represents the qualities of the observed phenomenon (i.e., that the subjective perception is identical to the objective essence). This stage, however, is usually followed (as a later stage of the chrono-architecture of the process) by a correction stage, which enforces its logic and provides interpretation that takes other points of view into consideration and examines the information in a broader context.

Childhood Memories
The Width of Memory Street—My Childhood Home

A description that might trigger a sense of familiarity in numerous brains is a description of a person who returns to the street on which he lived as a child. During the nostalgic visit, he feels as though the street is much narrower and shorter than he remembers. He is amazed to realize that the roof of the building in which he used to live, and which he remembers as reaching the clouds, is actually much, much closer to the ground. Perhaps the street itself did not shrivel, but it is his consciousness that shrank, and the height of the building's roof probably did not become shorter—it is his consciousness that lowered its stature.

Some researchers claim that half the children in kindergarten might have photographic memory, which enables them to preserve an accurate visual picture of a complex pattern in their brain. This natural talent fades away as years go by, and such a talent is hard to find after adolescence.

Two Constitutive Processes of Brain Maturation

Two main processes that take place during the development of a young brain are synaptogenesis—the constant creation and dismantling of synapses (with a high tide of creation up to the age of two or three, followed by a period of "supervised pruning")—

and the process of coating the axons in myelin (myelination), which takes place until the twenties and, in a more subtle manner, even beyond the twenties. These constitutive processes determine the outline of our brain, since it meets the world until it becomes a "mature citizen" in it.

The Maturity Touch of Myelin

Brain maturation, the process during which brain structures mature to a point at which they are capable of implementing their potential, is based, as a main component, on the myelination process. During this process, the exposed axons are coated in myelin, which is oily in essence. The coating of myelin allows an effective transmission of the bioelectrical action potentials and reduces the depreciation of the intensity of the signal along the axon. Its role is similar, in many senses, to the role of the coating of an insulating material around conductive electrical wires. The axons sometimes seem like tendrils of trees in a thick tropical forest. They beat according to changing melodic rhythms of the pattern of bioelectrical action potentials that go through them.

The axons, which are coated in myelin, are grouped in common clusters, and they are the creators of the areas of the brain core known as the "white matter," so called due to their pale appearance, as revealed in brain surgery.

The assessment of glucose consumption, the source of brain energy, serves as indirect testimony to the windows of acceleration and slowing down in brain development pace. The increase in glucose consumption is observed until the age of four, stabilizes around the age of ten, and decreases to the level of an adult brain from the ages of sixteen to eighteen.

It was found that the white material in the brain increases in volume in an almost linear, age-dependent pattern between the approximate ages of four and twenty. On the other hand, the gray matter increases in volume in a non-liner pattern, and an increase in volume is observed prior to adolescence, whereas reduction in its volume is observed after adolescence. This pattern also has unique regional characteristics, so that the pace of acceleration and slowing down is not uniform across the brain.

An Ethos of Maturation

The process of myelin coating takes place according to different rhythms at different brain areas. The last area in which myelin coating (myelination) takes place is the area of the frontal lobe and, particularly, its front part—the prefrontal area. This area matures structurally at the end of the second decade of life and even later. Due to this process, the volume of the frontal lobe increases gradually. The frontal lobe areas are the birthplace in which "managerial soul functions" are formed, such as judgment, long-term planning, abstract conceptualization ability, emotional regulation, and creation of inhibitions. It seems that partial maturation of the frontal lobes areas during adolescence, along a more complete maturation of other brain areas that are in charge of producing raw emotions and urges (such as the amygdala and structure of the limbic system), is at the basis of the high-amplitude emotional storms that characterize adolescence. During adolescence, in fact, there is structural imbalance between emotion-production areas whose structures have matured (undergone full myelination) and operate vigorously and the areas that are in charge of moderating and refining emotions (i.e., the prefrontal lobes), which have yet to undergo structural and functional maturation (they haven't completed the myelination process yet). Upon structural maturation, when functioning is at its best, the frontal lobes serve as a dam against the huge emotional waves, and their ability to refine and moderate the magnitude of raw urges betters behavioral products.

The prefrontal cortex matures a year or two sooner among female adolescents, compared to that of male adolescents. This finding suggests that, on average, the brain of female adolescents is wired to more reasonable behaviors earlier than the male adolescent brain.

On the other hand, the rising level of testosterone in male adolescents' brains also puts the partially built dam, positioned by the prefrontal lobe against the raging waves of emotional urges produced by the amygdala, at an inferior position.

Generally speaking, we can say that control of urges is structurally inferior among boys compared to girls during adolescence.

The waxing wave of testosterone dims the glow of social connections and leads the way to activities of a competitive, sexual nature.

With regard to the brain of the two genders, we can say that the linkage between maturation of functional potential in key areas of the brain, especially at the frontal lobes, close to the end of the second decade of life and defining this age as the age of "mature wisdom"— the age in which a person is considered qualified for social functions such as driving, and participating in elections—is not coincidental.

Memories' Period of Glory

The second and third decades of life constitute the period from which we preserve most of our memories. Memories from other periods are less preserved.

Chapter 16
Aspects of Mature Brain Functions

Maturation of the Social Brain

The Mirror Neurons

"Mirror, mirror in my skull, how does the person in front of me feel?" The magic mirror that can answer this question is composed of the unique neural networking of a cluster of cells called "mirror neurons." These are neurons whose activity reflects the sense of the other within us. Mirror neurons make emotions become "infectious," and their function serves our empathy skills. They create a sort of a "phantom sense" of other people's feelings and constitute a neural infrastructure for the ability to feel the feelings of the other and understand the motives of the other. This ability sometimes makes vague social situations look clearer and allows us to extract meaning from interpersonal situations that seem enigmatic.

An average female brain contains more mirror neurons compared to an average male brain, and some may add, "as expected."

A central ability that is enabled through mirror cells is imitation of skills, including cognitive skills.[50]

The act of imitating is a central layer with respect to acquiring world knowledge, particularly among children. Due to the mirror neurons, we are not required to conceptualize the emotions of the other in an indirect manner, since they allow us to sense them directly. They make our consciousness penetrable to emotional impact at any given time and turn the skull barrier into a membrane penetrable to the osmosis of emotions.

Aspects in Reading the Soul of the Other

The brain should not be viewed as a closed ecological system. It seems more appropriate to study it in a contextual pattern. Context carries us beyond the boundaries of the body of the person who owns the brain.

A human being, being a social creature, is defined, at least partially, according to the relationships he has (i.e., in terms of his connection with other people).

A human brain creates a representation of the person it is part of and contains the representation of the status of its owner in the social map (i.e., with respect to other people). Our brain creates models of other people's souls, as it creates models of the physical world. That is why we are embedded in the mental world of other people, no less than we are embedded in the physical world.

A person who is important to us often serves us as a guide for self-definition.

Life in groups led to the development of skills related to "the other's sense" of our brain. According to a common supposition, the reference group that constitutes the social shell significant to a person includes 150 men and women. These are the stars in the skies of our social universe, in the light of which we find our way. According to evolutionary psychology, this number is similar to the number of members of an average tribe that used to be the social universe of our ancestors.

Due to perception limitations built-in within our brain, we are only able to experience our own consciousness directly. With regard to others, we are forced to infer the qualities of their consciousness from external observations—i.e., indirectly. Thus the thought that is probably familiar to many of us: we will never be able to know whether other souls truly exist. In philosophy, this problem is referred to as "the problem of the other souls." Each of us knows for sure only the fact that he or she has a soul. As for the others—we rely on inferences and our ability to "read the other's soul."

The illusion of particularity is a universal experience experienced by our brain. Its extreme expression is solipsism (which means "self alone" in Latin), which is a view that acknowledges the existence of the "self," which is experienced in consciousness, and doubts the existence of other entities in the world of phenomena, which is external to us.

Altruism (behavior that prioritizes the needs of the other) is a behavior that attempts to bridge the tendency for a solipsistic gap to exist between two people. Perhaps certain types of autism are experienced by the ones who suffer from them as super-solipsism.

The 1998 film *The Truman Show* might be interpreted as being taken from solipsism's school of thought. In this film, Jim Carry plays Truman, who was raised from the beginning in an artificial world in which everybody, except for him, is an actor or actress. Truman is not aware of this—as far as he is concerned, this is "the real world." His acts in this false world are broadcast continuously on TV as a reality show in which Truman is the only authentic character. The world in which Truman lives presents to Truman a false presentation, which is supposedly as authentic as Truman is, until the painful moment of disillusionment.

Wearing the mental shoes of the other is a metaphor that relies on the understanding that, from the other's point of view, you are the other—i.e., you are the other in the eyes of someone else, and, in fact, you are the other in the eyes of any other person.

Seeing the soul of the other can be seen as the ability to look into somebody else's soul and to virtually walk the paths of life according to the walking pattern that is typical of this person. Mirror cells probably constitute a main biological layer of this skill.

The ability to look into the other's soul enables us to allegedly experience "how it is to be this other person," to feel this person's "whisperings of the heart" in our brain and in our "heart."

This ability results in the sense of empathy, the sense of identification with the other's suffering and distress. It seems that this ability is quite new in evolution's scale.

The inter-gender reading of the other's soul is most probably the most complex interpersonal psychology that serves as an inexhaustible source of jokes. For example, Bernard Shaw claimed, as a typical man, that attempting to understand a woman's brain is like reading the paper facing the wind. A common assumption is that a man's brain interprets the hidden tendency of another man's brain more easily, and, alternately, a woman's brain understands the secrets of another woman's brain more easily. The question is whether a woman will hit the secrets of a man's heart more accurately than a man when both the woman and the man know the other man quite at the same level. On the face of it, it seems that the "third eye," whose aperture is considered wider among women, will, in most cases, provide them with better resolution with regard to distinguishing between nuances.

The gradient of emotions turns the skull bone into a penetrable membrane. A person who is not a monk and secludes himself should acknowledge that his cognition is always in the midst of a voiceless dialogue with other cognitions.

Average and Uniqueness

Numerous studies have shown that the average hypothesis of an average person is that he is not average. In most cases, we tend to position ourselves above average with respect to various functional aspects. The need for differentiation and uniqueness is a powerful psychological motivation.

The sense of uniqueness is universal. Every person believes that he is different from the other and, in this spirit, the German statesman

Konrad Adenauer said that, although we all live under the same sky, each of us has his own horizon.

We also tend to exaggerate with respect to assessing the uniqueness of everybody else. We tend to believe that people are different from each other more than they really are.

In order to improve the understanding of the other and, by doing so, to contribute to our understanding of ourselves, the areas of our social brain try routinely to see things rationally from the other's point of view as "cognitive mindsight." We accept the fact that each brain has a different array of memes, a different array of insights, and different interpretive bias, which lead to a different worldview, to the creation of different creatures of thoughts, and to different behavioral tendencies.

On the other hand, this claim does not go as far as total relativism of insights and memes. Some memes are certainly better than others. A possible criterion for assessing the "moral" quality of a certain meme is the level of its contribution to the promotion of welfare and thriving of humanity and the biosphere in general.

Me and You

Self-focused considerations and empathy-driven considerations, in various mixes, guide our behavior.

Considerations that focus on our own benefit and considerations that focus on the benefit of others are a big part of the decisions we make throughout our life.

Once we become familiar to the owner of another brain, the mental representation that symbolizes us flickers on that person's screen of consciousness. The frequency and intensity according to which we appear on the screen of consciousness of that person depend on how important we are to that other person, the nature of the emotional relations between us, the frequency of our meetings, the extent to which we are able to assist this person at a given time ("the relevance coefficient"), etc. The mental representation that represents us is etched in the brain of the other as a passive representation when we are absent from the screen of this person's consciousness and becomes active once we flicker onto the screen of consciousness. In this sense, we "exist in this person's brain," and this person exists in our brain as present or absent.

Don't Judge a Person Until You Walk in Their (Mental) Shoes

Judging the insights and behavior of another person is fairer, in most cases, when the criticizing person has experienced similar circumstances. For example, the difficulties our parents faced when they raised us will be evaluated more reliably when we experience parenthood and deal with related difficulties.

When two people identify with each other, their mental maps undergo a sort of virtual merger and integrate. A conscious merger takes place, which sometimes leads to the question, where do I start and where do you end?

Failure in Seeing the Other's Soul

The pendulum of emotional identification with the other ranges from one end of complete defamiliarization—as was expressed by Jean-Paul Sartre, "Hell is other people"—to the other end of sweeping empathy—which puts the other's needs ahead of our own, sometimes to the point of nullifying the self and subjecting it to the needs of the other in the spirit of ultimate service offered to Aladdin by the genie: "Your wish is my command."

The mirror cells enable us to create an emotional portrait of the other's state of mind.

The spectrum starts from a hawk's view of the other's soul, among those who have the skill of empathy, through various levels of impaired sight, such as soft autism among those who suffer from

Asperger's syndrome, to total blindness with regard to the other's soul, which might exist among those who suffer from severe autism.

In absence of "mindsight" skills, our life is deprived of a key skill. Our relations with others will be of an alienated nature, and we might adapt a worldview that sees other people as objects ("objectivation of the other"—perceiving the other as an object) that lack emotions. In extreme cases, we might see them as worthless objects and become "mindblind"—a disorder that psychopaths suffer from.

Extreme difficulty with respect to mindsight of the other might lead to an infertile emotional dialogue between people; it does not rely on a common ground, similar to an argument between a person and a fish who are discussing whether it is more convenient to breath in water or on land.

Researchers of emotions distinguish between two types of empathy: warm empathy, which derives from feelings and is based on the brain areas that are in charge of producing emotions (primarily the limbic system), and, on the other hand, cold empathy: the ability to hypothesize the other's feelings by means of logic that does not involve any feelings. Psychopaths lack warm empathy, but they might be skilled in cold empathy, just as color-blind people are able to tell when the traffic light orders "Stop."

"The third eye"—which, according to various mystic theories, is in charge of mindsight of the other—matches, in terms of its traditional location in the forehead, the fact that mindsight skills are mostly produced in the frontal lobes. Damage to the frontal lobes severely impairs the ability to sense the other. In such a case, there is a retreat to an egocentric world in which the person, as an individual, feels as if he is the center of the universe.

The inner monologues in our brain are silent to other people, and vice versa. A human brain, however, is equipped with skills for indirect inference, which is rich in terms of its insights regarding the inner monologue within the other's brain. On the other hand, those who suffer from autism have difficulty in creating working assumptions regarding the monologues that take place in other people's brains.

One of the explanations for failures with regard to mindsight of the other among those who suffer from autism is that the intense inner monologue that takes place in their brain consumes all resources of attention. According to another hypothesis, they experience a constant flood of stimulations and find it hard to separate the wheat from the chaff. On the other hand, some autists demonstrate performances that are better than the average in terms of tasks that involve mental dismantling of complex inanimate systems and understanding the interrelations between the components.

The principles of game theory can be recognized in their inner-skull version. Often a zero-sum game takes place in the brain. In this game, focusing on one central processing task is performed at the expense of another processing task, which is ignored. Among those who suffer from various behavioral syndromes as a result of different wiring in their brain, the ordinary dynamic balance, which distributes attention and processing resources among the tasks routinely, is disrupted. Instead, there is a sort of dictatorship of a dominant functional system that regularly claims most of the processing resources. It might be that, in a similar pattern, among those who suffer from autism, one functional system takes control of the brain's agenda and abolishes the role of the "reading-the-other" system—a sort of dyslexia or alexia to the other's being. This is difficulty in reading to the point of a sweeping lack of ability to read the pages in the book of the other's mind.

"Mindsight of the other" requires attention resources. Whenever our attention resources are not focused on the other, there is a decrease in its function. One example is that when we walk in the streets of a busy city, we tend to introvert. A type of emotional barrier is created between us and the people around us. A sort of virtual glass, which is almost impenetrable, separates us from the ones around us. It is usually referred to as the "urban trance." The emotional echoing that a homeless person triggers among people walking nearby is sometimes vague and subdued, since our realty inputs undergo emotional inferiorization. We are at a state of depressed attention with respect to strangers around us and, therefore, tend to be "blind" to their emotions.

Unfortunately, our world is abundant in human suffering—every human being has an intimate correspondence with the suffering experience. Emotional wear that might affect people who are at the forefront facing human suffering, such as people who work at a hospice that treats terminal patients, might thin out their emotional resources and impair their function. They might be affected with "empathy failure" (a familiar term related to it is "compassion fatigue") and suffer from continuous depression if they do not take action to preserve their personal emotional reserves.

Though it is not a heuristic rule, and each brain owner has unique advantages, there is a common hypothesis according to which the tendency of the average male brain is to deal more with abstract, "emotion-free" systems; it is inferior to the average female brain with respect to analyzing emotional maps and reading the emotions of others. If we make a generalization, knowing that generalizations have built-in failures in terms of their application to specific people, we might say that the phenotype of the average male brain is inferior to the phenotype of the average female brain in terms of empathy skills.

Feeling Felt

The sense that the other uses mindsight skills while interacting with us is at the basis of the sense of "feeling felt." This is the sense that the echoes echoing in the cave of your skull create vibrations within the skull of the other as well.

The impact of our presence on the other is reflected in the other's looks and body language, like a ball bouncing from the walls of a billiard table. The lion's share of our self-image derives from the walls of the social billiard table.

We express a unique behavioral repertoire in the presence of different people.

A similar physiological state that takes place between us and the other at a given moment makes it easier for us to sense the other's emotions, and vice versa—a person who is satiated will have difficulties understanding the one who is hungry. Physiological compatibility increases the chance of mental and emotional compatibility.

Emotional compatibility between therapist and patient in therapeutic professions is a central layer with respect to the success of long-term therapy.

Emotional synchronization and minimization of the psychological distance between us and the addressee of the message intensify the impact of the message.

A human skill that is usually referred to as "emphatic accuracy" exists beyond basic identification with the emotions of the other (which is primarily mediated by "mirror neurons" at the subconscious level), and it allows cognitive understanding of the other's emotions. It is valuable in terms of predicting the intentions of the other. This skill is expressed in senior spouses who excel at decoding one another.

An elephant with silky skin – refinement of the built-in roughness of our interaction with the world is a high-level social skill. In order to create strong identification with the other, we must skip the "egocentric subject barrier" and rely on our natural ability to sense the other's senses within us, and translate this "mind reading" into gestures of generosity and emphatic "emotional climate" that is likely to create a spring-like emotional atmosphere at our work place, at home, etc.

The Truth? The Whole Truth? Nothing but the Truth?

Not speaking the truth is built into our communication with the other. The negative reputation related to this pattern should be restricted by understanding that telling a lie is context-dependent; it should not be considered as a negative thing just because it is a lie (we are all familiar with the term "white lies," which originate in good intentions and are usually intended to improve someone's mood or feelings). According to studies, if we were Pinocchio our nose would become longer between two to three times during each ten minutes of conversation. Studies suggest (taking into account the stigmatization trap that is part of sweeping claims) that women tend to lie more in order to improve the feeling of the other, while men do it to improve their own feeling.

When telling a lie, we tend not to use first person, emotionally charged words, whereas we sometimes provide superfluous details.

The skills related to lying and, at the other end, the ability to detect a lie told by another person are woven in the loom of evolution and wired in our brain as innate skills that are shaped in the light of the culture in which we live.

Thus, for instance, with respect to the "voice of truth," irregularity in the shades of the voice, its intonation or its rhythm—called dysprosody—is a vocal aspect we are highly sensitive to. Experts think that a telephone conversation is an inter-brain interface that allows us to sense when someone is not telling the truth (which is, in most cases, in correlation with dysprosody) more accurately than a face-to face conversation does, since most of us are more skilled in maneuvering our facial expressions and body language than we are in maneuvering our voice.

People who suffer from Parkinson's disease sometimes have difficulty in detecting prosodic clues (the tone of voice, the music of words) in other people's speech and, sometimes, also in inferring other people's emotions from their facial expression. Thus, they are more prone to err when assessing other people's intentions.

Lies are routinely woven into the texture of our social life, but they are not made of one piece. While some lies improve the durability of the social fabric that is being woven (such as white lies that are intended to minimize insult), others unweave the fabric.

It was found that men and women lie with similar frequency, but, as aforementioned, women tend not to tell the truth in order not to hurt other people, while men tend to make their nose longer in order to improve their own inner feeling. This is a generalization, with all it entails.

The pitted road of not telling the truth might turn into a slippery slope when new lies are often "required" in order to cover old lies that have already been told. Numerous energy resources are required in order to weave a lie, and a person who tends to lie frequently might enter a state of prolonged energetic deficiency. It is not that rare that the threads that weave the lie entangle into a ball that cannot be disentangled.

Important Brain Structures in the Cycle of Social Functioning

When we interact with the people around us, the abilities of synchronizing interaction, showing empathy, emotional accuracy, combining rationalistic dimensions in seeing the other, etc., are the criteria of social intelligence, which is a group of skills based on areas of the social brain.

In most cases, the more complex a thinking task is, the more scattered the involved nerves networks are.

The core cycles of the "social brain" (a functional definition from which an anatomic definition derives) are mostly the limbic system, the frontal lobes, brainstem structures and the cerebellum, which has an important role in the neural infrastructure at the basis of the ability to focus our attention, which is necessary for proper social functioning.

The mindsight glasses are mostly contributed from the front part of the brain. The mental representation of the other materializes mostly in the prefrontal cortex. The nonemotional layer of moral and ethical judgment is mostly contributed from the activity of the frontal lobes. Thus, the maturity of this brain area is at the basis of social maturity. The maturity of the frontal lobes does not take place at a specific magic age but, rather, during "age areas," and there is certain variance that reflects the individual brain. This maturity is the result of a graded process and, at any stage of the process, the question is related to quantity (what is the level of maturation?) and not to quality—of the dichotomous type of full maturation versus lack of maturation. The brain area called the amygdala is the main generator of tantrums—streams of emotions that erupt like geysers (a common expression of them is uncontrollable anger). When its function is more moderated, it is like a plate of colors that pours shades of emotions into the mental drawing of the experience.

Various studies show cases of functional impairment at the channels of the cingulate gyrus (a brain area in charge of contributing to outlining links between our consciousness and the emotional aspects of the experience), which is in correlation with a disorder that is characterized by poor, "flat" emotional expression (called alexithymia by mental health professionals), which, metaphorically, "pecks the eyes of emotions."

Unconscious synchronization, which is nonverbal, such as the complex choreography of body language that takes place during a conversation, depends on the activity of areas at the brainstem, the basal ganglia, and the cerebellum.

Chief ingredients of the potion that induces altruistic "trips" (focusing on the needs of other person) are oxytocin—the "female" socialization hormone—and vasopressin—the "male" socialization hormone.

A look is like a wireless connection between brains, similar to personal computers that create an interface between them without any physical connection. Eye contact—the visual input that is mediated through the eyes when we are in the midst of a dialogue with another person—actually connects the areas of the cortex above the eye sockets (which are called the orbito-frontal areas) that are in charge of producing information during face-to-face interaction. In this sense, the eyes are the window to our soul.

Paths of Trust and Doubt

Brain-imaging studies suggest that a central area in which the brain processes information it perceives as false or doubtful is the front insula, which is related to perceiving aversion and rejection, mostly with respect to smells and flavors. This can explain the saying "It does not smell good" with respect to lies.

The neural activation pattern that is correlated to the difficulty related to deciding whether a certain sentence is true is stimulation of the cingulate cortex, which is in charge of settling conflicts.

The ventral prefrontal cortex is a main brain area with respect to mediating between the output from brain areas that are in charge of factual-rationalistic assessment and the output produced by areas in charge of intuitive-emotional assessment. In other words, the prefrontal cortex is the place where the contents of the unconscious brain are being consciously doubted.

Emotional stimulations are mostly processed at the amygdala, from which the processed input is transferred as output to the hypothalamus (the bean-sized brain area located at the ventral area), which regulates physical reactions of various organs, and to the cortex—for more complex processing of the stimulation.

The brain area called "island" (Insula), which is hidden at the bottom of the lateral brain channel, processes sensory stimulations that are perceived as causing aversion and rejection.

Triggering the area of the insula might result in laconic, impulsive behavior of the type favored by comics artists due to its aphorismic nature, reflected in expressions such as "Yuck" and "Gross."

In a study during which subjects were asked to decide whether a certain statement was true or false, it was found that the classification of statements as true was made more quickly (which supports the philosopher Spinoza's claim that we tend to believe first, and start doubting our beliefs at a later stage).[51] Functional brain imaging studies show that the initial sense of trust is characterized by intensified activity at the prefrontal cortex, whereas the sense of doubt is characterized by intensified activity at the area of the amygdala.

Above the Eyes and Straight to the Heart

The cortex above the eye sockets (the orbitofrontal cortex) is also an important interface point between the courses of the upper pathways and the courses of the lower pathways. This area is the main evaluator of our social experiences (whether we had a really good time or, alternately, felt deep distress in a certain type of social interaction), and it is, in fact, the mint that coins hedonistic value currencies to our experiences. In the stock exchange where the value of different elements is determined according to their ability to please us, a place of honor is saved for the neural circuit of the cortex located above our eyes. These circuits are in charge of labeling our social world according to a scale of "hedonistic value," determining whose company we enjoy, and vice versa. The potential for the development of a romantic kiss is mostly determined in that area.

Personal and Interpersonal Synchronization

Synchronization of energies means matching the levels of energy among the people who interact. This synchronization increases the success chance of the inter-personal interface.

Synchronization during interaction between people is created by means of neural systems that are called "fluctuation generators." These systems determine the frequency of the "shooting rhythm" of information transmission among neurons and adapt it to the frequency of the perceived signal. These built-in pacers create the interface of our rhythm with the world and the people around us. They also create compatibility between various expression elements in our body and match, for instance, our body movement to the tone of speech.

Synchronization also exists in the sense of a subconscious merger with the rhythms of the other. The lower pathway is in possession of most of the neural oscillators that determine the shooting rhythm and the frequency of activity; thus, the lion's share of their activity derives from the nonconscious kingdom.

The synchronization is partially based on nonvocal language—the body language.

Thus, in a wireless neural network, emotional infection sometimes leads to energetic synchronization between the brains, related to cognitive and physiological functions, which crosses the skull barrier between the two brains.

A successful merger between two brains leads to a situation in which one brain's output becomes the other brain's input, and vice versa, in the sense of the reference point according to which they mutually "calibrate themselves" and metaphorically dance according to matching choreography.

The expiration of such synchronization usually leads to the expiration of the dialogue. When two people are having a conversation and one of them (or both) feels that the conversation has exhausted itself, they step out of synchronization (which is reflected, when it still exists, in nonverbal physical gestures that match the topic of conversation, timely looks, and facial expressions that fit the situation) and, by that, a silent signal that indicates that the conversation should come to an end is given.

Different levels of mental and physical energy among the participants of the interaction might constitute a significant obstacle to its success.

Many parents of young children are familiar with situations in which there is a lack of compatibility (desynchronization) of energy levels: when they return from work, exhausted, their little child has just woken up from his afternoon's nap and he is vital and full of energy—the energies of parent and child are desynchronized. Thus, the famous claim that all mothers are glad when their children, as cute and precious as they are, finally fall asleep. In situations of energetic synchronization, the duet of interaction between parent's brain and child's brain is much more melodic.

The sounds of silence: Music is based on single sounds, and the rhythmic organization (the intervals between sounds) determined by the internal metronome of the musician has a significant effect on the quality of the music. The famous violinist Isaac Stern once said that all the people who were certified in a school of music knew how to read notes, as he did. Stern added that real art resides in the silence between the notes, and that this is where emotions and talents are hidden.

Listening to music together creates a sort of "common metronome" that synchronizes the brains of the listeners, and, usually, emotional synchronization that crosses skulls' boundaries and matches the nature of the musical piece is created, as well.

Some find similarities between social behaviors that are composed of single activities of numerous individuals grouped together (i.e., social processes conducted by multiple participants) and the principles that derive from thermodynamics rules and the Brownian motion of molecules in a solution. And, similar to the different levels of solidarity in different human societies, it is possible to refer to the human texture of a society as existing in between a homogeneous solution and heterogeneous mix.

The Modality of Socialization

A famous study estimates the "people distance" that links any person to any other person on Earth as six or seven. In other words, a course of six to seven people, each one of whom knows a reference group that is closer and closer to the desirable person, leads from a housewife in the suburbs of Bogota, Colombia, to Queen Elizabeth in Buckingham Palace in Britain. This seemingly proximity, however, conceals the fact that sometimes the skip from one acquaintance to the other is like a skip from one cultural galaxy to another.

The rules of this social chess game, which are based on interpersonal psychology, are much more complex than the rules of an ordinary game of chess.

The mask of mannerism we wear is partially transparent with respect to other people who are familiar with the secrets of the algorithm of socialization, which is considered a norm in the society in which we function. Deeper interaction usually requires the creation of unique contents in our brain that do not derive from general, non-unique behavioral codes.

Each period changes the pattern of socialization among people's brains. Thus, for example, the communes that were popular across the western world in the 1960s were formed as a microcosmos in which the psychological climate was often characterized by sweeping juvenile joy, which served as a protective wall against the winds that were blowing outside. Socialization, then, was often based on holding a common ideological set of beliefs and served as a greenhouse in which additional insights grew, often in the image of the ones that preceded them. The period we live in is characterized by winds of change as, some people claim, are well described in the words of Lev Tolstoy in his book *War and Peace:* "…multitude of people in which no one is close, no one is far." In the era of the Internet, many people have replaced participating in "flesh and blood" community activities with alternative participation in activities of a virtual community. The possibilities of Internet socialization rewrite the experience of human socialization (for better or for worse).

The critics of online socialization claim that once we became "friends of everyone" on the social media, we forgot how to be a "friend of someone." We became "wholesalers," in the sense of proposing friendship that minimizes the individual value of each of our friends. We prefer "instant friendships" that are formed by a keyboard's clicking to building a friendship attachment as we did in the past, through face-to-face meetings and heart-to-heart talks that are not mediated by technology.

The range of socialization possibilities allows a higher degree of selectivity with respect to choosing our friends, in the spirit of the saying "Friends are the selected family" (as opposed to the biological reality, in which we were born to a family that was not chosen by us). This is, perhaps, the advantage of "socialization technology." Many of us, however, have a virtual mega-family, whose members have temporary, superficial relationships as a routine—but real friendship, which is based on long-term relationship, is rare. Some might say that, in this era, the more friends we have, the lonelier we are. The tribal fire was put out; the circle of people who used to sit around it, shoulder-to-shoulder, has evaporated, and now many of us are isolated within the walls of our electronic cave, trying to reinvent a "tribe" whose members are represented by pixels on the computer screen.

As technology progresses, the virtual world simulates reality in a way that becomes more and more reliable and blurs the borderline between truth and fiction even further, to a point at which it is hard to see through the screen that distinguishes between reality and virtual reality.

Culture, which is mediated by technology, shapes our brain more and more intensely as years go by. At the same time, the skills of inter-personal, direct relations are dying out. As the poet T.S. Eliot commented in 1963, when television started its campaign of conquering the hearts (and brains) of people as preferred mode of recreation, "Television is a medium of entertainment which permits millions of people to listen to the same joke at the same time, and yet remain lonesome." The impersonal nature of mass entertainment has contributed, and still contributes, to a decrease in the value of inter-personal, direct interaction and changes the activity patterns of our social brain.

The Journeys of the Brain in Terra Sexica (the Land of Sex)

Brain areas that mediate our sexual behavior are approximately two times bigger in the brain of an average man compared to those in an average brain of a woman.

The sexual desire (libido) is like a generator that generates tension, in a changing pattern, and is tuned to sexual behavior that results in relief and decrease of tension, and so on, in a repetitive pattern.

The libido is intensified by testosterone in both men and women, and among women by estrogen as well.

The potion in which our brain is immersed during an orgasm is composed mostly of a spurt of the neural transmitters dopamine, oxytocin, and endorphins. Immediately after an orgasm, the level of oxytocin rises among both men and women.

Stress as an Aphrodisiac or its Opposite

It seems that cortisol, the stress hormone, has an inverted effect on the brain of women and men in terms of the tendency to fall in love. Women in stress do not tend to establish a romantic relationship, whereas among men, stress and accompanying cortisol act as an aphrodisiac that increases their craving for female company.

Among men, reward-driven brain areas become alert at times of stress, whereas the same brain areas become subdued among women.

The Day Before the Cigarette After

The complexity of the female orgasmic process has led many researchers in the field to think that, often, a woman's ability to experience an orgasm depends on the collection of experiences she has undergone during the previous twenty-four hours. This is the average period of time of the female's pregame. The average pregame time required for men is about three minutes.

On a Journey to the Supernova

Studies have shown that a necessary (though not sufficient) state for female's orgasm is a sense of relaxation, whose main expression is a decrease in the activity of the amygdala and... warm feet.

An average woman requires three to ten times longer than the average man in order to reach orgasm.

A possible explanation for that lies in the fact that studies have shown that the chances of becoming pregnant increase when the woman reaches orgasm after the ejaculation of the man. Thus, if a woman wishes to get pregnant, the man should not act like a gentleman and finish last.

A Brain in Love—Under the Spell of Romantic Love

Love and attachment are basic layers originating in our evolutionary and cultural heritage.

Thus, for example, the first time the expression "not good" appears in the Bible, is in the verse "It is not good that the human being should be alone."

In Dante's *Divine Comedy*, love moves the sun and the stars. Love in its version of "romantic love" for the masses, in which people select the object of their love, is a young phenomenon in terms of historical review. In the absence of reliable historiography, some place the creation of romantic love in the Renaissance period, from where it embarked on a journey to conquer the heart of the world.

In brain-imaging studies, it was found that, during the formation of the process of falling in love, the most active areas in women's brain are those that process memory, attention, and intuition processes. On the other hand, among men, excessive activity was found in the areas in charge of processing visual input. Thus, we might claim that the pattern of "love at first sight" is mainly a male pattern.

Many evolutionary psychologists believe that, in the "casino of match-making," our brain directs us to gamble on the one it feels has the chance of yielding maximal "genetic yield."

About four hundred mature eggs are produced in a female body throughout the years of fertility. The male body ejaculates, during orgasm, about 250 million sperms. These great gaps, with regard to the availability of reproductive cells, have contributed to the development of different courting and reproduction strategies among men and women.

On the other hand, nowadays a big part of our repertoire of behaviors above the covers and under them does not obey the evolutionary dictate that is directed at conception and procreation.

The Secret Cocktail in Which Cupid's Arrows Are Immersed

The hormone cocktail in our brain is like a chief choreographer of the communication gestures between men and women.

The romantic dance takes part considerably according to the directions of hormonal choreography. The hormones are the "application assistants" of desire—the invisible threads that move the marionette of urges in our brain.

The intoxicating cocktail smeared over the arrows of Cupid is created from a mix of oxytocin, vasopressin, dopamine, phenylethylamine, testosterone, estrogen, etc., whose level within our brain increases at the time of falling in love.

The oxytocin and opioids come together and lead to the sense of pleasant relaxation we feel when we are close to our lovers and they are also attached to the orgasmic experience.

As for the level of testosterone at the time of falling in love, opposite changes take place in the female brain and the male brain: among men in love, the level of testosterone in the brain decreases, while the level of testosterone among women increases. In this sense, we might say that, at the time of falling in love, the female brain becomes a little "masculine" and the male brain becomes a little "feminine."

The level of serotonin in a brain in love is greatly decreased (thus, some think that antidepressants that are based on increasing the level of serotonin at the communication junctions between neurons decrease depression and the tendency to fall in love and, in this sense, serve as the opposite version of love potions).

It seems that there is not a single soul on Earth who does not oppose the replacing of the bill of intimate love with coins of neurotransmitters' mixture, though this intuitive opposition is not sufficient to blunt the sword of heartbreaking facts that cause us to act, at least partially, as marionettes that move according to the instructions of the threads of biochemistry. On the other hand, the claim that a complex phenomenon like love can be fully explained by the biochemistry discipline is haughtiness whose pretension is not supported by sufficient evidence.

The Nectar in the Absence of . . .

The period that comes right after separation from a significant person in our life is described as the time in which that person is present in our life more than ever. When a person who is important to us vanishes from our world, his mental representation often appears frequently on the screen of our consciousness close to the time of disappearance.

The absence of the loved one induces in the spouse's brain a "withdrawal syndrome" that also has a biochemical expression, although it is usually referred to only as a "mental phenomenon."

The Minus is Missing from Love Arithmetic

When we are in love, our brain is prone to avoid critical thinking about the weaknesses of our loved one. The minus sign (-) is missing from the arithmetic of romantic love. The areas that are in charge of anxiety—mostly the amygdala and the cingulate gyrus, which is thought to have a major role in the doubtful brain—are active, as if they are on a go-slow strike, and their output reduces significantly. Hidden from the burning sun of reality, in the dim shadow of romantic love, our uncompromising willingness to compromise thrives.

"The terror of the ideal" is a frequent companion of romantic love, and it dictates an outline that does not match the plane of reality.

Romantic love—a kingdom whose capital is a dream palace, whose anthem is "paradise now"—is a place desired by everybody, although its rules are illogical.

Some see it as a journey, during which people are looking for their own happiness—within the happiness of the loved one or, alternately, their self, within the self of others. When physical intimacy is accompanied by emotional intimacy, the separate being of one person is melted and merged into the being of the other person.

The duration of romantic love is limited. There are not any safety nets for the acrobats of the show of romantic love, and sooner or later the lack of a safety net becomes apparent.

The ingredients of the intoxicating cocktail, in which Cupid's arrow are immersed and that floods our brain during falling in love, change as time goes by, and their intoxicating effect vanishes. Numerous researchers indicate a period of three to four years, similar to the life expectancy of romantic love, before the burning thread of romance perishes. There is a teleological explanation to this phenomenon, according to which this is the estimated period of time during which the craving is realized by an act of lovemaking whose result—the fruit of love—the offspring—reaches partial independence.

And what comes after romantic love? On the "day after," the possible scenarios range from relationship characterized by deep affection and respect (a sort of quiet love—as is common among married couples who live in relative harmony for a long period of time) to losing interest and returning to the hunting fields for the next prey.

The Brain in an Intimate Relationship

Love and marriage—or was Frank Sinatra right in his song?

"Arranged marriages" are popular in traditional societies. The families of the designated bride and groom arrange the treaty of marriage, and the ones who are married accept the verdict.

Through a psychological questionnaire called "Rubin's questionnaire," which quantifies the illusive intensity of love, researchers have attempted to assess the height of the love flames in various marriage scenarios in different societies. The averaging of results revealed that the flames of love among Indian couples who married by choice are not stable, which is similar to the unstable pattern that is common among western couples who married by choice and out of love: at first, the flames are very high, and then they become shorter, as a reflection of the decrease in the intensity of love. On the other hand, in arranged marriages, the situation is inverted: the height of the love flames at the beginning is low, but the flames become higher and higher as time goes by. Within five years, the height of the flames in both types of marriages becomes even, but it follows two contradicting trends: in cases of marriage by choice, love loses power, and, on the other hand, in cases of arranged marriages, it becomes greater. Ten years after getting married, the "love dial" among those whose marriage was arranged points to a value two times higher than the value pointed to by the dial of those who married out of love and by choice. A possible explanation for that is that couples in arranged marriages, in which one of the spouses is not particularly underprivileged, "learn to love" their spouses differently, and they are less affected by the whims and limited time of romantic love.

The Biochemistry of Spouses' Loyalty

In the long-term routine of couples in intimate relationship, due to the long-lasting interaction, each one of the spouses shapes the other's brain. In this context, it is said that the loving person declares that he loves his loved one not only because of who she is but, also, because of who she makes him be when he is with her. In other words, it is the unique tune that the fingers of her presence play on the piano of his brain.

An interesting study found that, in a stable, happy relationship, the common ratio between experiences that are followed by positive emotional resonance and those that are followed by negative emotional resonance, in terms of interactions between the spouses, is about five to one (in favor of the positive ones). Five to one, in favor of good relationship.

Throughout the lifetime of successful relationship, the spouses move from the stage of excited infatuation to the stage of supportive, intimate relations as the level of dopamine decreases and, on the other hand, the levels of oxytocin and vasopressin increase.

The vasopressin is the male version of oxytocin, and it induces a need for social closeness and attachment in the brain of human males. There is a larger amount of vasopressin receptors in men's brains and a larger number of oxytocin receptors in women's brains. The male brain requires both the effect of oxytocin and vasopressin in order to adhere to a romantic relationship.

Testosterone increases the production of vasopressin and the intensity of orgasm.

Vasopressin increases the level of attention and vitality, but also the level of aggressiveness among men.

Among various types of mammals, including human beings, it seems that an increase in the level of the neurotransmitter oxytocin in the brain prepares the females' heart to select a spouse and adhere to it, whereas an increase in the level of the neurotransmitter vasopressin induces the same tendency among males.

An accurate, reality-compatible view of things shall not dismiss the great complexity of the structural bedding (i.e., the brain structure) on which this "simple chemistry" operates. Its "simplicity" is enabled due to the structural complexity. Nevertheless, the ability to affect a complex behavior by spraying some oxytocin or vasopressin around the nostrils of the owners of a human brain evokes some thought. Various commercial companies also recruit this esoteric knowledge—exactly the type of knowledge that might be used cynically and unfairly—and implement it on romantic dates (such as using a spray called "Liquid Trust," which is marketed commercially).

Oxytocin was named "the trust hormone." In a study conducted in Switzerland, it was found that participants who received nose spray that contained oxytocin tended to offer the other participants sums of money that were twice as big as the sums offered by the participants who received a spray that did not contain any active substances (placebo).

Among people who suffer from autism, a low level of oxytocin was found, and it is possible that this fact contributes to certain behavioral characteristics, including their difficulty in feeling empathy and trusting the other.

The monogamous party (a party of loyalty between spouses) represents a minority of five percent in the parliament that includes all types of mammals on the planet. Oxytocin is probably the ticket to entering this club, since it has a major role in inducing emotional relations between spouses.

Length matters: some look for the code of loyalty in the genes. It was found that the longer the gene version for the receptor for vasopressin among men is, the greater their tendency toward monogamy and active fatherhood; thus, it seems that the dream of any young bride should be that her groom is a carrier of the longer version of the gene.

The Brain and Parenthood

The Birth of Parents

During the last three months of pregnancy, the brain of the woman decreases a little in size. It does not occur in a unified pattern, since certain areas become thicker while others thin out. This condition starts to change about two weeks before giving birth and goes back to the original state within six months after labor, so it seems that this process does not involve loss of neurons, which is an irreversible process in most brain areas, but, rather, takes place due to general changes in the tissue, such as an increase in the volume of cells—a reversible process.

Neurohormones—hormones that affect the activity of neurons directly—have a major effect on inducing a parenthood state of mind among young parents.

The Birth of a Mother

Oxytocin induces maternal behavior in a manner of chemical imprinting.

Babies and the changes they induce in our brain, which are followed by changes in our behavior, sometimes seem like an evolutionary adaptation in a movie that might be called *The Invasion of the Brains Abductors.*

A newborn baby is equipped with powerful methods of psychological impact, in order to manipulate others to fulfill its needs, and which are mostly designed to trigger caring parental instincts.

According to certain studies, breastfeeding, which has numerous advantages, might induce a condition of lack of focus and attention deficit in the breastfeeding mother. It seems that the constant flow of prolactin and oxytocin might be the cause of that. These phenomena are reversible, and when the woman quits breastfeeding, and the amount of hormones decreases, these phenomena vanish.

Mothers' falling in love with their babies shares a common pattern with any other kind of falling in love, and the brain of mothers also experiences spurts of dopamine and oxytocin when they look at their babies or when they are close to them. This cocktail, as always, dims critical and judgmental thinking and is at the basis of maternal attachment, which, at its peak, is totally free of selfishness.

In most cases, the button of maternal pleasure is pressed time and time again, and the love for the offspring becomes a routine state of mind.

Stress containment, or serenity, induced by the people who take care of the baby, which is a type of "epigenetic imprinting"—one that is not imprinted in the genes themselves—is imprinted in the babies' state of mind and might pass from one generation to the next, in human beings and in other mammals.

It seems that there is a higher level of "internalization" of the neural surrounding of the parents among female babies. Parents who frequently express signs of stress cause an epigenetic imprinting of stressed-out dealing with the challenges of life among their babies, and girls are probably more sensitive, in this respect, than boys.

The Birth of a Father

In praise of the male's brain capacity for empathy, a domain in which there is a tendency to consider the man as a "deprived boy," one can indicate the couvade syndrome, in which future fathers experience pregnancy symptoms similar to those experienced by their pregnant spouses, such as severe morning nausea (various expressions of this syndrome were reported among two-thirds of future fathers!).

The brains of numerous future fathers showed an increase in the production level of prolactin up to 25 percent, and a reduction the level of testosterone in the weeks prior to the estimated date of birth. Some claim that pheromones (volatile molecules that can potentially change behavior) from the pregnant spouse's body contribute to the induction of this condition.

Aspects of Mother-Baby Communication
The Maternal Magic Touch

The heartbreaking experiments of researcher Harry Harlow on infant monkeys, who were separated from their parents and put in isolation, showed that they clearly preferred a dummy simulating the mother monkey, made of soft cloth but without a nourishing nipple filled with milk, to a dummy made of wires that had a nourishing nipple filled with milk. The little monkeys sucked quickly from the nipples of the wire mother and returned to the silent but soft bosom of the dummy mother made of soft cloth, where they spent hours. According to the words of the experimenters, "the touch variable was far more important than the feeding variable." The soft touch of the still, silent dummy, however, was not sufficient. The infant monkeys who grew up in the bosom of the cloth mother but were deprived of the company of their own kind during the first months of their life suffered from severe behavioral disorders at later stages of their lives (such as great hostility expressed during interactions with other monkeys, and lack of "parental skills" when rearing their own offspring), which prevented them from integrating in an environment of monkeys who were reared the natural way in the arms of their real monkey mothers.

We can infer from that that parental warmth, in general, and maternal warmth, in particular, is essential to the development of basic social skills and constitutes a role model for the mirror cells. It seems that these cruel experiments broke the little monkeys' hearts so that their "skills of the heart" were irrecoverably damaged. The implication for human infants is self-evident.

History has provided a tragic illustration of the validity of the hypothesis regarding the importance of critical, age-dependent windows of development in terms of social functioning skills among humans as well.

In 1966, Nicolae Ceausescu, former ruler of Romania, published a decree that prohibited abortions. The only exceptions were for women over the age of forty or women who had at least four children, who were allowed to have an abortion. Until 1989, when his rule ended, two million babies, called "the decree babies," were born in Romania, and many of them were sent to government orphanages. At the end of Ceausescu's rule, about 170 thousand children were living in more than six hundred huge institutions—the government orphanages in which there was an extreme shortage of caregivers. As a result, thousands of babies lay deserted all day long in their cradles—fed by a bottle attached to their beds, at the appropriate height for their mouths—without human touch, maternal voice, or caressing hand. This lasted for years. When the magnitude of the horror was revealed, the wrongs were partially corrected, but the scarred souls of many of the graduates of these orphanages could not be cured. Many of them suffered (and still suffer) from severe impairments in terms of social functioning. Many of them had a hard time integrating into the working world, building a family, and functioning as parents. In general, they had difficulties in integrating into cycles of normative social functioning. Many of them continued to live at the margins of society, suffering from a high percentage of mental impairments and functioning difficulties. The absence of a caressing voice and hand, touch and play, had irreversible effects on the development of social skills, and it also led to developmental delay, irreversible at times, at the intellectual level, as well.

Speaking "Motherese" means adjusting the voice's tone and melody to create a pleasant, reassuring echoing in babies' brains. Prosody (the manner of pronouncing the syllables—the melody of words) is similar in all cultures. "Motherese" is a universal language that crosses cultural borders. The main aspects of this language are high, cheerful vocal sounds (major) uttered in a rising-falling tone. The maternal body language is in harmony with the speech prosody.

Mothers who speak Motherese well are, in most cases, mothers who manage to create energetic compatibility with their babies.

Speech addressed to babies is rich in "prosodic clues," since the semantic contents do not constitute a significant component at this stage. This is how the emotional message of intentions is conveyed.

Babyese—the Esperanto of Babies

Language serves to convey information through prosody, syntax, and semantics (the meaning represented by words). Young babies have yet to master the secrets of language in the syntactic and semantic sense, but they are very attentive to the prosody and the emotional messages it represents. Thus, "Babyese" (the language of babies) is a sort of "Esperanto of world babies," and its characteristics are similar in different human cultures. The sounds of speech, addressed to babies, are higher, in most cases, and the pace of speech is usually slower. Thus, for example, words that mean prohibition are uttered in a low pitch, and vice versa: words that mean approval are "played" in a high pitch. "Babyese" stresses the vowels and increases the distinction between them in the scope of vowels (stretching the scope of vowels), which creates a sort of caricature-like speech.

An interesting fact is that adults who listen to speech in an unfamiliar language are, to some extent, in a position that is similar to babies' position, since there are not enough syntax and semantic clues to decipher the information, and they rely on prosodic clues, as babies do.

The crying of a human baby accelerates the amygdala in the parent's brain to a maximum level of activity within a shorter time than the time needed for a modern sports car to go from zero to a hundred kilometers an hour. Analysis of the magnetizing influence of a baby's crying showed that the unexpected fluctuation in terms of pitch (called "vibrato") is strumming the strings of emotions very powerfully. The crying sounds, to the parent, like the ultimate emergency call, and it was also found to be the most effective method of ordering "take away" food.

Faces That Capture the Heart

A structural characteristic of a human face that can be seen as a characteristic of universal beauty is its level of resemblance to a baby's face. This insight, first defined by ethologist Conrad Lorentz, was confirmed in fMRI studies. It was found that when we look at faces with a high "index of childishness" (round face, big eyes), we experience aesthetic pleasure, which is reflected in intensified activity in the area of the nucleus accumbens in the brain.

Pleasure and the One That is Bigger than It

Taking care of the offspring saturates the mother's brain in generous amounts of dopamine and oxytocin, which might be a possible (and partial) explanation for the fact that new mothers often do not feel the need for other relationships that (potentially) saturate the brain with the same cocktail. The decrease in new mothers' interest in sex is a well-known phenomenon. Taking care of the offspring sends virtual, though powerful, fingers to the maternal pleasure button and decreases the potential of other interactions to press it with similar force.

The offspring's needs become the biological commandment for the mother.

Blue Brain, Pink Brain

The saying "Anatomy is not destiny" is well known, and, in this context, the French philosopher Simone de Beauvoir claimed in her book *The Second Sex* that "One is not born, but rather becomes, a woman" (which, naturally, can also be implied for men). It seems that these sayings are partially true. The repertoire of gender-dependent behaviors is woven out of the dialogue between the innate (genes' dictate) and the acquired (cultural influences).

Language and Sound in the Mirror of a Woman's Brain and in the Mirror of a Man's Brain

Talking Facts

The number of neurons in the centers of language processing and hearing in the brain is 10 percent bigger, on average, among women compared to men. Typical differences in language formulation between girls and boys (gender-dependent dialect) are expressed in the fact that boys tend to use imperative or threatening tones, whereas girls tend to use wording that encourages cooperation. Though it is a generalization, which might not be correct with respect to an individual, truth can often be detected in it on the average.

There are observations that suggest that the lingual repertoire of the average man is, in most cases, poorer than a woman's repertoire, but a comprehensive study, which is considered very thorough, showed that both men and women "produce" a similar amount of words daily; both men and women said about sixteen thousand words a day on average.[52]

The range of prosody (distinction of nuances in the tone) among women is more diverse; women are capable of detecting vocal nuances, such as the range of frequencies, with better resolution, compared to men. On average, women's brains are more capable of detecting nuances that map the emotional state of others.

Feminine Brain and Masculine Brain

Although the origin of both sexes is the same planet (which is not Mars or Venus), their brains, on average, have different tendencies. A typical gender-dependent behavior is a result of innately different brain wiring and reciprocal relations with the environment.

According to common hypothesis, the feminine brain excels over the masculine brain, on average, in empathy skills and mindsight of the other. On the other hand, it is commonly assumed that the masculine brain is more prone to systemic understanding of basic rules at the basis of complex mechanisms. We might claim that women are stereotypically viewed as "wired to relationship" and men as "wired to competition."

According to a cynical cliché, women get lost in foreign cities and men get lost in relationships.

Many men and women have a "balanced brain," which represents a similar mix of systemic understanding, a tendency to compete, and empathy skills. According to a common supposition, the brains of about 5 percent of the population have extreme "feminine" characteristics or, alternately, extreme "masculine" characteristics. According to the famous autism researcher professor Baron-Cohen, people who suffer symptoms of soft autism (called "Asperger's syndrome") are, in fact, the owners of a brain characterized by extremely "masculine" characteristics.[53]

It seems that our gender does not necessarily determine the tendencies of our brain. The insights related to "gender-dependent brain" derive from averaging. It is certainly possible for women to own a brain that has a high tendency toward competitiveness, or for men to have a high tendency toward empathizing.

Brain Seasons in Pink and Blue

As aforementioned, among both men and women the hands of time in the clock of brain seasons indicate daily and monthly fluctuations, and fluctuations that are compatible with the various periods in life. This can be roughly compared to the hands of seconds, minutes, and hours.

The Gender Trip—Blue Trip and Pink Trip

Feminine brains and masculine brains tend to dip into different pools of neurotransmitters and thus experience different types of "trips."

To a certain extent, the owners of a masculine brain are forced to be subjected to the mischief of the androgenic hormones (testosterone and its derivatives), and owners of feminine brains usually find shelter in the shade of estrogens. This fact results in two brains that share numerous similarities but also significant differences.

The cocktail of hormones, which changes over the course of our life, greatly affects our desires and the way we perceive reality.

Among women, during the years of fertility, the brain changes a little every month due to the influence of hormones. The monthly period is also a brain cycle and not only a cycle of the endometrium.

Although the hormones do not lead to a certain behavioral pattern, in a deterministic manner, their level induces a convenient climate for the sprouting of certain behavioral (and thinking) patterns, and they induce a higher tendency or "general probability" of adopting such thinking or behavioral patterns.

The weather during the feminine seasons of the brain is greatly affected by the monthly period.

Thus, for instance, the premenstrual syndrome, which affects many women prior to the monthly period, might pour into the drink of reality perception a few drops of gloom that do not derive directly from external reality itself but, rather, from an emotional tendency induced by the hormonal cocktail prior to the monthly period.

Wave-Surfing as a Way of Life

The brain of a woman during the years of fertility sometimes seems as if carried by waves of progesterone–estrogen, whose changing levels lead to emotional fluctuations.

During the first two weeks of the female monthly period, estrogen is the prominent hormone, and one of its impacts is increasing the volume of the hippocampus in about 25 percent of women on average. It seems that the increased volume leads to improvement in memory performance. During the second part of the period, the main hormone is progesterone, and it seems its impact slows down, a little, the main thinking functions, as was reflected in studies that examined measurements of attention and concentration. The progesterone suppresses the influence of testosterone and decreases sexual desire.

During the last day of the period, while the level of progesterone is decreased, the brain becomes more sensitive to stimulations, and sometimes there are fluctuations with respect to its emotional conduct.

Premenstrual syndrome might lead to extreme emotional reactions, due to the impact of the balance of powers in the brain, induced by hormonal changes.

The estrogen improves memory capabilities. Its beneficial effects in this respect were observed in women whose monthly periods stopped and who took estrogen as an alternative hormonal treatment, although such treatment also involves some medical risks that require cautious consideration.

The monthly period induces a transition, similar to "musical chairs" in the brain parliament, which regulates thought and emotions, in the shape of changes in the activity mix at various brain areas.

In a simplistic, generalizing note, we might say that the amygdala has more seats at the brain parliament during the days that precede the monthly period, whereas the prefrontal lobe has more seats at the parliament mostly during the first two weeks of the period.

Advanced Maturation in Pink and Blue

The average age of cessation of the monthly period is fifty—at this age, along with the decrease in the level of estrogen, the level of oxytocin drops as well, and a biochemical profile that might lead to a tendency toward decreased interest in the more subtle shades of emotion is formed.

A few years prior to the cessation of the monthly period, at about the age of forty-three, the female brain becomes less sensitive to estrogen, and some see it as a sign of the beginning of perimenopause. The level of estrogen decreases, along with the level of testosterone, and interest in sex reduces.

The period that precedes premenopause is sometimes a period that, in certain senses, constitutes a mirror image of the hormone-dependent emotional chaos of adolescence—"just without the fun," as was defined by a woman of this age.

The male brain and the female brain age according to a different pattern; the male brain loses cells of the cortex sooner than the female brain does.

The stream of testosterone, which flows more intensively in the blood of alpha males of the primate society, grants them their excessive capabilities on the scale of apes' machoism. The implications for their human cousins are clear.

Although the level of testosterone decreases as men age, their brain does not experience a sharp drop of the levels of hormones, as a woman's brain does.

As men age and the level of testosterone decreases, the volume of activity in the amygdala decreases as well, and the vector of influence of the prefrontal lobe increases in the parallelogram of forces versus the amygdala. This increase moderates the tendency to feel anger. An aging man is less prone, on average, to behavior that is urge-motivated, and the influence of the vectors of moderation and restraining, in the behavioral parallelogram of forces, become more prominent in correlation with a decreased amplitude of the wave of testosterone in men's brains.

Anima and Animus

According to psychologist Carl Jung, Freud's student, behavioral characteristics that are stereotypically considered masculine and were merged into the cocktail of feminine behavior are called *animus*. Alternately, the behavioral components that are stereotypically considered feminine and were merged into the cocktail of masculine behavior are called *anima*. According to Jung, in midlife behavioral changes take place that expand the behavioral repertoire of gender-dependent behaviors, so that both men and women incorporate the anima and the animus, respectively, in their behavior, and are less prone to stereotypical, gender-dependent behavior. It seems that, as we grow older, our brain turns more purplish, as the "pink" female brain gets some of the "blue" shade that characterizes the male brain, and vice versa. We complement ourselves from within ourselves.

The Illuminated and Shadowed Areas of the Soul

According to a common view, as adults we accept with more understanding the "shadowed areas" of our personality (elements we considered unworthy in the past) as part of us; they are embedded in our consciousness and experienced as components of our personality, even if we still refer to them as "human weaknesses" and nothing to be proud of.

The hippocampus in women's brains is relatively larger than the one in men's brains, and some relate it to the better ability of women to retain emotional memories.

The volume of the hippocampus in a woman's brain deceases, as aforementioned, during pregnancy and regains its normal size a few months after giving birth.

The amygdala, which is, inter alia, the main generator of the sense of fear and emotion-derived urges such as aggressiveness, is larger in men's brains, and some relate it to men's tendency to act more competitively than woman. The amygdala in men's brain also contains more testosterone receptors, which makes the finger of men's amygdala more prone to pull the trigger of motivation for aggressiveness.

A Woman's Word and A Man's Word

Generally speaking, we might say that several studies show that women "document" social situations better and detect "emotional nuances" that accompany them better than men. On the other hand, men were more accurate regarding details with respect to violent events. It might be that the emotional barometer among women is more prone to failure in extreme violent situations, which compromises their ability to preserve the impressions of the experience as it is.

Philosophy of Gender-Dependent Risk Management

The different brain of men and women also results in a different behavioral tendency with respect to risk management philosophy in various aspects of life. When referring to five fields of coping with life—health-related, morals-related, financial-related, social-related and recreation-related—it was found that women are significantly more deterred by risks, compared to men, in four out of the five fields (except for the social-related field).

Chapter 17
The Aging Brain

Old Age—Advantages Along with Disadvantages

During the passage from childhood to old age, it seems that time turns from a friend to a rival. The birthday experience turns from a celebration of achievement to a painful reminder, which indicates that many of us turn from chronophiles to chronophobes.

Comical descriptions regarding the hardships and disadvantages of old age are common. "It's like playing billiards with a rope" was the response of comedian George Burns when he was asked about having sex at the age of ninety. One financial disadvantage is that, in old age, the candles cost more than the birthday cake.

"I am a person": Homo sum, with all of the weak and strong points that come from it. Even at an old age, recognizing the weaknesses and strengths that match our age will allow us to minimize their damages and maximize their strengths respectively.

Steps of Acquisition and Loss

The advantages of natural aging are acquired gradually and with hard work, and so are the disadvantages, which usually reveal themselves gradually.

It seems that, at an old age, we tend to encode less and less specific memories, such as memories of the episodic type that have personal significance. A real-life example can illustrate this: After a grandmother and her young granddaughter watch a children's play together, it is likely that the granddaughter remembers various details from the play better than her grandmother. The "average" grandmother is likely to have watched a number of plays in the course of her life, and the plot of this specific play is not that "special" to her, as opposed to the granddaughter to whom it is very special.

On the other hand, the grandmother's brain is filled with more inclusive (generic) memories that constitute a type of supercategory for more specific memories, which are considered the basic layer of wisdom. Wisdom is based, to a large extent, on a multitude of generic memories, which allows for rapid recognition of patterns that can be applied to a wide variety of phenomena.

Studies show that processing-related laterality (i.e., the different types of input that each of the hemispheres tends to process—for example, the natural tendency of the left hemisphere is to process verbal input, whereas the tendency of the right hemisphere is to process visual-spatial input) weakens in the aging brain.

The activity in the frontal lobes during performance of a cognitive task tends to be more bilateral among old people compared to the tendency to "choose sides," which is common among young people. As we become older, the asymmetric pattern of brain activity decreases. It seems that the growing tendency of cooperation between the two hemispheres of the brain among old people derives, at least partially, from the need to compensate for the decreased efficiency of processing in each half separately. Using the pattern recognition processing method, which is more typical of an old brain, compensates for the neural wear that takes place as we grow older.

The Complexity of Old Age

Many old people speak about the humiliation they experience when they feel that their image seems to be transparent in the eyes of other people. Their presence is not taken into consideration as part of the people who deserve proper consideration at a certain scene. Also known are the burning burns of the insult felt by the elderly who have lost their position in the family. The value of their share in the stock exchange of importance, as determined mostly by young "value traders," decreases continuously.

Society nowadays tends to emphasize the losses related to old age but almost totally ignores the accumulated insights—assets that are the hidden treasure of old age.

According to a common supposition, the bite of the teeth of time blunts the sharpness of our intellectual sword that fights the dragons of mystery in our world on a daily basis.

The deep streams of our life, however, shake, as years go by, the ship of our brain across streams and opposing streams, in terms of loss and acquisition of new abilities and skills. The directionality is not uniform. At an old age we experience loss of capabilities and thinking skills, on the one hand, but there are also deep and wide-perspective insights, on the other hand. Reinforcement versus extinguishing. The tough struggles of life are often our best mentors, and being old often means to learn a lesson from difficult crises.

A Harvard study on the development of old people is a long-term study (which follows the same subjects for years) that followed (and still follows) subjects from the last stages of adolescence until old age. One of the conclusions drawn from the study is the conclusion of psychiatrist George Valliant, who analyzed the findings and found that many old people acquire new skills at an old age, and most of them are more involved in society than they used to be as young people. The range of skills of many of the subjects is increased at an old age, and this challenges the common notion that sees aging as a monotonous process of retreat and fading out.

The solstice of our life (solstice is the day of change—the twenty-first of June in the northern half of Earth and the twenty-first of December in the southern part of Earth; this day is the longest day of the year, and its night is the shortest night of the year), or, metaphorically speaking, the time in which we are at our best and from that point our function gradually deteriorates, is not uniform with regard to our various skills, some of which will reach solstice at the height of the mountain of the years.

Old Age Wrinkles of an Exhausted Brain

During the advanced stage of adulthood, old age wrinkles appear in our brain as a result of age-dependent changes in structure and function.

These are some of the changes in structure and function that take place in the old brain: a decrease in the density of synapses, disconnection between adjacent neurons, a decrease in the amount of blood that reaches brain cells, a decrease in the amount of neurotransmitters that mediate communication between brain cells, accumulated damage to the cells' power stations (the mitochondria), and change in the structure of the proteins of brain cells, which disrupts their function. On the other hand stands the cognitive reserve—the ability to back up failing brain areas, to compensate for the wear of various skills, and using our finger (or more than one finger) to block the cracks in the dam.

When old age wrinkles appear in the brain, all of a sudden the sky seems lower. We feel that the limits of our ability are minimized. The slowing down of thinking processes might affect our brain when it looks from its position from the height on the mountain of years. A "decrease in memory" is a general term that commonly refers to the difficulty related to various cognitive skills and not necessarily to memory function impairments. Along with memory difficulties, the aging brain might experience a decrease in the span of attention, difficulty in focusing the beam of attention, a decrease in mental stamina, a decrease in the capacity of perceiving new perception stimulations, the speed of their processing, etc.

As we grow older, there is also a decrease in brain flexibility (neuroplasticity), and the maneuvering space for designing brain maps becomes smaller.

During childhood, brain maps that frequently encode neuropsychological skills are shaped in our brain. Structurally, the maps are networks of neurons that are connected in the three spatial dimensions in the brain, and in the fourth dimension as well—the timing of their activity (the aspect of chrono-architecture). There are many degrees of freedom in shaping these maps. These structures are the basic layer of the towers of perceptions and habits on which our personality is based.

As we grow older, and operational flexibility decreases, it is more difficult to deactivate maps of undesirable neuropsychological skills and more difficult to learn "new tricks."

In tests that were intended to check the ability to skip between conceptual groups, it was found that, during old age, our brain is more prone to thinking permanence—to a point of fixation—and, in a sort of mental inertia, we tend to stick to the conceptual group we are familiar with and hesitate more about skipping to another group.

The duration of a thinking beat in the old brain is longer, and it tends to scatter more, due to the difficulty of the prefrontal lobes in focusing the beam of attention. Our brain tends to be more democratic, in the sense that a larger number of brain areas are active while a task is being performed, but also more anarchic, and less task-oriented and focused on it.

Factors that affect the retreat in the efficiency of thinking skills, which is common in old age, are, among other things, a retreat in the quality of sensory input, such as the abilities related to smelling, hearing, and eyesight, which become weaker at the level of the sensory organs. "Turbid" input results in "turbidity" of the memory that is encoded from it.

There is also a retreat related to brain processing abilities and the speed of brain activity—also due to thinning out of neural networks that operate the brain. It is reflected in brain scans as a slowing down of metabolism.

Among old people who suffer from dementia, deterioration processes in thinking functions usually take place gradually and do not occur at once. Just as the sun does not set at once, the transition from the light of clearheadedness and clarity of thought to the twilight of mental decline is mostly gradual.

An increased tendency to suffer from dementia relies on genetic tendency, vulnerability that increases as the brain ages, and a lifestyle that increases brain morbidity.

Memories with Aging

There is a common saying that "a moment of old age fell on me" with respect to memory failure, and indeed the synapses do not "synapse" as in the past.

Old people usually remember fewer details compared to young people when witnessing the same event. Some of the main explanations for this phenomenon: decreased sharpness of sensory organs (decreased quality of hearing, etc.) and difficulty with division of attention, which increases as we climb the ladder of the years.

Memory functions are not homogeneous, and some of them are prone to failure during old age more than others.

As we get older, our brain stores a large number of patterns that share similarities. The reason for that is that our memory tends to be built from modular parts, a sort of repertoire of subunits, like Lego blocks that might be suitable for building numerous memory towers.

Over time, many memories lose their uniqueness. Memories that share perception impressions in which there is "overlapping" with similar memories tend to merge with each other, which makes it difficult to attribute them to a specific frame of time and place.

According to a harsh, painful description related to those who suffer from dementia in old age, an aging person loses the memory of this world, just as a baby loses his angelic memory once he come to this world (Bronson Alcott, quoted by Ralph Waldo Emerson).

The Last to Mature—the First to be Damaged

The brain areas that mature last, in terms of their structural and functional competence throughout a person's life, were the last to appear on the stage of evolution. In other words, these are the newest structures in the human brain, from a phylogenetic perspective. In a noncoincidental manner, these structures are the most sensitive to the harm of time and the first to sense the negative aspects of old age. The ones that are most prone to damage as we grow older are the areas of the prefrontal lobes, which are the newest purchased players in the dream team of the human brain in terms of ontogenetics (development axis at the individual level) and phylogenetics (evolutionary development axis at the species' level).

Wrinkles in time: The age wrinkles of the brain, which are formed throughout the years, are expressed, inter alia, in the fact that as we grow older our ability to preserve episodic memories in our brain decreases. We are prone to having memories that are more and more general (generic). The generic memories tend to take control and color everything in a uniform shade. For example, young spectators of a sports game tend to remember more details more accurately compared to old people who watch the same game. It is likely that the older fans have watched many games in the course of their life, so they will not encode the memory of an average game as unique enough. For them, this game shares many similarities with previous games they have watched, and the lack of uniqueness results in encoding details in a more superficial, less detailed manner.

From a Tsunami to Bath Waves—the Level of the Waves of Emotion

The waves of emotion are mostly motivated by the amygdala, which blows winds in higher intensity at a young age.

The level of the waves of emotion and their intensity are at their peak among young people. As we grow older, it seems that the average level of the waves of emotion decreases. *Romeo and Juliet* scenarios probably belong to the young ones, with respect to the romantic storm of emotions.

An Intellectual Challenge

The mental life expectancy is sometimes tragically shorter than the physical life expectancy. We all know people whose general health is considered to be good, but their brain has betrayed them and they live in mental twilight, although their body, except for their spirit, is fit and functions well. Thus, it is of utmost importance to preserve and even reinforce our cognitive abilities as we grow older.

Using Our Brain in a "Learning" Mode Versus Using Our Brain Based on "Past Learning"

As the years are piled on top of each other, we tend to place ourselves in conceptual comfort zones and act less and less as students do to use the skills that were previously learned. We usually tend to avoid tasks that require focused attention and concentration.

An attempt to cope with a threatening challenge, such as learning a foreign language at an old age, encourages the aging brain to act contrary to the comforting trend of decreasing "learning" mode and relying mostly on "past learning."

Don't Close the Right Door

Findings that reviewed processing of cognitive, emotional, and sensory-motor information show that the right half of our brain tends to lose its functions, in old age, at a higher speed compared to the left half of our brain.

Learning as a way of life, and initiated coping with the new, might slow down the degeneration of the right hemisphere.

Functional flexibility is one of the prominent characteristics of our brain.

The ability to wander from the comfort zones at the core of thinking functions to the peripheral zones is a central layer of the term "brain flexibility."

The brain is not a stagnated structure—it shows flexibility and liquidity, in accordance with its owner's richness of experiencing.

Cognitive encouragement by means of taking part in brain-challenging activities, particularly according to a graded pattern of increased levels of difficulty, strengthens the "brain's muscles" and increases its reserves that are kept for rainy days.

The level of cognitive reserve, which is mostly reflected in the ability of the brain to compensate for damage by means of "delegating" functions of the damaged area to brain areas that are still healthy, might derive from a fortunate array of genes, from an adequate environmental exposure, and from practice. The correlation between brain pathology at the structural level and cognitive retreat is not linear. The effect of the cognitive reserve might significantly blunt the level of damage to the structure, even with respect to the darkest diseases, such as Alzheimer's.

Education and literacy are risk-decreasing factors in relation to sweeping brain retreat.

Intellectual challenging is a lifetime project. It is an attempt to preserve mental flexibility in an active manner, which leads to the formation of new threads (new webs of neurons' networks) constantly.

Preserving the flames of thought is the recipe for "mental longevity," hoping that, as old people, we will be the proud owners of a "used brain in great condition."

Intellectual challenge is also reflected in structural change in the brain. As aforementioned, words and thoughts also have a "chemical manifestation" in the brain (i.e., they are reflected at the material level of molecules). Brain cells continue to multiply at the brain structure of the dentate nucleus in the hippocampus, which has an important role in assimilating new information in advanced adulthood, as well.

Brain Actuary

Actuarial calculations that rely on studies in which multiple subjects participate support the approach according to which mental vigor throughout the course of life can be compared to purchasing an insurance policy. Like any insurance policy, it is not free of breaches, but, not as with any policy, its range of coverage is rather wide. Maintaining intellectual vigor was proved to guarantee the preserving of our cognitive skills and increase their durability in front of time's harms. The Notre Dame "Nun Study" supports these finding as well. In this study, which began in 1986, researchers from the University of Kentucky followed about seven hundred nuns from a Notre Dame convent in Minnesota. One of the most prominent findings from this longitudinal study was that the nuns who excelled in lingual richness and verbal creativity as young women, as well as in expressing positive emotions, as reflected in the autobiographic reviews they wrote in their early twenties, maintained their cognitive skills better and were less prone to dementia compared to their friends who did not demonstrate such characteristics at a young age.

An Expert Brain and a Novice Brain

Simultanagnosia—difficulty in merging single impressions into an inclusive impression—characterizes the brain of a novice who starts to study a new discipline. The brain of a novice tends to perceive forest trees as separate entities and has difficulty in identifying the holistic pattern of the forest that is created out of the individual trees.

Over time, if the novice's brain becomes thoroughly familiar with the specific discipline, it turns into an expert's brain, and, thus, the synthesis of individual impressions might take place in a blink of an eye.

There are numerous possibilities in a beginner's cognition, though there are only a few possibilities in an expert's cognition.

"The magic wand of expertise" is expressed in the fact that an expert's brain enables him to solve problems related to his field of expertise, as if by a magic wand, without the blood, sweat, and tears that are sometimes part of the task when performed by a novice in the field.

As we climb the different steps of the ladder of expertise, clusters of neurons create new webs among them that serve as infrastructure for information encoding, according to new aspects or higher levels of resolution.

A treasure of implicit knowledge that derives from a lifetime of being you is imbedded in the perception of our experiences and the products of our activities during the various periods of our life.

Senior, Economical Brain

Senior brains in our world are more prudent consumers of brain power.

As we grow older, our skills, experience, and "life wisdom" turn our insights into thinner ones, in terms of metabolism (consuming a relatively small amount of brain power), but muscular ones, in terms of their contents.

The energy saving takes place courtesy of the expertise and seniority of pacing in the paths of life.

Practice and repetitive experiencing, with respect to performance of a familiar task, decrease the metabolic burden involved in it. The main cause for that is related to the different information processing approach that is usually used by an expert's brain. Such an information processing approach, which derives from thorough familiarity, relies on identification of patterns and requires the involvement of fewer neurons at the different phases of performance, compared to information processing according to other approaches, such as tiresome serial and logical processing of a task.

The combined effect of expansion of patterns' representation areas and transferring to the pattern recognition approach leads to a decrease in metabolic consumption. This effect turns the brain into a more sensible and reasonable energy consumer, and, during challenging times, this economical approach is of great value to it. When, at an old age, the blood supply to the brain decreases, which is accompanied by a reduction in oxygen and various nutrients' supply, the brain is well prepared and performs its task with decreased energy consumption.

Senior Brain—Song of Degrees

Our culture tends to present old age as an inevitable sequence of losses. When our hair turns white and thin, we assume that the same process is taking place underneath it, in our brain, and that the flame of our brain is slowly extinguishing.

Is our brain, indeed, slipping down an inevitable slope as we climb the height of the mountain of years and grow older? New findings do not support this "inverted topography."

The brain at an old age is an oxymoron. On the one hand, it is slower in terms of processing information and tends to scatter the beam of attention, but, on the other hand, its gaze penetrates more layers of reality.

The speed of information processing, in which a young brain has a built-in advantage, does not guarantee the quality of performance with respect to many tasks of life.

The ability to "see beyond the matrix"—to understand the essence of things or the core of an idea—is a skill that is related to a senior brain more than to its younger version.

The "Seattle Longitudinal Study" is a study that has followed subjects for many years. This study began in 1956 and, from then, has traced the cognitive functions of six thousand people over a period of forty years. The subjects' population was divided equally between men and women whose age ranged from twenty to ninety. Analysis of the results has yielded some surprising findings: "middle-aged" subjects, from the age of forty to the midsixties, yielded better results in most of the thinking-function tasks examined, compared to the group of subjects who were in their midtwenties. It was found that, in four out of the six examined categories—logical inference ability, spatial orientation, verbal memory, and vocabulary—the results of the middle-aged subjects were better than the results achieved by the subjects who were in their twenties. In two categories—speed of calculation of numbers and speed of perception (the time gap between stimuli and response)—the younger subjects showed better performance. In terms of gender, male brains were better at spatial performance and female brains were better at verbal functions.[54] It seems that the "expert in life" brain owned by middle-aged people is a well-lubricated machine in the various senses of brain functioning in which it shows its superiority (contrary to the common prejudice regarding its fading competence) compared to young adults' brains.

Thinking Skills Over Time:
Changes over time in six thinking skills from the age of 25 to the age of 65.
Some of the skills fade out significantly over time.
Others, such as vocabulary, inference ability, and verbal memory, are preserved and even improve over time.

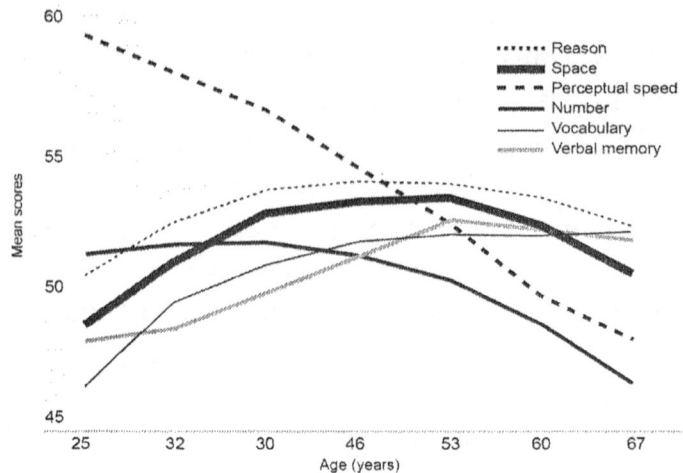

When we tell a person on his birthday "May you live to be a hundred and twenty," we are usually infected with mathematical overoptimism, although it seems that in the near future we will not be accused of that.

As life expectancy grows, gerontologists (researchers of aging) have started to divide old age into more specific sections. They do so with respect to all periods of life. A period that became more specific is "middle age," which is referred to by many as the period between the ages of forty and the midsixties.

The mature brain during middle age is the preferred residence of expertise. An expert brain performs tasks related to its field of expertise in an elegant manner, which is free of disruptions, in its "left hand," as they say in Spanish.

It seems that, during middle age, our parenting skills also improve; we act more as an "expert parent" and tend to create a better "situation-compatible interface" with our children. Our social skills are also at the peak during this period.

The Age of Good Advice

At different stages in our life we act as both students and teachers.

The older ones among us, who are rich in terms of mileage on the path of life and its accompanying insights, are a promising source of good advice, which is mainly based on shortcuts based on pattern recognition.

The saying "The wisdom of old age resembles winter sun—illuminating, but no longer warming" is only partially true. Indeed, numerous insights that arrive in later stages of life can no longer be implemented in our personal life, but they might be passed to the next generations and be preserved in a manner of memetic transfer.

As we grow older, the insight regarding our ability to understand the world only partially (in the sense of "I am not young enough to know everything") is often established.

Old age brings along the "roughly," the insights of "both," and fuzzy logic, which are suited to many situations in life.

In brain-imaging studies it was found that the reaction of "senior brains" to sensory stimulations with negative contents (as was reflected in increased activity at the amygdala) was more moderate compared to the more stormy reaction of the amygdala among young brains.

Although our brain is programmed to focus on the negative aspect, due to the survival-related advantage that directs us to deal with danger first, at an old age we tend toward a softer interpretation of reality and toward increasing the dosage of rosy shades in the picture of our life. We tend to be more optimistic—a mirror image of our tendency at a young age. It seems that part of the tilting of the pendulum toward the end of positive interpretation derives from an improved ability to regulate our emotions.

Some might see it as evidence that supports the hypothesis that emotional sensibility is often intensified as we grow older. In this sense, old people are more positive than young ones. It seems that they are capable of producing a more "subtle" and reasonable emotional reaction in situations that induce emotional stress.

Wisdom and Expertise as the Fruit of Maturity

A common view sees wisdom as a threshold passage of a critical number of core insights. The threshold refers both to the quality of insights and to their quantity. Genius resembles the eruption of a volcano. It bubbles and comes up against world orders. The strength of wisdom is usually like the strength of still waters that run deep. One can choose to remain at a place of naivety, but life unfolds as a place of experience in the battles of life (also known as "life's wisdom"), and the experience relies on memories and is usually in correlation with an older age.

There is an ancient saying that experience is a flashlight that illuminates backward, to the road we have already traveled. Using experience in a manner of extrapolation, however, turns the future into something partially predictable and enables us to change the direction of the beam of experience flashlight toward the dark depths of the future.

One of the definitions of wisdom is a correct mix of heart and brain (in a neurological tone: between the amygdala, as the main representative of emotion, and the frontal lobe, as the main representative of rationalism).

Some ascribe the wisdom of middle age—the age of "matured adulthood"—to the amount of white matter in the brain, which reflects a more efficient transmission of signals between brain cells. In contemporary jargon, it may be referred to as a wider "band width." It was found that, at certain areas of the temporal and frontal lobes, the lining of myelin—due to its light shade, brain areas that contain a large amount of it are called "white matter"—reaches its peak at about the age of fifty.

Life is like a novel whose ironic truths are hidden from the eyes of those who have just started reading the book. A "non-ideal" by-product might be added as we grow older, when idealism tends to be covered by the scaled armor of cynicism.

Wisdom does not only mean familiarity with the facts of life, but also the ability to apply these insights, in practice, by means of real actions. The guiding power of wisdom is reflected in outlining a practical course of action.

Snow on the roof does not mean that the fireplace inside the house does not spread heat, just as white hair and the burden of years do not prevent the intellectual burning in the brain of an old person.

Young and Old Hemispheric Action Profile

The hypothesis according to which the center of the cognitive gravity moves toward the left side as we grow older, which means a reduction in the volume of activity on the right side, is supported, inter alia, by brain-imaging studies. These studies show that among young people, the basic activity at the prefrontal cortex on the right is more significant from the prefrontal activity on the left side, almost regardless of the nature of the cognitive task. An inverted mirror image was found among old people, in whom a larger volume of basic activity was detected at the prefrontal cortex on the left side. At a young age, we tend to look at the world through the lens of pursuing innovation and coping with it, whereas at an older age we tend to view the world through the lens of accumulated past experience.

The Personal Pattern of Aging

Each person experiences the crises of life in a personal, unique way that is based on the circumstances of his life, his beliefs, and his views. The fact that time has limits is known to every child, but it seems that only adults, in the second part of their life, really sense it. As in many other aspects of life, the pace of brain aging changes from one person to another. "The wisdom of the individual" includes the personal insights of a person, which are stored in his brain throughout the years. The repertoire of insights becomes larger as we grow older, and the insights are piled in the cellars of our brain.

The number of volumes of information patterns on the brain shelves predicts the cognitive durability against the moth of forgetfulness, which nibbles at the volumes in the neural library.

Every one of us has weaknesses and strengths in our mental armor, which, to a large extent, derive from the patterns we have acquired in the course of our life. The level of vulnerability with respect to the fading of various thinking skills depends on the depth of the backup and the richness of the patterns that are stored in our brain and relate to these thinking skills. If an animation painter and a literary critic are affected by dementia, it is likely that the literary critic will suffer more damage related to spatial orientation, while the animation painter will suffer more damage related to language skills.

The Indented Shoreline of the Sea of Forgetfulness

Even when dementia is developing, the frontline of cognitive deterioration is not uniform.

The shoreline of mental deterioration's frontline is indented, full of islands and peninsulas, and does not have a uniform interface with the sea of forgetfulness. Islands of cognitive excellence might be preserved in people who suffer deterioration in specific aspects of their cognitive skills.

Each one of us has a personal map of mental strengths and weaknesses. Cognitive strengths and cognitive weaknesses might be compared to peaks and valleys in the personal map of skills. This map is a significant actuarial marker with regard to the skills that are more prone to be damaged as a result of cognitive deterioration processes. The valleys—areas of weaknesses—are more susceptible to calamity and to being flooded by the wastewater of loss and extinguishing. The cliffs of strengths will still sprout from the boggy swamp.

As we grow older, the birds of thought and emotion migrate from the "lands of daringness," at the right side of the brain, to the "lands of safety," at the left side of the brain (although they still nest at the right side as well, their spreading pattern becomes more bilateral as we grow older)—this migration process is not uniform in terms of pace and contents.

The process of universal change in the human brain that takes place as we grow older involves tilting of the information-processing pendulum from a unilateral tendency toward the right, to a growing collaboration of the left brain (and a more cooperative action pattern based on both hemispheres). Nevertheless, this tendency also depends on personal aspects.

At an old age, people sometimes feel as if there is no excitement left—as if the circus has left town.

Sometimes it seems that, as we grow older, the glow in our heart becomes dimmer from winter to winter, but the ability to "enjoy differently," such as enjoying a quiet musical piece, might compensate for losing our ability to enjoy jumpy computer games. We can enjoy in different ways.

Part C:
Aspects of Body and Soul

Chapter 18
Body and Soul—Are They the Same?

Domain of the Individual and Domain of the Numerous

The issue of the connection between body and soul is a central branch in the Tree of Knowledge of philosophy, and there are many who walk the intellectual track that attempts to understand the descent of spirit from substance.

According to a common supposition, since we do not have the ability to rise above the limits of our senses we are not able, by means of direct observation, to grasp the body–soul essence, although we do have some perceptional images that represent it. All we can do is infer inferences according to the impressions of the perception that conceptualizes the inner aspects of our being (the world of phenomena restricted by our epidermis).

This is not a unique case. Other basic entities in science cannot be perceived or directly observed, either. Inferring the existence of these entities derives only from their perceived results. An example of that is gravity. Each apple that falls from the tree to the ground attests to its existence, but for a long time, no gravitational radiation or gravitational particles, which confirm its existence, have been registered. We inferred its existence only by means of results that are perceived by our senses.

Many soul researchers claim that when the object of investigation is ourselves, as conscious entities, the main barrier of knowledge derives from the fact that our subjectivity sometimes seems as if it cannot be reduced.

We cannot observe the double-headed creature called body–soul. We gain insights about it out of inferences that result from cold, objective, scientific observation, such as during brain surgery, and out of a warm sense—the subjective awareness.

Today there is growing awareness of the incapacity of a single discipline to explain the complexity of the soul. Relying on a single discipline in the field of mental research might lead to interpretive failure similar to that in the fable about the six blind men who tried to assess the shape of an elephant. Each of them touched a different part of the elephant's body but could not grasp the overall being of the elephant.

It seems that the language that describes the soul of human beings must be "multilingual" and include concepts that are taken from different disciplines. Understanding the human soul requires hermeneutic thinking, which combines insights taken from numerous disciplines: from philosophy and psychology to biochemistry and neuropsychology.

The mental mechanism can be observed from two points of view simultaneously: as subjective consciousness that can only be peeked into by its legal owner (i.e., every owner of a soul can observe only his own soul from within), or as material entity. A combination of these two modes of investigation creates an added value, compared to each of the modes separately. In this way, we get a more fitting picture of the mental reality. The combination of the world of subjectiveness, which is invisible to the eye of an outsider, and the study of visible brain tissues might greatly improve our understanding of mental life.

The two observable aspects of this entity—the material aspect of the brain and subjective awareness—greatly vary in terms of their availability to be observed, in the sense that the human brain can be observed scientifically by other brain owners, due to the fact that it is tangible and measurable and can be quantified (available to the domain of numerous) while, on the other hand, subjective awareness belongs to private territory and is unavailable for observation except by its owner (the domain of the individual).

The situation in which, despite the great differences between our brains and our unique perception, we share numerous conventions in common might be termed "objective subjectiveness." Some believe that our common perception of reality is like a collective mirage, similar to the one shown in the movie *Matrix*.

A common definition of the "me" is a subject who is aware of himself. The qualitative aspects of our experiences are hard to explain (how it is to "be me from within"), similar to the confusion we sense when watching films like *Being John Malkovich or Fight Club*.

Many scientists tend to focus their work on the "domain of the numerous." They prefer to focus on phenomena that can be observed by a multitude of spectators. The great number of observations increases, as they see it, the reliability of the findings of observation, and the risk of bias, which might exist with respect to observations whose findings are only reported by a single spectator, decreases.

As part of the "domain of the numerous" approach, some use evolutionary models of cognition, in the spirit of the work of Darwin, who searched for the element that unifies life manifestations, as opposed to the trend of searching for the element that differentiates, which was common in his time. The tendency of evolutionary models of cognition is to search for the unifying, universal elements and to focus less on the cognition of the individual.

In this sense, the tendency of these models is mythic—they reveal the almost-"timeless" causes (as representatives of evolutionary eras) that define the human condition.

It seems that the combination of the domain of the numerous observations, which can be refuted or confirmed by means of contemporary scientific methods, and those that are related to the domain of the individual and which rely on a testimony of a single witness who "owns the soul" and cannot be refuted or confirmed by means of contemporary scientific methods, offers us panoramic observation as part of the attempt to better understand the mysteries of body–soul.

The researcher Alexander Luria stood out in his attempt to reflect the elements that rise on opposite sides of the barricade. Along with the objective information, he tried to reflect the inner experience of the investigator. According to Luria, "When the observation is performed appropriately, it realizes its traditional purpose—to explain the facts, and at the same time, it is focused on the romantic purpose of maintaining the varied richness of the subject."

The Weight of the Soul

Does our soul, the refined core of our being, become as thick as a cloud at the moment of death and wander off to an invisible heavenly plane—as in the common, enchanting image?

The attempt to extract the honey of the soul from the brain beehive and to grant scientific acknowledgement to the existence of the soul led Dr. Duncan MacDougall to carry out a study whose results were published in 1907. As part of the study, the weight of patients was checked carefully at the moments of dying and right after death. The findings of the weighing led Dr. MacDougall to conclude that the weight of the human soul is twenty-one grams. His findings were not confirmed in other studies. As part of an interview with the *New York Times*, Dr. MacDougall told of his ambition to document the human soul in an x-ray scan.

Studies that involve innovative technologies in brain imaging, such as functional MRI, try to detect correlations between neural activities and mental phenomena, but the portrait of the soul has not been captured in a photograph yet.

The Ghost in the Machine

Artificial intelligence constitutes, in the eyes of certain brain researchers, an alternative to the complicated approach of finding a correlation between conditions of the soul (which are subjective conditions in essence) and aspects related to structure (anatomy) and function (physiology) of the brain. Certain researchers of artificial intelligence aspire to reach a point in which they will be able to create consciousness out of substance in practice.

Is the creation of artificial consciousness the key to understanding consciousness, in the spirit of the words of physicist Richard Feynman, who said that we cannot understand what we cannot create?

Is it possible to exchange human intelligence with digital intelligence, or is it as the physicist Roger Penrose claimed in his controversial book, *The Emperor's New Mind*: brain activity cannot be imitated by a digital machine, since cognition and thought are not algorithmic actions but rather actions that are subject to the laws of quantum mechanics?

People who suffered brain damage and afterward experienced changes of perception and exceptional insights and behaviors were described by neurologist Oliver Sacks as "giants who walk in lands beyond imagination—lands we would not have heard of if it weren't for them." These people were forced to embark on a discovery journey into themselves, during which they exposed hidden talents and abilities within themselves and extracted from their brain new insights about the reality inside and outside.

The birth date of the self, in its common meaning, does not overlap with the date of our birth. The sense of self, in the sense of our being as an entity that is separated from others, is probably born at the age of eighteen to twenty-four months, and a manifestation of that is that this is the time at which infants are able to identify themselves in photos.

The Self—A Multi-Entities Entity

The self is multifaced and is not a unified entity whose boundaries are well defined. The "cluster" of the aspects of the "self" get together and create the overall sense of "self." In this sense, the "self" is like a multi-aspects cocktail in which each aspect of the self has presence and weight that constantly change. Since the weight and the level of presence of each aspect of the self is weighted in a sequential pattern within the overall self, this self actually changes constantly according to the mixing changes in the subgroups of the self that compose it, but within it exist a preservation of the "preserved core" of the self representations, which are like a guardian angel for the continuum of our being.

The "self" includes many "selves"—for example, the self that places us in the sense of time; the self that places us in the sense of the physical location; the self that places us in the sense of social sphere; the self that looks inside (the introspective); the physical self that monitors body senses; the emotional self that is attentive to emotional waves; the autobiographic self that preserves the core memories of our identity; the reflexive self, which is like a mirror that reflects ourselves to ourselves; the self of self-esteem, which determines the value of our self-evaluation in changing circumstances; and the perspective self, which sets us in a unique square in the chess board of the situation.

The *Fata Morgana* of the "Self"

Is our consciousness a *fata morgana*—a mirage that derives its liveliness from the shadow theater within our brain?

In philosophical terms, we might define this approach regarding the building of the "self" as "naive realism"; we refer to the representation built by our brain as if it were reality itself.

David Hume, the eighteenth-century Scottish philosopher, wrote, "Selves are nothing but a bundle or collection of different perceptions which succeed each other with an incomparable rapidity.... The mind is a kind of theater where several perceptions successively make their appearance."

And, contrary to the common approach of cause and effect, it is possible that our thoughts and emotions generate us and not the other way around.

A similar approach sees a human being as a collection of "selected representatives" (with respect to the perceptions, memes, and memories that reside in his brain, in particular) whose composition might change every now and then, and, in this sense, it is constantly changing.

Just as a collection of wooden boards that are organized according to a certain pattern is termed "a ship," so is the collection of personal impressions termed the "self." The self is like a turning wheel whose spokes are our memories and the core memes that nest within us.

The perception according to which the "self" is an illusion and the sense of self is nothing but a cluster of changing impressions is common in Buddhism as well.

According to an interpretation of the view of the British philosopher George Berkeley, there are not any entities that exist in themselves in our world—the existence of these entities depends on someone else who perceives them in his senses. This view is illustrated in the saying that in the absence of ears that can hear, a tree that falls down in the forest makes no sound.

People are used to telling stories to themselves and to others, and some believe that the self is the core story, a narrative whose plot is woven like a braid from the threads of our memories, our memes, and senses' impressions and secondary plots branch out from it.

The philosopher Daniel Dennett emphasized language as a tool in the service of the "continuous self" due to its ability to grant narrative consistency to our life. Some might say that, more than we tell the story, the story tells us.

Among those who have a deterministic approach regarding the human soul, some believe that, just as one should not be upset with a falling stone, since it is doomed to fall due to gravity, we behave in the shadow of certain behavioral motivation laws that allow only limited maneuvering toward free choice. Thus, when we judge people's conduct, we must take into account that their behavior derives from coercion.

It seems that the circumstances that are formed under the influence of randomality are the invisible forces that motivate many of our decisions. "Free will" is often a secondary motivation at best. It seems that we are often the victim of the oppression of random circumstances, which lead to a behavioral outcome that is preordained by means of various psychological biases that are invisible to the eye of our consciousness.

Are We Biochemical Marionettes?

The belief in the existence of the soul is a top psychological tool in the service of survival. According to this view, our consciousness invents the soul, like Baron Munchausen manages to get out of a boggy swamp by holding on to his braids.

According to an approach that is becoming popular among many mind researchers of various disciplines, our brain builds in an automatic manner, and hidden from our consciousness, a model of the self that relies on self-mapping of the systems and processes within our body. The activity of the neural network of the default system greatly contributes to the formation of this model. The processes of building the model are invisible, like transparent windows that are piled on top of each other on the computer screen while a software is running. Our perception penetrates them unnoticeably, and we only get to know the final virtual outcome of the process. This is the process of inner mapping, from which the model of the self is created in a manner that is invisible to our consciousness's eyes. The self-model, which is perceived as the "self," is a delusional entity in nature, since it does not constitute an entity in itself but, rather, an inclusive representation of the condition of our body. The abstract term "soul" loses its mysterious aura and is referred to as flesh and blood that must obey the rules of time and space. Thus, we end up with an essence that is not metaphysical.

When a researcher looks into the skull cave of another person, all he sees are brain cells and a dance of information (a dance that is reflected in the pattern of shooting potentials along the axons) that takes place between them. In this spirit, some might say that the brain cells and the dance of information that is formed between them are like the props used by the brain, the master of illusion, and the "self" is the illusion itself.

Some compare the "self" to an (automatic?) pilot who sits in the cockpit and flies the bird of our soul. Others might claim, in a nihilistic tone, that there is nothing at the cockpit and that all of us are passengers captured in a ghost flight that is headed nowhere. The ones who believe that the model of the self is a genetic dictate ascribe a teleological explanation to it. According to them, the model of the self serves the organism and grants it a survival-related advantage by creating distinction, which enables the collection of genes that is the "self" to manage itself as an autonomic unit. This reductive, gloomy approach might suggest that we are nothing but a passive vessel that contains genes, soulless couriers who row and carry the genes through the waterfall of generations.

Others see the model of the self as a reflection of the excessive ability of the human brain, which derives from its complexity, without any "agenda" on the part of the genes. On the other hand, some might claim that interpreting the brain in Darwinian terms is biological determinism, a charged concept that sometimes triggers negative connotations.

In modern human history there have been a number of "insight decrees" that challenged the unique charm of our species; thus, for instance, Copernicus "stole" the centrality of our home planet in the solar system; Charles Darwin "removed" the crown of divine creation from our head and discovered our relatives among the monkeys. Some see the materialistic decree related to body–soul (whose validity is yet to be confirmed) as the cruelest one—one that "banishes" our souls from us.

The "Self" Avatar?

The word avatar comes from Sanskrit. In Hindi religious literature it means the embodiment of God on Earth. In today's virtual world, the term refers to a virtual image that represents a human user in the cyberspace universe.

Is our sense of self also a type of avatar—a virtual entity that our brain fabricates and grants an aura of real materialism to? For instance, brain researcher Michael Gazzaniga believes that the left hemisphere is the one that creates the narrative of the one consistent "self," irons the wrinkles in the fabric of our reality perception and unfurls it as smooth and coherent, and, in fact, is the producer of the mirage of the unified self in practice, as Gazzaniga sees it.[55]

Is our "self" the producer of thoughts or is it our thoughts that produce it? Is the "self" a mere lingual creature rather than a realistic being?

Like a flourishing oasis that looms as a *fata morgana* when we explore its nature, so is the soul, in the eyes of the skeptics, the "desert of the actual"—a desolate sand dune rather than the magical oasis we tend to perceive. This view is the result of the de-mythologization of the mythological concept of soul.

The "self" represents, as a common human convention, our essence as unique and independent human beings. But there are people who are forced to deviate from the course of the consolidated "self," such as those who suffer from psychotic episodes or neurological diseases that crumble the "self." In the absence of a control agent within us, which demands control over our emotions, the self dissolves, but still we will not claim that these patients are not human beings.

It seems that many brain researchers assume that we are watching a play during which a disheveled scientist is about to approach the front of the stage, in the role of a little boy who will declare that the royal garments of the soul are nothing but an illusion, and just after him the temporary similitude of Nietzsche will step forward and pronounce the death of the soul.

Knowing the self thoroughly might turn out to be very disappointing in the emotional sense. It is difficult to desert the comforting view related to the existence of the soul, just as the Copernican world slapped the ones who believed in the centrality of Earth.

Moreover, the intellectual killing of the self might shake the foundations of our moral perception. The core of ethics and the principles of the good and worthy are based on viewing human society as a collection of autonomic agents with unique characteristics who enjoy freedom of choice and the ability to build railways for the train of their destiny. As a result of that, we are all responsible for our choices, and out of this basic hypothesis, the layers of personal responsibility, and morality- as a whole, sprout.

If we pronounce that the "self" that is related to the ability of understanding and freedom of choice is nothing but an illusion, we endanger ourselves with a chain reaction that might lead to the splitting of social atom and activating forces that are invisible to the contemporary horizon of our vision.

Up and Down

The nesting nature of phenomena enables us to go up and down the scale of resolution and abolishes the need to start each thorough discussion with the "story of genesis" in which the developer of the idea believes.

The nesting nature of things is well known, and it is possible to start from a reasonable point in the middle of the scale, assuming that the more accurate levels of detail it contains are already known to the addressee of information.

Due to the nesting pattern of knowledge, according to which old knowledge nests within newer knowledge and the latter finds its nest in the most up-to-date knowledge, we tend to prefer new information to old information due to the implicit supposition according to which new information contains old information and adds to it, or processes it in a pattern that elaborates on the conclusions that can be derived from it.

The direction of the arrow of reductionism is opposite to the direction of the arrow of holism and might reach the end of the conceptual horizon.

A familiar claim that opposes reductional explanations is that, in the interspace that is created between the level of reducing and simplifying explanation and the phenomenon as it is perceived in our daily life, essential characteristics of the phenomenon trickle out of the mental field of vision.

Like a divine *fata morgana*, a rainbow looks spectacular and colorful from a distance. As we come closer to it, however, its image starts to vanish gradually. The level of resolution, according to which we observe a phenomenon, dictates the way we perceive it, which means that a phenomenon that is perceived as a certain entity at a certain level of resolution might be perceived as a totally different entity when the level of resolution is changed.

Some claim that we cannot use reductionism with respect to subjective situations because substantial aspects of the interpreted experience tend to evaporate during the reduction.

Reductionist explanations have always stirred up resentment. For example, the nineteenth-century English poet John Keats claimed that Isaac Newton "destroyed the poetry of the rainbow by unweaving it to the prismatic colors."

The difficulty in applying a reductive approach in the interface between matter and spirit has led to the implementation of the softer approaches of "supervenience" and "emergence." The supervenience approach is applied when we do not have enough knowledge to reduce a single discipline and contain it, as a whole, in another discipline. An example is the claim that mental conditions rely on the structural, functional infrastructure of our brain. A relation of supervenience is common in assumptions regarding the type of correlation between the material infrastructure of the brain and its function and the sphere of our mental life.

When there is difficulty in translating concepts from a certain discipline into another discipline in a way that no substantial feature of the described manifestation is lost in translation, the "emergence" approach is implemented. Thus, for example, the quality of wetness, which is essential to the essence of water, derives from the pattern of molecular joining of water molecules to each other (a chemical bonding pattern called "hydrogen bond"), which results in a liquid state of aggregation. The wetness feature, which does not exist when the components of water are separated, as separate molecules of water, emerges from this state of aggregation.

An additional example of emergence is the temperature that emerges out of the average of kinetic energy of a certain substance, or the weather, which is also formed as a result of changes in the movement of molecules and the changes in temperature that follow that.

Emergence relations seem suitable to the description of the reciprocal relations between the biological bedding of our brain and the spiritual life that results from it. The thoughts sprout from the material infrastructure of the brain according to its unique pattern of organization. Organization is not a material quality; thus it constitutes an additional possible explanation for supervenience and emergence of spirit out of substance.

Freudian Mind

The relationship between brain scientists and Freud's ideas were bipolar. On one pole, there were those who adapted Freud's theory wholeheartedly, and, on the other end, there were those who rejected the theory and claimed it was anachronistic and lacking solid factual basis. The biological approach has been acquiring a foothold (and hearts) among many contemporary mental doctors, however, at the same time, the pendulum that tilted in the past toward Freud's supporters, and later moved sharply toward the other end, has become more moderated in its tilt recently. The moderation of the dichotomous tendency of the pendulum derives from the forming recognition that Freud was, indeed, responsible for significant conceptual revolutions, even if some of his hypotheses were incorrect.

Interdoctrine Conceptualization

Some compare the terms "id," "ego," and "superego," created in Freud's conceptual mint, to the terms "unconscious," "conscious," and "preconscious."

According to Freud, the ego is the servant of two masters—the superego and the id—and, by means of possible extrapolation, the conscious serves the unconscious layers.

An additional correlation between terms taken from Feud's theory and terms taken from another conceptual doctrine about the brain is the one that matches terms of phylogenetic evolution of the brain with Freud's doctrine; the topographic correlation to the structure of the mind in Freud's doctrine is embodied in the model that conceptualized the human brain as a hierarchic, multilayered model. This model, proposed by brain researcher Paul McLean, is based on the assumption that the brain develops from lower core structures toward higher structures of the brain. It may be applied both in the functional sense and the topographic sense. The brain stem (the reptilian brain), as the basic phase in the evolution of our brain, can be correlated to the id (the generator of raw urges). The next stage, which represents the transition from reptile brain to mammal brain, is the development of the limbic system (whose main structures are the hippocampus, the amygdala, and the hypothalamus), which can be correlated to the ego (the main pathways of emotional motivation). An additional stage in development is the transition from the mammal brain to the primate brain, which is characterized by the development of the cortex, and the "transition-to-human" stage is characterized by the "excessive" development of the frontal lobes, and their front areas in particular, where the superego—the main generator of judgment and ethical perception—resides.

The Language of the Land and the Language of the Sky— Materialism Versus Idealism

Some believe that the body is the temporary residence of the mind (a very temporary residence—more like a motel), while others think that the destruction of the body equals the destruction of the mind.

By means of an aphorism that may oversimplify the complexity of the subject matter, we could say that there are three main approaches related to the question of body–soul, from which numerous subapproaches derive.

Materialism: According to this approach, the entire universe, including human beings' consciousness, is composed only from matter.

Idealism: According to this approach, the entire universe, including human beings' consciousness, is composed only from spirit.

Dualism: This is the "mediating approach," according to which the world is composed of materialistic aspects and spiritual aspects and both reside side by side.

The concept of the separate, existence of the soul has deep evolutional roots. The natural sense of many of us is that we carry within us a mental core that cannot be minimized.

The view regarding the separate existence of the soul was the only show in town during various periods in history (in the sense of the ruling meme).

According to the view of the philosopher Rene Descartes, which was dominant for many years, the body has a material manifestation—flesh and blood—that obeys the rules of physics, and, at the same time, it is also the residence of the mind. According to Descartes' theory, the mind is not material, does not take up space, and does not obey the rules of physics and the principle of causality. A human being is the junction in which materialism meets nonmaterialism, and from this combination the human being is formed.

"I think, therefore I am," said Descartes. The question is, who is this "I" that "think"?

It seems that the tendency to adapt the concept of separate entities of body and soul is built into our brain in a pattern similar to the collective unconscious. Descartes related to this heart's wish and built his philosophical theory around it.

The failures that are typical of this dualistic view derive from the difficulty in explaining the way in which nonmaterial thought makes the neurons that are imbedded in the material world operate.

Is Our Mind the Result of Physical Rules?

According to a contradictory approach called the "physicalism approach," or "reducing materialism," all aspects of our soul life can be reduced and explained in terms of material and energy. The effect vector is unidirectional: the body influences the spirit and not the other way around. The spirit is activated by the matter but does not affect it.

Consciousness resides in the thick forest of neurons and, more specifically, among them. The spirit is formed out of an interactive process. In this sense, the self as the creature of the spirit exists in the gap between the things.

The mind is formed out of the "discourse of neurons," and cognition, as a main aspect of it, derives from this activity.

The technical specification of the human mind was written throughout ages of evolution. The daily challenges in the survival journey of life have brought about more updated versions.

"Complexity that is formed out of simplicity" is a known phenomenon in our world. Fractals and other recursive processes are an example. Some claim that it is also true for the brain, which is composed of about a hundred billion similar "simple" cells and produces products of great complexity, the most amazing of which is the human mind.

According to an approach that sees the manifestation of the mind as an aspect that derives from pure material processes, the mind is a huge production of brain models that have different fields of expertise.

Is our mind a mirage—a phantasmagoric result of perception tricks? Loops of perceptional representations that are formed within our brain, one inside the other, and that mutually feed each other? Some believe that the conception that sees our inner part as the "mind" derives from this highly complex pattern of weaving.

Thus, for example, the delivery room of the sense of self, according to the theory of brain researcher Rodolfo Llinás, is formed in the space of mutual connections between the thalamus, basal ganglia, and cortex.

The marvelous exchange act of the brain from material to spirit echoes in the words of mathematician Paul Erdös, who once defined a mathematician as a machine whose purpose was transforming coffee into algorithms.

In the eye of Buddhists, the issue is reflected in the argument that, as a flower that grows in the garden reflects the continuation of the matter and the atoms that compose it have existed from the very first days of Earth (though they have undergone different incarnations in the evolutionary process, to the point where they incarnated as a flower), so, similar to entities in the materialistic world, the cognition of an individual must also derive from a continuum that originated in the past.

Skeptic Mind

Some might say that the attempt to use the self to see "beyond the self" is doomed to failure, and, despite the scientific craving that has lasted for years, it seems that it is impossible to separate the process of observation from the observer. Possible support for that can be found in quantum mechanics as well. A poetic metaphor for that might be that we are both the poet and the poem, and the poem defines our nature as poets.

In a similar sense to the meaning used by Carl Sagan when he said that we are stardust looking at the stars and exploring their secrets,[56] our mind is a creature of our brain, and we use it to explore the brain as being the interrogator and the interrogee simultaneously; some might claim that there is built-in failure in that.

Some think that one of the substantial characteristics of cognition is that, by its nature, it is incapable of explaining itself—the limitation of self-knowledge is built into our cognition. In other words, our nature prevents us from fully comprehending our nature, including understanding the manner in which the flower of spirit blossoms on the bedding of materialism.

Philosopher Gilbert Ryle claimed that body and soul do not exist on the same reference plane, thus their conceptualization and investigation in similar tools of analysis are doomed to failure.

The personal experience is formed in the first-person universe; science is a resident of the third-person universe. Robert Frost once said that poetry is what is lost in translation. Some claim that, in a similar manner, the "poetry of cognition," which is experienced in the first person, is lost in the attempt to translate it into a third-person description in the language of anatomy and biochemistry.

Is our mind an immeasurable aspect of the physical universe?

Some believe that among brain functions there are immeasurable qualities, not only with respect to measurement modes known to us today, but also in a more comprehensive sense, and that the understanding of cognition exceeds the ability of conceptual understanding of the human brain, exactly as we do not expect a lizard to understand grammar rules or a goldfish to understand the principles of thermodynamics. Mysterianism is a philosophical approach that represents a softer version of the anti-romantic saying about the limited ability of human intelligence. Mysterianism suggests that the human mind is not at the evolutionary stage that enables it to understand how cognition grows from neural activity, though it is not a supernatural phenomenon. This approach, as opposed to the one that preceded it, does not totally deny the future potential of human intelligence to cope with the riddle.

Will We Be Able to Know Our Mind?

Metaphysics—the kingdom of the things we cannot get to know by means of observation and experiments—is the space in which numerous hypotheses regarding the essence of the mind and the body–soul alchemy that is formed in the workshop of the brain take place.

Science is empirical (based on experiments) and quantitative, whereas metaphysical approaches are qualitative and, though they are sometimes characterized by a great power of explanation, they usually cannot be quantified or proven by means of an experiment.

Science is suspicious of metaphysical hypotheses, and often justifiably so. But metaphysical "axiomatic" hypotheses are hidden, like corpses of the unlucky rivals of Mafia members, in the "concrete pillars" that constitute a basis for the scientific method, even if only implicitly. Thus, for example, there is a hypothesis according to which reality manifestations are real in and of themselves, regardless of human acknowledgement. Thus, trees in the forest will make a sound when falling down, even if nobody hears this sound. So also is the assumption that reality manifestations have some kind of orderly regularity that can be traced.

The dispute that is based on the question of whether the ontological existence of phenomena depends on or does not depend on human consciousness that experiences it seems like a metaphysical dispute. In this spirit, for example, there are contradictory arguments that claim that mathematical entities are only expositions (ideas) of physical entities in human thought and, on the other hand, the notion that one plus one will remain two also after the death of the last person on Earth.

Insights that were perceived as metaphysical and unexplainable empirically in certain periods of time might be either confirmed or refuted as time goes by—for example, the insights regarding the essence of time and space. Contradictory positions were introduced by Newton, who claimed that time and space are absolute, and Leibniz, who claimed that time and space are relative.[57] The issue that seemed irresolvable, irrefutable, and unconfirmable was classified at the time as a metaphysical issue. This issue has undergone conceptual metamorphosis with the emergence of the theory of relativity, whose common inferences are compatible with Leibniz's position. Will the riddle of cognition, characterized nowadays as a metaphysical issue, be empirically illuminated in a pattern similar to the question of time and space?

Some define the constant search for the secret of the mind as "spiritual pilgrimage."

According to William James, a philosopher resembles a blind person who looks in a dark room for a black cat that is not there. Some might cynically say the same thing about "soul hunters": they search our brain for an abstract entity that is not really there.

Is our mind—a non-material entity—the production of a material organ—our brain?

Is there an "autonomous, perceiving entity" in our brain that operates the constant channel of perception in our consciousness? And does this entity include a unique component that contributes the essence of consciousness? And who perceives it? And who perceives the perceiver?

Is the pattern of perception a type of matryoshka doll, which contains smaller and smaller matryoshka dolls that are contained within one another? Containment in a pattern that gets smaller and smaller is referred to as endless regression, and it is one of the familiar logical failures that researchers of consciousness are dealing with.

The discussion that relates to the issue of the mind sometimes seems like a semantic, undecidable argument that relates to the meaning of the concepts rather than the facts.

There are situations in which cracks are revealed in our worldview, which usually seems whole. Even the most rationalistic people among us have difficulty with applying cold, scientific methods when it comes to researching the high spheres of the human mind.

The body–soul unity approach often meets fierce internal resentment, which usually derives its strength from the lands of emotion.

It seems that the mysteries of the mind will remain a "terra incognita" (unknown land) for years to come. Some say, "We will never know," while others make do with saying "We do not know yet."

Epilogue

For those who have forded the river of words and made it to the rear bank of this book, I truly hope you have also found some insights that might be useful for your daily life.

In the spirit of the philosophical, existential view, we should create a purpose for ourselves for our short cadence on the planet. Our brain enables us to design this purpose and pursue it. A complete understanding of the human brain's pattern of action is a remote objective, and some claim it is unachievable, like an asymptote to which one can draw closer and closer but is never able to reach completely, though additional knowledge and information about this magnificent organ bring us closer to the objective. And I do hope you are, indeed, closer to it now.

Reference

1. **Hippocrates about the Brain**
   In *The Genuine Works of Hippocrates*. Translated by Francis Adams. Vol. 2, 344–5. 1886.
2. **John Locke-Self Identity**
   Uzgalis, William. 2012. "John Locke—Self Identity." In *The Stanford Encyclopedia of Philosophy*. Edited by Edward N. Zalta. Fall edition.
3. **Franz Kafka – Metamorphosis of Identity**
   In *The Tremendous World I Have Inside My Head, Franz Kafka: A Biographical Essay*. Atlas & Co., 2008.
4. **Richard Dawkins – The evolution of genes**
   Dawkins, Richard. 1976. *The Selfish Gene*. Oxford University Press.
5. **Paul Bach-y-Rita- To see with the tongue**
   Abrams, Michael, and Dan Winters. 2003. "Can You See with Your Tongue?" *Discover*, June 1. Retrieved October 3, 2009.
   Bach-y-Rita, Paul, and Y. P. Danilov, M. E. Tyler, R. J. Grimm. 2005. "Late Human Brain Plasticity: Vestibular Substitution with a Tongue BrainPort Human-Machine Interface." *Intellectica* 40:115–122.
6. **Jose Delgado -Electrical Conditioning of Bull's Behavior**
   Delgado, Jose M. 1969. *Physical Control of the Mind: Toward a Psychocivilized Society*. New York: Harper & Row.
7. **Arthur Schopenhauer-Between Dream and Reality**
   Schopenhauer, Arthur In *The World as Will and Presentation*. Translated by Richard E. Aquila in collaboration with David Carus. New York: Longman, 2008.
8. **The Bouba-Kiki Effect**
   Gómez Milán, E., and O. Iborra, M. J. de Córdoba, V. Juárez-Ramos, M. A. Rodríguez Artacho, J. L. Rubio. 2013. "The Kiki-Bouba Effect: A Case of Personification and Ideaesthesia." In *The Journal of Consciousness Studies*. 20(1–2): pp. 84–102.

9. **Daniel Kahnemanand Amos Tversky– The Framing Effect**
   Kahneman, D., and A. Tversky, eds. 2000. *Choices, Values, and Frames.* New York: Cambridge University Press.
10. **John Edensor Littlewood- about Miracles**
    Littlewood, John Edensor. 1986. *A Mathematician's Miscellany.* Cambridge University Press.
11. **Karl Popper- Knowing and Not Knowing**
    Popper, Karl. 1978. "Natural Selection and the Emergence of Mind."
12. **Stephen Hawking – Soft Determinism**
    Hawking, Stephen, and Leonard Mlodinow. 2010. *The Grand Design.* Bantam Books.
13. **Friedrich Nietzsche – Life in the Second Round**
    Young, Julian. 2010. *Friedrich Nietzsche: A Philosophical Biography.* Cambridge University Press.
14. **Attentional Blink**
    Chun, Marvin, M., and René Marois. 2002. "The dark side of visual attention." *Current Opinion in Neurobiology* 12(2): 184–189.
15. **Marco Polo in the Land of Smell**
    Arzi, A., and N. Sobel. 2011. "Olfactory Perception as a Compass for Olfactory Neural Maps." *Trends in Cognitive Science* 15(11): 537–45.
16. **The Zeigarnik Effect – The Effect of "Uncompleted Matters"**
    Baumeister, R. F., and B. J. Bushman. 2008. *Social Psychology and Human Nature.* United States: Thompson Wadsworth.
17. **Benjamin Libet – who is in charge of making decisions**
    Libet, Benjamin. 2004. *Mind Time: The Temporal Factor in Consciousness, (Perspectives in Cognitive Neuroscience).* Cambridge, MA: Harvard University Press.
18. **The Rubber Hand Illusion**
    Botvinick, M., and J. D. Cohen. 1998. "Rubber Hands 'Feel' Touch What Eyes See." *Nature Magazine* 391: 756.
    Makin T. R., and N. P. Holmes, H. H. Ehrsson. 2008. "On the Other Hand: Dummy Hands and Peripersonal Space." *Behav Brain Res.* 191: 1–10.
    Out-of-body experience
19. **Out-of-body experience**
    Blanke O., and T. Landis, L. Spinelli, M. Seeck. 2004. "Out-of-Body Experience and Autoscopy of Neurological Origin." *Brain Journal* Feb; 127 (Pt 2): 243–58.

20. **Frederik Skinner – Behaviorism**
    Smith, D. L. 2002. "On Prediction and Control: B. F. Skinner and the Technological Ideal of Science." In *Evolving Perspectives on the History of Psychology.* Edited by W. E. Pickren and D. A. Dewsbury. Washington, D.C.: American Psychological.
21. **James Olds and Peter Milner- Reward Cycle in the Brain**
    Olds, J. 1958. *"Self-Stimulation of the Brain."* Science 127:315–32.
22. **Wolfram Schultz-Dopamine in the service of happiness**
    Schultz, W. 2013. "Updating of Dopamine Reward Signals." *Curr Op Neurobiol* 23: 229–238.
23. **Alexander Shulgin- Mendeleev of the Psychoactive Elements**
    Shulgin, Alexander, and Tania Manning, Paul Daley. 2011. The Shulgin Index Vol 1: *Psychedelic Phenethylamines and Related Compounds.* Berkeley: Transform Press.
24. **Aldous Huxley - Soma Pill**
    Huxley, Aldous Leonard. 1932. *Brave New World.* Huxley, Aldous Leonard. 1954. The Doors of Perception.
25. **Eric Kandel- How Memory Remembers**
    Antonov, Igor, and Irina Antonova, Eric R. Kandel, Robert D. Hawkins. 2003. "Activity-Dependent Presynaptic Facilitation and Hebbian LTP Are Both Required and Interact during Classical Conditioning in Aplysia." *Neuron* 37 (1): 135–147.
    Dreifus, Claudia. 2012. "A Quest to Understand How Memory Works." *New York Times*, March 5.
26. **Hermann Ebbinghaus- The Curve of Memory Fading**
    Thorne, B., Henley, T. 2005. "Hermann Ebbinghaus." In *Connections in the History and Systems of Psychology.* 3rd ed., 211–216. Belmont, CA: Wadsworth Cengage Learning.
27. **HM (Henry Molaison) – Prisoner of Eternal Present**
    Scoville, Beecher W., and B. Milner. 1957. "Loss of Recent Memory after Bilateral Hippocampal Lesions." *J. Neurol. Neurosurg. Psychiat.* 20(11), 11–21.
28. **Elizabeth Loftus-False Memories**
    Loftus, E. F., and J. M. Doyle, J. Dysert. 2008. *Eyewitness Testimony: Civil and Criminal.* 4th ed. Charlottesville, VA: Lexis Law Publishing.
    Loftus, E., and G. Geis. 2009. "Taus v. Loftus: Determining the Legal Ground Rules for Scholarly Inquiry." *Journal of Forensic Psychology Practice* 9(2): 147–62.

29. **Alexander Luria- Memory and Head Injuries**
    Luria, A. R. 1987. *The Mind of a Mnemonist: A Little Book About A Vast Memory*. Cambridge, MA: Harvard University Press.
    Luria, A. R., and Lynn Solotaroff. 1987. *The Man with a Shattered World: The History of a Brain Wound*. Cambridge, MA: Harvard University Press.
30. **Arthur Koestler - The Theory of Relativity in the Sunflower Field**
    Scammell, Michael. 2009. *Koestler: The Literary and Political Odyssey of a Twentieth-Century Skeptic*. Random House.
31. **Solomon Asch - Conformity Experiment**
    Asch, S. E. 1955. "Opinions and Social Pressure." *Scientific American* 193, 35–35.
32. **Jean Jacques Rousseau - On Human Nature**
    Zalta, Edward N., ed. 2012. "Jean Jacques Rousseau." In *The Stanford Encyclopedia of Philosophy*. Winter ed.
33. **Thomas Hobbes - On Human Nature**
    Zalta, Edward N., ed. 2013. "Hobbes's Moral and Political Philosophy." In *The Stanford Encyclopedia of Philosophy*. Summer ed.
34. **Stanley Milgram- Obedience to Authority Test**
    Milgram, S. 1974. *Obedience to Authority; An Experimental View*.
35. **Jorge Luis Borges—Books and Darkness**
    The Garden of Forking Paths. An Internet site on life and work of Jorge Luis Borges.
36. **Ray Kurzweil- The Singular Point**
    Kurzweil, Ray. 2005. *The Singularity Is Near*. Viking Penguin
37. **Richard Haier and Rex Jung – Men's & Women's Intelligence**
    Haier, R. J., and R. Jung, R. Yeo, K. Head, M. T. Alkire. 2005. "Structural brain Variation, Age and Response Time." *Cognitive, Affective, and Behavioral Neuroscience* 5(2), 246–251.
38. **The Effect of Physical Activity on the Brain**
    Voss, M., and L. Nagamatsu, T. Liu-Ambrose, A. Kramer. 2011. "Exercise, Brain, and Cognition across the Life Span." *Journal of Applied Physiology* 5: 1505–1513.
39. **Isaac Asimov- Exceptional event and mental chaos**
    Asimov, Isaac. 1941. *Nightfall*.
40. **Edwin Abbott- Transdimensional Meme**
    Abbott, Edwin. 1884. *Flatland: A Romance of Many Dimensions*.

41. **Sigmund Freud- The Super-memes of the mind apparatus**
Tauber, Alfred I. 2010. *Freud, the Reluctant Philosopher*. Princeton, NJ: Princeton University Press.

42. **Rosalind Franklin – Basic steps on the ladder of DNA**
Maddox, B. 2002. *Rosalind Franklin: The Dark Lady of DNA*. London: Harper Collins.

43. **Immanuel Kant – Categorical Imperative**
Hanna, Robert. 2006. *Kant, Science, and Human Nature*. Clarendon Press.

44. **Moral Conflict- The Railway Scenario**
Mikhail, John. 2007. "Universal Moral Grammar: Theory, Evidence, and the Future." *Trends in Cognitive Sciences*. 11, 143–152.

45. **The U- Curve of Happiness**
Blanchflower, David G., and Andrew J. Oswald. 2008. "Is Well-Being U-shaped over the Life Cycle?" *Social Science & Medicine*, 66(8): 1733–1749.

46. **Akinetopsia- freezing of movement perception**
Gazzaniga, Michael S., and Richard B. Ivry, George R. Mangun. 2009. *Cognitive Neuroscience: The Biology of the Mind*. 3rd ed. New York: W.W. Norton.

47. **Lorenz Konrad- imprinting**
Lorenz, Konrad, and Paul Leyhausen. 1973. *Motivation of Human and Animal Behavior: An Ethological View*. Van Nostrand Reinhold Co.

48. **Synapsis' Creation (Synaptogenesis) during Infancy**
Gilmore, J. H., et al. 2007. "Regional Gray Matter Growth, Sexual Dimorphism, and Cerebral Asymmetry in the Neonatal Brain." *Journal of Neuroscience* 27(6): 1255–1260.
Rakic, P. 2006. "No More Cortical Neurons for You." *Science* 313: 928–929.

49. **The Sally-Anne Test- to concept the other**
Wimmer, H., and J. Perner. 1983. "Beliefs about Beliefs: Representation and Constraining Function of Wrong Beliefs in Young Children's Understanding of Deception." *Cognition* 13(1): 103–128.

50. **Mirror Cells- To reflect the other**
Oberman, L. M., and V. S. Ramachandran. 2008. "Preliminary Evidence for Deficits in Multisensory Integration in Autism Spectrum Disorders: The Mirror Neuron Hypothesis." *Social Neuroscience* 3 (3-4): 348–55.

51. **Baruch Spinoza – Trust and Doubt**
    LeBuffe, Michael. 2010. *Spinoza and Human Freedom*. Oxford University Press.
    Yovel, Yirmiyahu. 1989. *Spinoza and Other Heretics, Vol. 1: The Marrano of Reason*. Princeton, NJ: Princeton University Press.
52. **The number of words spoken by women and men throughout a day**
    Matthias, R. Mehl, and Simine Vazire, Nairán Ramírez-Esparza, Richard B. Slatcher, James W. Pennebaker. 2007. "Are Women Really More Talkative Than Men?" *Science Magazine*: 317(5834) 82.
53. **Simon Baron-Cohen – Seeing the Other's Soul**
    Baron-Cohen, S., and H. Tager-Flusberg, D. J. Cohen, eds. 2007. *Understanding Other Minds: Perspectives from Developmental Cognitive Neuroscience*. 2nd ed. Oxford University Press.
54. **Seattle Longitudinal Study of Cognitive Development**
    Warner Schaie, K., and Sherry L. Willis. 2010. "The Seattle Longitudinal Study of Cognitive Development." *Bulletin of International Society for the Study of Behavioural Development* 57(1): 24–29.
55. **Michael Gazzaniga - on the creation of the sense of "self"**
    Gazzaniga, Michael S. 2011. *Who's in Charge?: Free Will and the Science of the Brain*. Harper Collins Publishers.
56. **Carl Sagan– Star stuff**
    Sagan, Carl. 1998. *Billions & Billions: Thoughts on Life and Death at the Brink of the Millennium*. Random house.
57. **Gottfried Wilhelm Leibniz - Time and Space Relativity**
    Zalta, Edward N., ed. 2013. "Gottfried Wilhelm Leibniz." In *The Stanford Encyclopedia of Philosophy*. Fall ed.

www.ingramcontent.com/pod-product-compliance
Lightning Source LLC
Chambersburg PA
CBHW071408180526
45170CB00001B/10